International REVIEW OF Neurobiology
Volume 72

International REVIEW OF Neurobiology

Volume 72

SERIES EDITORS

RONALD J. BRADLEY
*Department of Psychiatry, College of Medicine
The University of Tennessee Health Science Center
Memphis, Tennessee, USA*

R. ADRON HARRIS
*Waggoner Center for Alcohol and Drug Addiction Research
The University of Texas at Austin
Austin, Texas, USA*

PETER JENNER
*Division of Pharmacology and Therapeutics
GKT School of Biomedical Sciences
King's College, London, UK*

EDITORIAL BOARD

ERIC AAMODT	HUDA AKIL
PHILIPPE ASCHER	MATTHEW J. DURING
DONARD DWYER	DAVID FINK
MARTIN GIURFA	MICHAEL F. GLABUS
PAUL GREENGARD	BARRY HALLIWELL
NOBU HATTORI	JON KAAS
DARCY KELLEY	LEAH KRUBITZER
BEAU LOTTO	KEVIN MCNAUGHT
MICAELA MORELLI	JOSÉ A. OBESO
JUDITH PRATT	CATHY J. PRICE
EVAN SNYDER	SOLOMON H. SNYDER
JOHN WADDINGTON	STEPHEN G. WAXMAN

Catatonia in Autism Spectrum Disorders

EDITED BY

DIRK MARCEL DHOSSCHE
University of Mississippi Medical Center
Jackson, Mississippi, USA

LORNA WING
Centre for Social and Communication Disorders
Bromley, Kent, United Kingdom

MASATAKA OHTA
Center for the Research and Support of Educational Practice
Tokyo Gakugei University
Tokyo, Japan

KLAUS-JÜRGEN NEUMÄRKER
Charité, Humboldt-University Berlin
Department of Child and Adolescent Psychiatry and Psychotherapy
DRK Hospital, Berlin-Westend, Berlin, Germany

AMSTERDAM • BOSTON • HEIDELBERG • LONDON
NEW YORK • OXFORD • PARIS • SAN DIEGO
SAN FRANCISCO • SINGAPORE • SYDNEY • TOKYO
Academic Press is an imprint of Elsevier

Cover image courtesy of Dirk Marcel Dhossche
University of Mississippi Medical Center, Jackson

Academic Press is an imprint of Elsevier
525 B Street, Suite 1900, San Diego, California 92101-4495, USA
84 Theobald's Road, London WC1X 8RR, UK

This book is printed on acid-free paper.

Copyright © 2006, Elsevier Inc. All Rights Reserved.

No part of this publication may be reproduced or transmitted in any form or by any means, electronic or mechanical, including photocopy, recording, or any information storage and retrieval system, without permission in writing from the Publisher.

The appearance of the code at the bottom of the first page of a chapter in this book indicates the Publisher's consent that copies of the chapter may be made for personal or internal use of specific clients. This consent is given on the condition, however, that the copier pay the stated per copy fee through the Copyright Clearance Center, Inc. (www.copyright.com), for copying beyond that permitted by Sections 107 or 108 of the U.S. Copyright Law. This consent does not extend to other kinds of copying, such as copying for general distribution, for advertising or promotional purposes, for creating new collective works, or for resale. Copy fees for pre-2006 chapters are as shown on the title pages. If no fee code appears on the title page, the copy fee is the same as for current chapters.
0074-7742/2006 $35.00

Permissions may be sought directly from Elsevier's Science & Technology Rights Department in Oxford, UK: phone: (+44) 1865 843830, fax: (+44) 1865 853333, E-mail: permissions@elsevier.com. You may also complete your request online via the Elsevier homepage (http://elsevier.com), by selecting "Support & Contact" then "Copyright and Permission" and then "Obtaining Permissions."

For information on all Academic Press publications
visit our Web site at www.books.elsevier.com

ISBN-13: 978-0-12-366873-8
ISBN-10: 0-12-366873-5

PRINTED IN THE UNITED STATES OF AMERICA
06 07 08 09 9 8 7 6 5 4 3 2 1

Working together to grow libraries in developing countries

www.elsevier.com | www.bookaid.org | www.sabre.org

ELSEVIER   BOOK AID International   Sabre Foundation

Susie Wing (1956–2005)

This book is dedicated to Susie Wing, beloved daughter of Lorna and John Wing. Susie had severe autism and life was not always easy for her. However, Susie had a tremendous capacity for living in the present moment and derived a lot of pleasure and happiness from her special interests and activities. These included collecting teapots, travelling by train/plane, horse-riding, swimming, going on holiday, art and craft, music, opera, and eating her favourite foods.

Susie is missed and remembered by all who knew her and of her, especially for her special significance and contribution to our understanding of autism and catatonia.

CONTENTS

Contributors xiii
Acknowledgments xv
Preface xvii

SECTION I
CLASSIFICATION

Classification Matters for Catatonia and Autism in Children
Klaus-Jürgen Neumärker

I.	Introduction	3
II.	Catatonia	4
III.	Autism	10
IV.	Autism and Catatonia	12
	References	14

A Systematic Examination of Catatonia-Like Clinical Pictures in Autism Spectrum Disorders
Lorna Wing and Amitta Shah

I.	Introduction	22
II.	Clinical Features of ASDs and Catatonia	23
III.	Manifestations of Catatonic Features in ASDs	23
IV.	Three Studies of Catatonic Features in ASD	24
V.	Discussion	34
VI.	Conclusions	37
	References	37

Catatonia in Individuals with Autism Spectrum Disorders in Adolescence and Early Adulthood: A Long-Term Prospective Study

MASATAKA OHTA, YUKIKO KANO, AND YOKO NAGAI

I.	Introduction	42
II.	Subjects	43
III.	Methods	43
IV.	Presentation of Cases	44
V.	Results	49
VI.	Discussion	51
VII.	Suggestions on Treatment	52
VIII.	Limitation of this Study	53
IX.	Conclusions	53
	References	53

Are Autistic and Catatonic Regression Related? A Few Working Hypotheses Involving GABA, Purkinje Cell Survival, Neurogenesis, and ECT

DIRK MARCEL DHOSSCHE AND UJJWAL ROUT

I.	Introduction	56
II.	Early Autistic Regression	57
III.	Late Autistic Regression	58
IV.	Catatonic Regression in Autism	59
V.	Are Autistic and Catatonic Regression Related?	61
VI.	Theoretical Treatment Implications	62
VII.	History of ECT in Childhood Psychoses	63
VIII.	GABA in Autistic Regression	64
IX.	GABA in Catatonic Regression	66
X.	GABA and the Mechanism of Action of ECT	68
XI.	ECS in GABA Models of Autism	69
XII.	ECS and Purkinje Cell Survival	70
XIII.	ECS and Neurogenesis	71
	References	72

SECTION II
ASSESSMENT

Psychomotor Development and Psychopathology in Childhood

DIRK M. J. DE RAEYMAECKER

| I. | Three Vital Steps in Motor Development | 84 |
| II. | Human Motility: A View From Psychoanalysis | 88 |

| III. | Developmental Psychopathology and Motor Impairment | 92 |
| | References | 99 |

The Importance of Catatonia and Stereotypies in Autistic Spectrum Disorders

LAURA STOPPELBEIN, LEILANI GREENING, AND ANGELINA KAKOOZA

I.	Introduction	104
II.	Catatonia and ASD	104
III.	Stereotypic Movement Disorder	110
	References	115

Prader–Willi Syndrome: Atypical Psychoses and Motor Dysfunctions

WILLEM M. A. VERHOEVEN AND SIEGFRIED TUINIER

I.	Introduction	119
II.	Development and Behavior	120
III.	Psychotic Disorders	121
IV.	Psychomotor Symptoms	126
V.	Conclusions	127
	References	128

Towards a Valid Nosography and Psychopathology of Catatonia in Children and Adolescents

DAVID COHEN

I.	Introduction	132
II.	Phenomenology of Catatonia in Young People	132
III.	Towards a Broader Nosography for Child, Adolescent, and Adult Catatonia	135
IV.	Models of Catatonia	136
V.	Psychopathological Model for Catatonia	138
VI.	Discussion	143
VII.	Conclusions	145
	References	146

SECTION III
BIOLOGY

Is There a Common Neuronal Basis for Autism and Catatonia?

DIRK MARCEL DHOSSCHE, BRENDAN T. CARROLL, AND TRESSA D. CARROLL

| I. | Introduction | 151 |
| II. | Motor Theory of Language and Autism | 152 |

III.	Are Autism and Catatonia Neurobiological Syndromes?	153
IV.	Catatonia as a Neurobiological Syndrome	154
V.	Autism as a Neurobiological Syndrome	156
VI.	Is There a Common Neuronal Basis for Autism and Catatonia?	157
VII.	Future Directions: Parallel Family-Based Studies in Autism and Catatonia	159
VIII.	Implications	161
	References	162

Shared Susceptibility Region on Chromosome 15 Between Autism and Catatonia

Yvon C. Chagnon

I.	Introduction	165
II.	Methods	166
III.	Results	167
IV.	Conclusions	173
	References	173

SECTION IV
TREATMENT

Current Trends in Behavioral Interventions for Children with Autism

Dorothy Scattone and Kimberly R. Knight

I.	Current Trends in Behavioral Interventions for Children with Autism in the United States	181
II.	Language	182
III.	Social Skills	185
IV.	Self-Management	189
V.	Conclusions	191
	References	191

Case Reports with a Child Psychiatric Exploration of Catatonia, Autism, and Delirium

Jan N. M. Schieveld

I.	Introduction	195
II.	Case Histories	197
III.	Methods	199
IV.	Results	200
V.	Discussion	201

VI.	Conclusions	203
VII.	Epicrisis	203
	References	204

ECT and the Youth: Catatonia in Context
Frank K. M. Zaw

I.	Introduction	208
II.	ECT and the Youth	210
III.	Indications for ECT	213
IV.	Pre-ECT Consideration	217
V.	Administration of ECT	219
VI.	Adverse Effects of ECT	221
VII.	Concurrent Medication	223
VIII.	Mortality	224
IX.	Patients' Satisfaction and Attitude to ECT	224
X.	What Follows ECT?	225
XI.	Conclusions	226
XII.	Conflict of Interest	227
	References	227

Catatonia in Autistic Spectrum Disorders: A Medical Treatment Algorithm
Max Fink, Michael A. Taylor, and Neera Ghaziuddin

I.	Introduction	233
II.	Catatonia	237
III.	Treatment Algorithm	239
IV.	Theory	241
	References	242

Psychological Approaches to Chronic Catatonia-Like Deterioration in Autism Spectrum Disorders
Amitta Shah and Lorna Wing

I.	Introduction	246
II.	Effects of Stress	247
III.	Effect of Medical Treatments	250
IV.	Problems with Assessing the Effects of Medical Treatments	251
V.	Psychological Methods of Intervention	252
VI.	General Principles of Psychological Treatment and Management	253
VII.	Management of Specific Problems	257
VIII.	Implications for Services and Staffing Levels	260

IX. Conclusions 260
References 263

SECTION V
BLUEPRINTS

Blueprints for the Assessment, Treatment, and Future Study of Catatonia in Autism Spectrum Disorders

DIRK MARCEL DHOSSCHE, AMITTA SHAH, AND LORNA WING

I. Introduction 268
II. Blueprint for the Assessment of Catatonia in ASDs 269
III. Blueprint for the Treatment of Catatonia in ASDs 273
IV. Blueprint for the Future Study of Catatonia in ASDs 280
References 283

INDEX 285

CONTENTS OF RECENT VOLUMES 297

CONTRIBUTORS

Numbers in parentheses indicate the pages on which the authors' contributions begin.

Brendan T. Carroll (151), Department of Psychiatry, University of Cincinnati, Cincinnati, Ohio 05267, USA; Psychiatry Service, VA Medical Center, Chillicothe, Ohio 43162, USA

Tressa D. Carroll (151), The Neuroscience Alliance, West Jefferson, Ohio 43162, USA

Yvon C. Chagnon (165), Laval University Robert-Giffard Research Center, Beauport, Québec, Canada

David Cohen (131), Department of Child and Adolescent Psychiatry, Hôpital Pitié-Salpétrière, AP-HP, Université Pierre et Marie Curie, 47 bd de l'Hôpital, 75013 Paris, France

Dirk Marcel Dhossche (55, 151, 267), Department of Psychiatry and Human Behavior, University of Mississippi Medical Center, Jackson, Mississippi 39216, USA

Max Fink (233), School of Medicine, State University of New York, Stony Brook, New York 11794, USA

Neera Ghaziuddin (233), University of Michigan School of Medicine, Ann Arbor, Michigan 48109, USA

Leilani Greening (103), Department of Psychiatry and Human Behavior, Center for Psychiatric Neuroscience, University of Mississippi Medical Center, Jackson, Mississippi 39216, USA

Angelina Kakooza (103), Makerere University, Kampala, Uganda

Yukiko Kano (41), Graduate School of Medical Sciences, Kitasato University, Sagamihara, Kanagawa, Japan

Kimberly R. Knight (181), Department of Pediatrics, Mississippi Child Development Institute, The University of Mississippi Medical Center, Jackson, Mississippi 39216, USA

Yoko Nagai (41), School of Nursing, University of Shizuoka, Suruga-Ku, Shizuoka, Japan

Klaus-Jürgen Neumärker (3), Charité, Humboldt-University Berlin, Department of Child and Adolescent Psychiatry and Psychotherapy, DRK Hospital, Berlin-Westend, Berlin, Germany

Masataka Ohta (41), Center for the Research and Support of Educational Practice, Tokyo Gakugei University, Koganei-Shi, Tokyo, Japan

Dirk M. J. de Raeymaecker (83), Department of Child Psychology and Psychiatry, "Kores" GGZ Group-Europoort, Rotterdam, The Netherlands

Ujjwal Rout (55), Department of Surgery, University of Mississippi Medical Center, Jackson, Mississippi 39216, USA

Dorothy Scattone (181), Department of Pediatrics, Division of Child Development and Behavioral Medicine, The University of Mississippi Medical Center, Jackson, Mississippi 39216, USA

Jan N. M. Schieveld (195), Department of Psychiatry and Neuropsychology, University Hospital Maastricht, The Netherlands

Amitta Shah (21, 245, 267), Leading Edge Psychology, Purley CR8 2EA, United Kingdom

Laura Stoppelbein (103), Department of Psychiatry and Human Behavior, Center for Psychiatric Neuroscience, University of Mississippi Medical Center, Jackson, Mississippi 39216, USA

Michael A. Taylor (233), Rosalind Franklin University of Medicine and Science, North Chicago, Illinois 60064, USA; University of Michigan School of Medicine, Ann Arbor, Michigan 48109, USA

Siegfried Tuinier (119), Vincent van Gogh Institute for Psychiatry, Venray, Stationsweg 46, 5803 AC Venray, The Netherlands

Willem M. A. Verhoeven (119), Vincent van Gogh Institute for Psychiatry, Venray, Stationsweg 46, 5803 AC Venray, The Netherlands; Department of Psychiatry, Erasmus University Medical Center, Rotterdam, The Netherlands

Lorna Wing (21, 245, 267), Centre for Social and Communication Disorders, Bromley, Kent BR2 9HT, United Kingdom

Frank K. M. Zaw* (207), Senior Clinical Lecturer (Hon.), Division of Neurosciences, Department of Psychiatry, University of Birmingham, United Kingdom

---

*Corresponding address: Huntercombe Stafford Hospital, Ivetsky Bank, Wheaton Aston, Staffordshire ST19 9QT, United Kingdom.

ACKNOWLEDGMENTS

I would like to thank the series editors, particularly Ron Bradley, and the editorial board of International Review of Neurobiology, for giving us this opportunity to focus on this novel topic. Surely, without the bold vision of Johannes Menzel, Senior Publishing Editor, Elsevier Academic Press, London, this book would not have seen the daylight.

Lorna Wing and Max Fink have made seminal contributions to this book and have inspired to persist. Lorna Wing has taught me many lessons about autism for which I am very grateful. Max Fink has taught me many lessons about catatonia for which I am equally grateful. I would like to thank Masataka Ohta and Klaus-Jürgen Neumärker for their help in completing the editorial task.

The September 2001 Mac Keith Meeting on childhood catatonia, organized by Michael Prendergast and Gregory O'Brien at the Royal Society of Medicine in London, presaged this book. Several of us met there for the first time. We were reminded of the seminal observations of Karl Leonhard on childhood catatonia. Max Fink introduced his upcoming book on catatonia. David Cohen described catatonia in adolescents. It was an eye-opening experience when Lorna Wing and Amitta Shah spoke about catatonia in autism.

At the August 2004 IACAPAP meeting in Berlin, Klaus-Jürgen Neumärker, Masataka Ohta, Yukiko Kano, Yoko Nagai, Frank Zaw, Dirk de Raeymaecker, and Jan Schieveld joined this project. Laura Stoppelbein, Leilani Greening, Angelina Kakooza, Willem Verhoeven, Siegfried Tuinier, Brendan and Tressa Carroll, Yvon Chagnon, Ujjwal Rout, Dorothy Scattone, Kimberly Knight, Michael Taylor, and Neera Ghaziuddin followed later.

Jan Schieveld discovered the relevance for this book of Elsevier's motto and a novel aspect of the Rosetta stone metaphor as applied to autism and catatonia. Allison East, Sara Stanfield, and Veronnica Taylor have helped tremendously with proofreading throughout this journey. Cindy Minor, Senior Developmental Editor, Elsevier Academic Press, San Diego, has skillfully kept this project moving forward. Ongoing research support from the Thrasher Research Fund, Salt Lake City in Utah, is gratefully acknowledged.

The work is dedicated to Susie Wing (1956–2005), beloved daughter of Lorna and John Wing, and all families who have entrusted us with the care of their ill children and relatives. We have been rewarded with a deeper appreciation of the overlapping psychopathology of autism and catatonia and with new hope for improved treatments.

Dirk Marcel Dhossche

I would like to thank my colleague Amitta Shah, who has, through her clinical work with children and adults, been the prime mover in developing the practical psychological approaches described in one of the chapters. I would also like to thank the care staff in specialist homes for adolescents and adults with autism spectrum disorders in UK, who have worked with us to implement and improve these psychological approaches.

Finally, the thanks of all contributors must go to Dirk Marcel Dhossche, who has had the courage and foresight to edit the first book in this highly controversial field.

Lorna Wing

PREFACE

"The most valuable lesson that knowledge can teach us is that its creation depends upon a continuous line of human relationships and traditions that go far back into the past. That continuity is an unbroken thread. It links cultures and peoples; it brings tolerance and understanding; it delivers hope and compassion."

Richard Horton, BSc, MB, FRCP, FMedScie
Editor-in-Chief, The Lancet
The 2004 Elsevier Library Connect Medical Library Lecture

Autism and catatonia have increasingly come into the limelight. Autism afflicts our young at a much higher rate than the early prevalence studies suggested. The prevalence of the whole spectrum is now considered to be in the region of 6 per 1000, but some studies suggest it may be as high as 9 or 10 per 1000. Catatonia, once thought to have disappeared, is increasingly recognized again with the development of rating scales, diagnostic tests, and effective treatments. Catatonia has lacked attention; autism lacks treatments.

Catatonia is a motor disorder characterized by stereotypy, rigidity, mutism, and posturing. These motor signs are also characteristic of autism. Catatonia has been almost completely absent in autism research, and vice versa. The similarity of behaviors that characterize autism and catatonia raises the question of the nature of the overlap between these two seemingly disparate disorders with divergent historical roots. Some authors in this book have much experience with catatonia in general psychiatry, some with catatonia in autism, and others with both. The Rosetta stone was a small stone tablet containing the same message written in Greek and two different Egyptian scripts. The stone was essential for deciphering Egyptian hieroglyphics. Can the group of patients who are diagnosed with both autism and catatonia be like the Rosetta stone, providing the complete sentences required for deciphering the meaning of some autistic symptoms? Do the syndromes have a common pathophysiology? Can the successful treatment of catatonia be applied to patients with both autism and catatonia? These themes are examined in this book.

All authors in this book point out that there is a lot more to learn about autism and catatonia and their commonalties. Despite (or because of) these uncertainties, they have been carefully restrained and cautious in their observations, hypotheses, and recommendations for research and treatment, and are to be commended for it. Their compositions, grouped into four sections, reflect the collective experience of a cross-disciplinary group of clinicians actively engaged in the treatment of patients of all ages, with autism and/or catatonia.

The section *Classification* consists of four chapters. In the first chapter, Klaus-Jürgen Neumärker paints a rich historical perspective of the concepts of catatonia and autism, as described by the old European masters like Bleuler and Kahlbaum. Next, Lorna Wing and Amitta Shah, and Masataka Ohta, Yukiko Kano, and Yoko Nagai, in their respective Chapters, describe the application of systematic criteria to dissect the catatonic syndrome out of the myriad of autistic symptoms. The chapter by Dirk Marcel Dhossche and Ujjwal Rout, compares autistic and catatonic regression and outlines focused research to test the hypothesis that, in some instances, autistic regression is an early form of catatonic regression.

Dirk de Raeymaecker, expertly cognizant of contemporary major psychological theories, starts the *Assessment* section by emphasizing the importance of psychomotor function for normal and abnormal development. Laura Stoppelbein, Leilani Greening, and Angelina Kakooza provide a comprehensive review of motor abnormalities in autism. Prader–Willi syndrome is another enigmatic neuropsychiatric syndrome that may provide useful insights for autism and catatonia research as pointed out by Willem Verhoeven and Siegfried Tuinier. David Cohen, presents a new psychopathological model of catatonia at the level of patients' intrapsychic experience and as such rooted in a long tradition of excellence of French psychiatry.

In the *Biology* section, neurobiological (chapter by Dirk Marcel Dhossche, Brendan, and Tressa Carroll) and genetic (chapter by Yvon Chagnon) risk factors of autism and catatonia, and their overlap, are discussed.

The *Treatment* section starts with the chapter in which Dorothy Scattone and Kimberly Knight describes current behavioral interventions and their indications in children with autism. Jan Schieveld, describes two illustrative case-reports of catatonia in youngsters with autism. He conceptualizes catatonia and delirium as reaction types of the brain and describes their similarities and differences. Severe catatonic stupor is an important indication for ECT but there is as yet little evidence for or against its use for catatonia in autism. Frank Zaw describes the use of ECT for catatonia in childhood and adolescence and discusses the controversies currently raging around this subject. Next, Max Fink, Michael Taylor, and Neera Ghaziuddin propose a medical treatment algorithm for catatonia in autism. Amitta Shah and Lorna Wing discuss the problems of medical treatment of chronic catatonia-like conditions in autism and offer a psychological approach that has proved helpful in clinical practice, especially when medical treatment is ineffective.

In the last section, Dirk Marcel Dhossche, Amitta Shah, and Lorna Wing submit *Blueprints* for the assessment, treatment, and future study of catatonia in autism. The authors aim to increase early recognition and treatment of catatonia in patients with autism, show the urgency of controlled treatment trials and increased collaborative and interdisciplinary research, both in the preclinical and clinical arenas, into the co-occurrence of these two enigmatic disorders.

Catatonia in Autism Spectrum Disorders is in effect a new field. "Autism + Catatonia" is also a novel permutation for research funding. There are no hits for this key word combination in CRISP.[1] There are hundreds for autism; none for catatonia. Funded catatonia research is nonexistent in the U.S., or merely a by-product. The history of autism and psychosis (including catatonia) has diverged since Kanner isolated autism from other early-onset disorders in 1943. Valuable autism research has been done albeit at the expense of exploring links between autism and childhood psychoses, including catatonia. Arguably, psychiatric interest in autism research has suffered for this reason, resulting in meager contributions and slow progress in finding effective treatments for autism. In the meantime, the uncanny resemblance between autistic and catatonic symptoms and co-occurrence of autism and catatonia in patients remains unaccounted. Treatments are often applied haphazardly.

This book has started to collect clues for solving the mystery of the overlap and co-occurrence of autism and catatonia. Connecting autism and catatonia research can only benefit our patients, their families and caregivers, as well as their treating physicians and health professionals. Our hope is that, by means of this new volume in the International Review of Neurobiology series, the stage is set for great progress so that Elsevier's motto can be honored: "Building insights. Breaking Boundaries."

Dirk Marcel Dhossche
Klaus-Jürgen Neumärker
Masataka Ohta
Lorna Wing

---

[1]CRISP stands for Computer Retrieval of Information on Scientific Projects, a searchable database of US federally funded biomedical research projects conducted at universities, hospitals, and other research institutions.

SECTION I
CLASSIFICATION

CLASSIFICATION MATTERS FOR CATATONIA AND AUTISM IN CHILDREN

Klaus-Jürgen Neumärker

Charité, Humboldt–University Berlin, Department of Child and Adolescent Psychiatry and Psychotherapy, DRK Hospital, Berlin–Westend, Berlin, Germany

I. Introduction
II. Catatonia
III. Autism
IV. Autism and Catatonia
References

Despite its chequered history, Kahlbaum's 1874 description of catatonia (tension insanity) and its categorization as a clinical illness is in outline still valid. Kahlbaum also acknowledged the existence of catatonia in children. Corresponding case studies have also been analyzed. The originators and disciples of the Wernicke-Kleist-Leonhard school proved catatonia in early childhood as a discrete entity with specific psychopathology. This does not mean that catatonic symptoms do not occur in other illnesses and in particular in organic psychoses. These are, however, of a totally different nature. Autism, as first described in connection with schizophrenic negativism by Bleuler in 1910, is one of the key symptoms of schizophrenia. As identified by Kanner in 1943, abnormal social interaction and communication, together with retarded development, are the main characteristics of autism in early childhood. Asperger's concept of autistic disorder (1944), although based on psychopathological theory, did not include retardation in development as an aspect. Consequently, autistic behavior can occur in a variety of mental disorders. Research into possible etiological and pathogenetic factors has been undertaken, but no clear link found as yet.

I. Introduction

Catatonia and autism are not only major mental disorders, but they also have archetypal characteristics that help explain mental disorders. This chapter investigates the specific characteristics of both disorders and examines the extent to which they can be considered illnesses in their own right, rather than mere symptoms of other underlying illnesses. It also considers their ubiquitousness. Understanding these illnesses today is also aided by adopting a historical perspective not only in regards to the classification of mental disorders but also because the early descriptions can be regarded as descriptions of unspoilt pathological conditions in which the patients had not been treated with psychopharmaceuticals. Analysis of catatonia and autism in children provides an insight into their developmental aspects.

II. Catatonia

It is the generally held view in psychiatry—historical research that it was Karl Ludwig Kahlbaum (1828–1899), who first coined the term catatonia in his book *Die Katatonie oder das Spannungsirresein* ("Catatonia or Tension Insanity") of 1874. He defined the following symptoms: "*Flexibilitas cerea, chorea–like cramps above all in the face and extremities, head, masticatory, logo–, laryngeal, and eye muscle spasms, crooked posture of the upper part of the body. Stereotypic motions and postures, involuntary tension of the limbs, monotonous and pointless movements, crying and laughing fits, mouth cramps, stiffness of the whole body, taking up and maintaining uncomfortable postures and grimaces.*" He also introduced another special term—verbigeration—by which he understood: "*the patient repeating meaningless and incoherent words and sentences, which take on the character of speech.*" Kahlbaum stated that catatonia took the following course: "*melancholia, mania, stupor, confusion, imbecility*" (Kahlbaum, 1874).

Kahlbaum had observed and examined 26 patients (18 males, 7 females, and 1 unspecified) with an average age 27.9 years for the males and 36.4 years for the females. He stated that catatonia had occurred "*at any age, even in children.*" Subsequently, the prevalence of catatonia in children was often described in German-speaking psychiatric literature. Of course, since child and adolescent psychiatry had not yet become a discipline in its own right, these were descriptions of psychiatrists usually working with adults in psychiatric institutions, which also provided mental health-care services for children suffering from mental diseases. The beginnings of institutionalized child and adolescent psychiatry, however, lie in the first decades of the twentieth century, when independent wards for children and adolescents were established—as in Heidelberg in 1917 by August Homburger (1873–1930); in Tübingen in 1920 by Werner Villinger (1887–1961); in Berlin in 1921 by Karl Bonhoeffer (1868–1948); and in Leipzig in 1926 by Paul Schröder (1873–1941)—to name just the first of their kind in Germany.

Among the studies which discussed catatonia as an illness in its own right and its relations to "*idiocy*" was that of Wilhelm Weygandt (1870–1939) involving seven children (five girls and two boys) for whom he pointed to "*abnormal postures and positions, abnormal motions, abnormal mimic expressions, grimaces, abnormal [i.e. incomprehensible] speech – verbigeration, order execution automatism, negativism*" (Weygandt, 1907). He considered these as characteristic symptoms of what he referred to as "catatonia in children" and contrasted with Emil Kraepelin's (1856–1926) description of adolescents suffering from "catatonic idiocy" in the 6th edition of his famous *Textbook of Clinical Psychiatry* (Kraepelin, 1899). At around that time Vogt (1909) and Plaskuda (1911) too dealt with the problem of infantile catatonia.

In 1909, Julius Raecke (1872–1930) working alongside Ernst Simmerling (1857–1931) at the Psychiatric and Neurological Hospital of Kiel University that the latter headed from 1901 to 1925, gave an account of 10 cases of *catatonia in children* (8 boys, 2 girls), which showed the same symptoms as Kahlbaum had described. In his paper, he clearly maintained that "*catatonia also occurred in children, especially between the ages of 12 and 15. As regards their main characteristics there are no major differences between catatonia in children and adults.*" He pointed out that although only four patients showed a hereditary background "*certain early detectible peculiarities*" should—and here he was in full agreement with Kraepelin (1899)—be regarded as "*first symptoms/indications for this disease.*" For Raecke, most characteristic of catatonia were an "*abrupt alternation between inhibition and agitation with patients tending to stereotypies and bizarre urges, impulsive motor eruptions and blind antipathy, although not showing abnormal affect or opacity of consciousness*" (Raecke, 1909). Other symptoms Raecke found included stupor with mutism (the refusal to speak), refusal to eat, uncleanliness, first indications of flexibilitas cerea, unmotivated eccentricity, childish demeanor, and retarded mental development in contrast to children of the same age. As regards the interrelationship of catatonia and imbecility, Raecke shared Kraepelin's view that in cases where "*imbecility or idiocy*" was accompanied by catatonia "*the first indications of acquired catatonia had been overlooked*" (Kraepelin, 1899).

Karl Pönitz (1888–1973) worked at Leipzig's Psychiatric and Neurological Hospital headed by Paul Flechsig (1847–1929) before taking on his study at Halle University in 1913 (Pönitz, 1913). Based on his observations made at Leipzig, he published a study on "*early catatonia*" in 1913. This reported the case of a boy who had been suffering from "*typical catatonia*" from the age of 12, and the boy from his symptoms showed the "*clear picture of schizophrenia.*"

Although providing a comprehensive bibliographic account of the subject, Theodor Ziehen (1862–1950) made no mention of Kahlbaum's original study in his monograph of 1915–1917 on mental illnesses in children. However, the symptoms he described coincided with those stated by Kahlbaum—starting with catatonic tension ("catatonic stupor") and ending with motor perseveration, which however, he assigned to "*dementia hebephrenicas. praecox (hebephrenia)*" defining it as "*acquired defect psychosis occurring in most cases during adolescence.*" As for the term hebephrenia, Ziehen acknowledged Kahlbaum's coinage, adding that it was the latter's *"student Hecker who gave a first exact scientific description"* (Ziehen, 1915–1917).[1]

Similarly, Hermann Emminghaus (1845–1904), in his comprehensive review of psychoses in children titled *Die psychischen Störungen des Kindesalters* ("Mental Disorders in Children") 30 years earlier, made mention of Kahlbaum in the chapter on hebephrenia (Emminghaus, 1887). In International Classification of Diseases, 10th revision (ICD-10, 2000), the syndrome described here is classified under F20.1 hebephrenic schizophrenia; in Diagnostic and Statistical Manual for Mental Disorder, 4th ed. (DSM-IV) (APA, 1995) under 295.10 as disorganized type.

In 1906 and 1908, Sante de Sanctis' (1862–1935) *variety of dementia praecox, namely dementia praecocissima* and Theodor Heller's (1869–1938) syndrome-defining study *on dementia infantilis* of 1909 (Heller, 1909), gave rise to a renewed discussion of this topic, above all in respect of classification. Whether these syndromes represented an early stage of schizophrenia was a major issue in this debate. In his later study summarizing his observations on the course of these illnesses, Heller (1930) concluded that dementia praecocissima and dementia infantilis constituted *"two separate and different forms."* As a result, he classified the latter as "a process of regression not belonging to the group of schizophrenias" but for which he could not identify any possible cause. However, for Heller these two forms had no connections to catatonia in children. In ICD-10, Heller-syndrome is classified among *"other childhood disintegrative disorders"* comprising dementia infantilis, disintegrative psychosis, Heller-syndrome, and symbiotic psychosis. DSM-IV, however, does not include Heller-syndrome or similar syndromes under Childhood Disintegrative Disorder 299.10.

One of those who opposed Heller's hypotheses was de Sanctis. Following his first description of three children (aged 6, 7, and 10 years) (de Sanctis, 1906), the Italian child psychiatrist gave a report on a similar case—namely a 3-year-old girl showing *"symptoms typical of catatonia: cataleptic phenomena, negativism, expression-less face, and mild stupor."* This new case as well as the follow-up study of the three children described earlier maintained that *"The diagnosis 'catatonia' cannot be contested. The question, however, is: Are there catatonias that are independent of dementia praecox? I am inclined to think so. However, for most psychiatrists following Kraepelin's approach, each*

---

[1] In his study of 1871, Ewald Hecker (1843–1909) described seven patients (five males and two females) aged an average 21.4 years who suffered from the disease during the ages of 18 and 22. Over the course of 4 years, Hecker observed "14 cases of hebephrenia out of a total of some 500 patients" and found "melancholia, mania, confusion, and even imbecility" in a certain succession the most common conditions. As major symptoms, he described "sadness, dejection of the disposition" "contrasted by cheerful mood, an inclination to laugh, silly jokes, silly, and precocious twaddle." He found "departure from logical sentence structures when writing letters, nonrefined language, a fondness to use foreign words, hallucinations, and excitement." Hecker concluded that the illness ended in "hebephrenic imbecility ... with defects" and was "incurable."

catatonia is inseparably connected with dementia praecox." The apparent contradiction led him to the conclusion that the symptoms he described could best be classified as an "*infantile catatonic form of dementia praecox ... i.e., dementia praecocissima catatonica*" (de Sanctis, 1908).

In the following years, clinical psychiatric research began to question the independence of catatonia as described by Kahlbaum (catatonia *sui generis*). In his *Textbook of Clinical Psychiatry*, psychiatry's German protagonist Emil Kraepelin (1904) had embedded a systematic classification of mental disorders. He identified "*hebephrenic forms*" and "*catatonic forms*" as well as "*paranoid forms*" of dementia praecox, differentiating this group of disorders from the group of manic–depressive illnesses. In 1908, Eugen Bleuler (1858–1936) replaced Kraepelin's term dementia praecox by his newly coined term schizophrenia that has been used ever since that time. Bleuler chose this term since "*the tearing apart or splitting of mental functions is an outstanding and characteristic symptom of the whole group*" (Bleuler, 1908). Three years later Bleuler provided a comprehensive account of *Dementia Praecox or the Group of Schizophrenias* (Bleuler, 1911) in the world's first *Handbook of Psychiatry*, edited by Gustav Aschaffenburg (1866–1944).[2]

As early as 1898, Heinrich Schüle (1840–1916) had discussed, in his extensive clinical study on catatonia, the problem of the specificity and unspecificity of symptoms. He maintained: "*Now we no longer have catatonias as such but catatonically modified states of dementia, paranoia etc.*" In particular, Schüle also found out that "*many periodic manias ... and among those specifically the juvenile forms (the so-called hebephrenious type ... brought forth acute catatonic episodes*" (Schüle, 1898). Similar to Schüle, catatonia in children was not an issue of primary interest to Karl Bonhoeffer. Yet, he touched the subject in both his 1908 study of symptomatic psychoses and his monograph of the whole topic of 2 years later. In the latter, he pointed out the fact that above all, juvenile patients "*suffering from an amentia-like febrile psychosis produced catatonic symptoms.*" An initial severe motor excitement accompanied by "*catatonic symptoms of negativism, grimacing, verbigeration*" could lead "*either to a severe catatonic psychosis or a febrile psychosis.*" From the differential diagnostic point of view, however, Bonhoeffer came to the conclusion that "*as a matter of fact there is not a single catatonic symptom that could not also be found in infectious psychoses*" (Bonhoeffer, 1908), that is, in symptomatic and organic psychoses. It states that the "*great diversity in the basic illness is opposed by a great uniformity of mental conditions. So we can conclude that what we face here are typical reactions that seem to a large degree to be independent of the specific noxae ... which we are entitled to call exogenous mental reactions*" (Neumärker, 2001). Having analyzed the cases of hospital patients treated for dementia praecox, one of Bonhoeffer's employees at the Berlin

---

[2]Gustav Aschaffenburg was appointed professor of psychiatry in Cologne. Because he was Jewish, he was forced to resign from his office in 1934. Thanks to Adolf Meyer (1866–1950), he was able to emigrate in 1939 first to Switzerland and then to the United States of America.

hospital, Ernst Grünthal (1894–1972) came to the conclusion that *"all kinds of psychoses also occur in children"* (Grünthal, 1919).

In 1914, Kurt Schneider (1887–1967) at that time still assistant psychiatrist under professor Aschaffenburg at the Psychiatric Hospital of Cologne University, contributed to the discussion of catatonia with his account on the *Nature and Importance of Catatonic Symptoms*. First and foremost he voiced his criticism on two major aspects: (1) first, that since Kahlbaum's first description the *"complex of catatonic symptoms had been watered down,"* as *"nowadays it comprises almost every outer sign patients suffering from schizophrenia exhibit"*! (2) Second, that *"recent textbooks of pediatrics barely mention this."* Yet, Schneider in his study also suggested that *"a tendency to catatonic states can occur whenever there is insufficient development of associative powers and a relative poverty of imagination and that consequently catatonic states can occur when there is disruption of the development of imagination"* (Schneider, 1914).

In 1928, August Bostroem (1886–1944) published his thorough clinical/pathological diagnostic investigation of *catatonic disorders* in the influential *Handbuch der Geisteskrankheiten*, edited by Oswald Bumke (1877–1950). Bostroem had worked under Bumke, while the latter was professor of psychiatry at Leipzig University. When in 1924, Bumke was appointed Kraepelin's successor in Munich, he took his most reliable consultant with him and arranged for him being appointed associate professor of psychiatry and neurology in 1926. Hitherto Bostroem's study has remained unsurpassed in German-speaking psychiatric literature.

Bostroem's (1928) comprehensive account of catatonic symptoms has been summarized in Table I.

TABLE I
BOSTROEM'S (1928) COMPREHENSIVE ACCOUNT OF CATATONIC SYMPTOMS

- Pure akinesia
- Akinesia with blockings and muscle tensions
- Pose perseverance (catalepsy)
- Compulsive automatism (mimetic automatism and echo phenomena: Echolalia and echopraxia)
- Negativism (senseless resistance to external influences affecting movement, speech, and thought processes)
- Mannerisms (*"a deliberate and necessary movement . . . is executed in a strange, unnaturally twisted way, or as Kraepelin put it with an exaggerated flourish, and is repeated the same way whenever the same situation arises"*) and abnormal postures (*"grotesque and complicated movements and postures, for which no purpose or aim can be identified"*)
- Motor excitement (hyperkinesia: *"aimless, scratching innervations often contradicting one another, but justified in isolation . . .* [which do] *not lead to purposeful motor activities"*); can be limited to particular parts of the body
- Impulsive acts

Anglo-American publications too dealt with the topic of catatonia and dementia praecox as exemplified by Leland E. Hinsie's (b. 1895) study of 1932 (Hinsie, 1932a,b).

In the 1950s and 1960s, adult psychiatry saw a decline in the number of patients in the industrialized countries, suffering from catatonic schizophrenia. This prompted Mahendra (1981) to ask: "*Where have all the catatonics gone?*" On the other hand, Karl Kleist (1879–1960) and Karl Leonhard (1904–1988), as well as disciples of the Wernicke-Kleist-Leonhard school, continued to provide clear descriptions of such cases. (For Kleist and the literature, see Beckmann *et al.*, 2000; Leonhard, 1957; Neumärker and Bartsch, 2003; Ungvari, 1993.) It was, however, Karl Leonhard, who reanimated the scientific discussion of the subject by his study of catatonia in children (Leonhard, 1960). Three years earlier, in the first edition of his book on the *Classification of Endogenous Psychoses* (Leonhard, 1957; 1st English edition by Russell Berman in 1979), he had already given a review of the different forms of catatonic disorders in adults. In the 7th German edition (Leonhard, 1995), he dedicated a whole chapter to the *Treatment and Etiology of Infant Catatonia* based on the description of 117 cases of catatonia that had begun in the first years of life. The differentiation of these cases is shown in Table II.

In summary, he drew the following conclusions.

Simple systematic catatonias in children do not differ from those in adults. However, the symptoms appear to be more distinct in children. *Combined systematic catatonias* can also appear in early childhood.

Frequency: There are twice as many cases of negativistic and mannerism-type catatonia as there are of speech-prompt, speech-inactive, and proskinetic and parakinetic catatonias.

Prognosis: Rather poor as in adults.

Periodic catatonia (*unsystematic schizophrenia*): Appears after the age of 6 only and seems to be highly hereditary. A dominant heretic origin is suggested.

TABLE II
CASES OF INFANT CATATONIA DIFFERENTIATED INTO GROUPS

Simple systematic catatonias	Male	Female	Total
Speech-prompt/voluble catatonia	3	6	9
Speech-inactive/sluggish catatonia	5	6	11
Proskinetic catatonia	9	2	11
Negativistic catatonia	14	6	20
Parakinetic catatonia	7	3	10
Mannerism-type catatonia	10	10	20
SUBTOTAL	48	33	81
Combined systematic catatonias	18	18	36
TOTAL	66	51	117

Etiology: Lack of communicative actions (psychiatric hospitalization, institutional upbringing, and separation from psychological parent(s)), and neurodevelopmental factors. In systematic schizophrenia the hereditary factor is of little significance.

According to Leonhard, infant catatonia can be differentiated from autism and *"states of feeblemindedness"* (*Schwachsinnszustände*) since *"idiotic children"* react to being approached by showing emotions, smiling, or even affection, whereas catatonic children are never *"emotionally"* engaged; they always *"remain distant."* *"This schizophrenic distance so strange to those in normal health . . . is something different from autistic seclusion"* as described by Hans Asperger (1906–1980). Leonard is of the view that his *"portrayal of schizophrenic children"* corresponds *"to many of the autistic children studied by Kanner."* Leonhard's differentiated psychopathological diagnostics and classification stands, in strong contrast to the uniform classification of negative and positive symptoms of schizophrenia of American psychiatry, both for adult (Andreasen and Olsen, 1982) and infant patients—the so-called Kiddie-PANSS (Fields *et al.*, 1994). In Germany, Remschmidt *et al.* (1992) adopted a similar approach and differentiation that, however, did not gain high currency.

On the positive side it is to be noted that throughout the twentieth century, schizophrenia in children has become a topic of international debate—for example, Lutz (1937–1938) in Switzerland; Bender (1947), Fish (1979), and Kanner (1954) in the United States of America; Kothe (1957) and Wieck (1965) in Germany, or Spiel (1961) in Austria. The debate continues even today, when one thinks of the recognition of earning warning signals (prodomi) and early stages of infantile schizophrenia. In both child and adult psychiatry, the debate is motivated by attempts to improve both preventative and therapeutic interventions (Eggers and Bunk, 1999; Gordon *et al.*, 1994; Häfner, 1995; Schaeffer and Ross, 2002). As a result of these studies on infantile schizophrenia, the number of publications, both on catatonia in general (Caroff *et al.*, 2004; Fink, 1994; Fink and Taylor, 2003; Johnson, 1993; Rosebush and Mazurek, 1999; Taylor, 1990; Taylor and Fink, 2003) and catatonia in childhood, have also increased (Ainsworth, 1987; Annell, 1963; Cohen *et al.*, 1999; Dhossche and Bouman, 1997a,b; Neumärker, 1995a,b; Neumärker *et al.*, 1994; Shah and Kaplan, 1980). Catatonia, as first described by Kahlbaum, and catatonic disorders are again topical (Barnes *et al.*, 1986; Carroll, 2000; Stöber and Ungvari, 2001).

III. Autism

In July 1910, Eugen Bleuler, head of the Burghölzli Psychiatric University Hospital in Zurich, published his study on *The Theory of Schizophrenic Negativism* in

Psychiatrisch-Neurologische Wochenschrift. This was followed by three further articles on the same subject (Bleuler, 1910–1911). In these articles he defined negativism as a "*complicated symptom with several causes which in the individual case often coincided or interacted.*" Going deeper into the matter he identified "*ambitendency, ambivalence, a schizophrenic splitting of the psyche, unclear or insufficiently logical thinking*" as predisposing causes of negativism. As regards "*the negativistic reaction ... common negativism, rejection of all kinds of external influence,*" Bleuler maintained that it was "*the patient's autistic withdrawal into his own fantasies, against which any influence from outside will be regarded as an unbearable disturbance. This seems to be the most important point. In more severe cases this alone can cause negativism.*" It is here that Bleuler used the term *autistic* or *autism* for the first time. He continued by providing an explanation of what he meant: "*By autism I understand roughly the same as Freud does by autoerotism. However, I would rather this term to be avoided since it will certainly be misunderstood by anyone not familiar with Freud's work.*" An even more comprehensive discussion of the term *autism* is given by him in his book *Dementia Praecox or the Group of Schizophrenias* (Bleuler, 1911). However, the journal paper had appeared 1 year earlier! A year after the book, Bleuler published yet another study on *Autistic Thinking* in the *Jahrbuch für Psychoanalytische und Psychopathologische Forschungen*, edited by himself together with Sigmund Freud (1856–1939) and proofread by Carl Gustav Jung (1875–1961). In this contribution, he repeated that "*One of the most important symptoms of schizophrenia is the predominance of the inner life combined with an active withdrawal from the outer world. More severe cases withdraw themselves totally into their own shell and live a dream; in less severe cases we can find minor degrees of the same phenomenon. I have called this symptom autism.*" It is remarkable from the historical perspective that in discussing the symptoms of schizophrenia, it was Bleuler, who made judgment-free reference to psychoanalytic thinking.

In the same article of 1912, he acknowledged that "to a large degree my term autism coincides with Jung's term introversion by which he understood the libido's reference to the inner world, whereas it should normally seek its objectives in reality" (Bleuler, 1912). In all Bleuler's publications quoted here, he also considered autism in children. And, as stated earlier, for Bleuler autism was one of the most important symptoms of the whole group of different schizophrenias.

In 1943, Leo Kanner (1894–1981) published his account on infantile autism based on observations and psychopathological findings of 11 children (8 boys and 3 girls) aged 2–8 years. In both the title and text of his publication, Kanner only used the adjective form *autistic* and never the noun. Even more remarkably, he did not make any reference to Bleuler's studies of the 1910s. As for Vienna professor Hans Asperger's study on *Autistic Infantile Psychopaths* (1944), Kanner had no pre-knowledge since his own article came out 1 year earlier (Kanner, 1943). That Kanner in his follow-up study of 27 years later included neither Bleuler's nor Asperger's findings is, however, beyond understanding (Kanner, 1971; Neumärker, 2003; Ritvo, 1976).

Contrary to Kanner (1943), Asperger (1944), in his study based on observations of four male patients aged between 6 and 8.5 years, does refer extensively to Bleuler's description of autism. In particular, he discusses the "*autistic particularity*" of one of his patients regarding mathematical capabilities as well as "*autistic intelligence* . . . *autistics' instinctive and emotional life*" as well as the "*social value of autistic psychopaths*." Like Bleuler, he suggested the similarities with the "*introverted way of thinking*" as defined by Jung, since "*introversion means nothing but restriction to the individual self (autism) and a limitation of the bonds with the environment.*" However, neither Asperger believed that a systematic typology could be applied to his "*autistic psychopaths,*" nor did he produce one. For him, only a long-term study of the course of this illness from childhood to adulthood could produce any such results (Asperger, 1944).

To date, there has been great controversy as to whether the autistic disorder (Kanner syndrome) or the Asperger disorder are valid diagnostic categories, disorders, or entities (Macintosh and Dissanayake, 2004). Both ICD-10 and DSM-IV list them, together with Rett's disorder as well-established categories within the group of pervasive developmental disorders. The similarities and coincidences between Asperger syndrome and autism have prompted the Gillberg team (Gillberg and Billstedt, 2000) to search for coexistences and comorbidities with other clinical disorders (Volkmar and Cohen, 1991). The spectrum of autistic disorders grows ever wider. Their research forms part of the attempts to explain mental disorders and illnesses with the help of continuity–discontinuity models (Gleaves *et al.*, 2004; Spencer *et al.*, 2000). Incidentally, it was Gillberg, who proved that autistic symptoms can be diagnosed in children even younger than 3 years old (Gillberg *et al.*, 1990; Stone *et al.*, 1999), and autism is not an "*extremely rare disorder*" (Gillberg and Wing, 1999), as has also been acknowledged in Fombonne's (1999) review of the epidemiology of autism. A historical review of both the dimensions and conceptualizations of the disorder, and how they have changed throughout history has been provided by Wolff (2004), who also took into consideration the sociopolitical consequences these changes produced.

IV. Autism and Catatonia

Research has been undertaken to evaluate the interrelationships of catatonia, or catatonic symptoms and autism (Dhossche, 1998; Realmuto and August, 1991; Wing and Shah, 2000). Dhossche (2004) characterized "*autism as an early expression of catatonia.*" As a result of his studies on both the relevant literature as well as case vignettes, the author concludes that *"the effects of anticatatonic treatments on autistic symptoms are unknown, abnormal gamma-aminobutyric acid (GABA) function has been implicated in both disorders, neuroimaging studies show small cerebellar structures in both disorders. There is genetic evidence that susceptibility genes for autism and catatonia are located on the long arm of chromosome 15. Differences between autism and catatonic of age-of-onset, symptoms, and illness course, do not exclude a common genetic etiology"* (Dhossche, 2004). Stöber *et al.* (2002) have undertaken systematic research on periodic catatonia and linked this with chromosome 15. Its clinical classification is based on theories established by Karl Leonhard, as a result of his research on families suffering from periodic catatonia. The authors' team (Neumärker *et al.*, 1995b) succeeded in providing molecular—genetic evidence on those families. Concentrating particularly on the children and adolescents suffering from systematic catatonias, periodic catatonia, and motility psychoses, as described by Leonhard and the authors' team, concluded that these three disorders could be put together in one group, for which the authors suggested the name, as *psychomotor psychoses* (Neumärker, 1990). Leonhard always emphasized that children suffering from catatonias often also showed autistic behavior. However, this schizophrenic alienation is different from autistic seclusion as described by Asperger: *"For healthy children, kids suffering from schizophrenia don't seem to have a soul at all, whereas autistic children only seem to have a different soul. Consequently, one should not mix autism, which children can empathize with, with schizophrenic unapproachability"* (Leonhard, 1986). Besides autism, *"autistic hebephrenia,"* as described by Leonhard, is characterized by affective hebetude, loss of drive, impulsive aggressive outbursts, and thinking disorders.

If one follows the phenomenological descriptions of Bleuler and Leonhard, autism is a part of *"schizophrenic negativism"* or in particular of *"autistic hebephrenia."* The latter, however, is different qualitatively from autistic withdrawal as described by Kanner and Asperger from autistic behavior and autistic spectrum. On the other hand, *"catatonic disorders"* (Bostroem, 1928) can also be found in several other diseases, not least in symptomatic psychoses (Bonhoeffer, 1910). What Kahlbaum (1874) described as *"catatonia or tension insanity,"* Kraepelin (1899) described as *"Catatonic Forms"* of dementia praecox besides *"Hebephrenic Forms"* and *"Paranoid Forms."* Accordingly, current classifications—ICD-10 and DSM-IV feature catatonia as an independent disorder.

Throughout history, the attempts of neuropsychiatrists to verify and substantiate nosological entities by pathogenetical or etiological evidence through morphological, neurotransmission, genetic research, or modern imaging methods have not been successful. On the other hand, different mental disorders produce the same pathological changes (e.g., alterations in dendritic spine density in schizophrenia and depression) (Glantz and Lewis, 2000; Rosoklija *et al.*, 2000), or abnormalities in the dendrites or hypoperfusion in the prefrontal cortex in schizophrenia and anorexia (Neumärker *et al.*, 1997; Selemon *et al.*, 1995; Takano *et al.*, 2001). All these changes are accompanied by reduced synaptic connectivity, and they result in network malfunctions. In this regard, the proliferation and decline in synaptic connections in children (density of D_2 receptor) as described by Seeman *et al.* (1987), as well as the process of axon pruning throughout the first 10 years and the "*minicolumn hypothesis*" as proposed by Buxhoeveden and Casanova (2002), and its practical application to autism (Casanova and Buxhoeveden, 2002) should also be taken into consideration. These and other hypotheses have been linked with the debate regarding the correlations between structure and function. It remains the role of the neurosciences to find answers to these questions.

Acknowledgments

The author wishes to thank Ms Irene Schramm (Berlin) for her most valuable help in procuring the relevant literature and typing this manuscript. Sincere thanks also to Dirk Carius (Leipzig) for the adequate translation into English.

Appendix A

Images in Psychiatry

The Burghölzli Clinic (1996). *Am. J. Psychiatry* **153,** 816.
Karl Ludwig Kahlbaum, M. D. 1828–1899 (1999). *Am. J. Psychiatry* **156,** 989.
Ewald Hecker, 1843–1909 (2000). *Am. J. Psychiatry* **157,** 1220.
Theodor Ziehen, M. D., Ph. D., 1862–1950 (2004). *Am. J. Psychiatry* **161,** 1369.
Kurt Schneider, 1887–1967 (1994). *Am. J. Psychiatry* **151,** 1492.
Karl Kleist, 1879–1960 (2000). *Am. J. Psychiatry* **157,** 703.
Karl Leonard, 1904–1988 (1998). *Am. J. Psychiatry* **155,** 1309.

References

Ainsworth, P. (1987). A case of lethal catatonia in a 14-year-old girl. *Br. J. Psychiatry* **150,** 110–112.
Andreasen, N., and Olsen, S. (1982). Negative v. positive Schizophrenia: Definition and validation. *Arch. Gen. Psychiatry* **39,** 789–794.

Annell, A.-L. (1963). Periodic catatonia in a boy of 7 years. *Acta Paedopsychiatrica* **30,** 48–58.
American Psychiatric Association (APA) (1995). "Diagnostic and Statistical Manual of Mental Disorders," 4th ed. (DSM-IV). APA Press, Washington, DC.
Asperger, H. (1944). Die "Autistischen Psychopathen" im Kindesalter. *Archiv für Psychiatrie* **117,** 76–136.
Barnes, M. P., Saunders, M., Walls, T. J., Saunders, I., and Kirk, C. A. (1986). The syndrome of Karl Ludwig Kahlbaum. *J. Neurol. Neurosurg. Psychiatry* **49,** 991–996.
Beckmann, H., Bartsch, A. J., Neumärker, K.-J., Pfuhlmann, B., Verdaguer, M. F., and Franzek, E. (2000). Schizophrenias in the Wernicke-Kleist-Leonhard school. *Am. J. Psychiatry* **157,** 1024–1025.
Bender, L. (1947). Childhood schizophrenia. Clinical study of 100 schizophrenic children. *Am. J. Orthopsychiatry* **17,** 40–56.
Bleuler, E. (1908). Die Prognose der Dementia praecox (Schizophreniegruppe). *Allgemeine Zeitschrift für Psychiatrie und Psychisch-Gerichtliche Medizin* **65,** 436–464.
Bleuler, E. (1910–1911). Zur Theorie des schizophrenen Negativismus. *Psychiatrisch-Neurologische Wochenschrift* **9,** 171–176, 184–187, 189–191, 195–198.
Bleuler, E. (1911). Dementia praecox oder Gruppe der Schizophrenien. *In* "Handbuch der Psychiatrie" (G. Aschaffenburg, Ed.), Spezieller Teil, 4. Abteilung, 1. Hälfte. Franz Deuticke, Leipzig und Wien; English translation Bleuler, E. (1950). "Dementia Praecox or the Group of Schizophrenias." Universities Press, New York.
Bleuler, E. (1912). Das autistische Denken. *In* "Jahrbuch für Psychoanalytische und Psychopathologische Forschungen" (E. Bleuler and S. Freud, Eds.), IV. Band, 1. Hälfte, pp. 1–39. Leipzig und Wien.
Bonhoeffer, K. (1908). Zur Frage der Klassifikation der symptomatischen Psychosen. *Berliner klinische Wochenschrift* **45,** 2257–2260.
Bonhoeffer, K. (1910). "Die symptomatischen Psychosen im Gefolge von akuten Infektionen und inneren Erkrankungen." Franz Deuticke, Leipzig und Wien.
Bostroem, A. (1928). Katatone Störungen. *In* "Handbuch der Geisteskrankheiten" (O. Bumke, Ed.), Vol. II, pp. 134–205. Springer, Berlin.
Buxhoeveden, D. P., and Casanova, M. F. (2002). The minicolumn hypothesis in neuroscience. *Brain* **125,** 935–951.
Caroff, S. N., Mann, S. C., Francis, A., and Fricchione, G. L. (Eds.) (2004). "Catatonia: From Psychopathology to Neurobiology." American Psychiatric Publishing Inc., Washington, DC.
Carroll, B. T. (2000). Mechanisms and Presentations of Catatonia. *CNS Spectrums. Int. J. Neuropsychiatr. Med.* **5,** 25.
Casanova, M. F., and Buxhoeveden, D. P. (2002). Minicolumnar pathology in autism. *Neurology* **58,** 428–432.
Cohen, D., Flament, M., Dubos, P. F., and Basquin, M. (1999). Case series: Catatonic syndrome in young people. *J. Am. Acad. Child Adolesc. Psychiatry* **38,** 1040–1046.
De Sanctis, S. (1906). Dementia praecocissima. *Revista Sperimentale di Freniatria* **32,** Fasc. I, II.
De Sanctis, S. (1908–1909). Dementia praecocissima catatonica oder Katatonie des frühen Kindesalters? *Folia Neurobiologica* **2,** 9–12.
Dhtossche, D. M., and Bouman, N. H. (1997). Catatonia in an adolescent with Prader–Willi syndrome. *Ann. Clin. Psychiatry* **9,** 247–253.
Dhossche, D., and Bouman, N. (1997). Catatonia in children and adolescents. *J. Am. Acad. Child Adolesc. Psychiatry* **36,** 870–871.
Dhossche, D. (1998). Brief report: Catatonia in autistic disorders. *J. Autism Dev. Disord.* **28,** 329–331.
Dhossche, D. M. (2004). Autism as early expression of catatonia. *Med. Sci. Monit.* **10,** 31–39.
Eggers, C., and Bunk, D. (Eds.) (1999). Early onset schizophrenia: Phenomenology, course, and outcome. *Eur. Child Adolesc. Psychiatry* **8**(Suppl. 1), 1–35.

Emminghaus, H. (1887). "Die psychischen Störungen des Kindesalters." Verlag der H. Laupp'schen Buchhandlung, Tübingen.
Fields, J. H., Grochowski, S., Lindenmayer, J. P., Kay, S. R., Grosz, D., Hyman, R. B., and Alexander, G. (1994). Assessing positive and negative symptoms in children and adolescents. *Am. J. Psychiatry* **151,** 249–253.
Fink, M. (1994). Catatonia in DSM-IV. *Biol. Psychiatry* **36,** 431–433.
Fink, M., and Taylor, M. A. (2003). "Catatonia: A Clinician's Guide to Diagnosis and Treatment." Cambridge University Press, Cambridge, UK.
Fish, B. (1979). The recognition of infantile psychosis. *In* "Modern Perspectives in the Psychiatry of Infancy" (J. G. Howells, Ed.), pp. 450–474. Brunner/Mazel, New York.
Fombonne, E. (1999). The epidemiology of autism: A review. *Psychol. Med.* **29,** 769–786.
Gillberg, C., Ehlers, S., Schaumann, H., Jakobsson, G., Dahlgren, S. O., Lindblom, R., Bagenholm, A., Tjuus, T., and Blidner, E. (1990). Autism under age 3 years: A clinical study of 28 cases referred for autistic symptoms in infancy. *J. Child Psychol. Psychiatry* **31,** 921–934.
Gillberg, C., and Wing, L. (1999). Autism: Not an extremely rare disorder. *Acta Psychiatr. Scand.* **99,** 399–406.
Gillberg, C., and Billstedt, E. (2000). Autism and asperger syndrome: Coexistence with other clinical disorders. *Acta Psychiatr. Scand.* **102,** 321–330.
Glantz, L. A., and Lewis, D. A. (2000). Decreased dendritic spine density on prefrontal cortical pyramidal neurons in schizophrenia. *Arch. Gen. Psychiatry* **57,** 65–73.
Gleaves, D. H., Brown, J. D., and Warren, C. S. (2004). The continuity/discontinuity models of eating disorders. *Behav. Modif.* **26,** 739–762.
Gordon, C. T., Frazier, J. A., McKenna, K., Giedd, J., Zametkin, A., Zahn, T., Hommer, D., Hong, W., Kaysen, D., Albus, K. E., and Rapoport, J. L. (1994). Childhood-onset schizophrenia: A NIMH study in progress. *Schizophr. Bull.* **20,** 697–712.
Grünthal, M. (1919). Über Schizophrenie im Kindesalter. *Monatsschrift für Psychiatrie und Neurologie* **46,** 206–240.
Häfner, H. Special Issue (1995). Schizophrenia in childhood and adolescence. *Eur. Arch. Psychiatry Clin. Neurosci. Special Issue* **245,** 57–100.
Hecker, E. (1871). Die Hebephrenie. Ein Beitrag zur klinischen Psychiatrie. *Archiv für pathologische Anatomie und Physiologie und für klinische Medicin* **52,** 394–429.
Heller, T. (1909). Über Dementia infantilis (Verblödungsprozeß im Kindesalter). *Zeitschrift für die Erforschung und Behandlung des jugendlichen Schwachsinns auf wissenschaftlicher Grundlage* **2,** 17–28.
Heller, T. (1930). Über Dementia infantilis. *Zeitschrift für die Kinderforschung* **37,** 661–667.
Hinsie, L. E. (1932). The catatonic syndrome in dementia praecox. *Psychiatr. Q.* **6,** 457–468.
Hinsie, L. E. (1932). Clinical manifestations of the catatonic form of dementia praecox. *Psychiatr. Q.* **6,** 469–474.
ICD-10 (2000). "Rantionale Klassifikation psychischer Störungen." Kapitel V (F) Klinisch-diagnostische Leitlinien (H. Dilling, W. Mombour, and M. H. Schmidt, Hrsg.), 4th ed. Huber, Bern.
Johnson, J. (1993). Catatonia: The tension insanity. *Br. J. Psychiatry* **162,** 733–738.
Kahlbaum, K. L. (1874). "Katatonie oder das Spannungsirresein. Eine klinische Form psychischer Krankheit." Hirschwald, Berlin. English Translation: Kahlbaum, K. L. (1973). "Catatonia" (Y. Levi and T. Pridou, Trans.). Johns Hopkins University Press, Baltimore.
Kanner, L. (1943). Autistic disturbances of affective contact. *Nerv. Child* **2,** 217–250.
Kanner, L. (1954). Childhood schizophrenia. *Am. J. Orthopsychiatry* **24,** 526–528.
Kanner, L. (1971). Follow-up study of eleven autistic children originally reported in 1943. *J. Autism Childhood Schizophr.* **1,** 119–145.
Kothe, B. (1957). "Über kindliche Schizophrenie." Carl Marhold Verlag, Halle (Saale).

Kraepelin, E. (1899). "Psychiatrie. Ein Lehrbuch für Studierende und Ärzte," 6th completely revised ed., Vol. II. Klinische Psychiatrie, Barth, Leipzig.

Leonhard, K. (1957). "Aufteilung endogenern psychosen." Akademie-Verlag, Berlin Thieme, Stuttgart, New York, 7th ed. (1995). Aufteilung der endogenern Psychosen and ihre differenzierte Ätiologie; 8th ed. (2003) Thieme, Stuttgart, New York. English translations: 5th ed. by R. Berman: "Classification of Endogenous Psychoses." Irvington, New York (1979); enlarged re-edition "Classification of Endogenous Psychoses and Their Differentiated Etiology" (H. Beckmann, Ed.), Springer, Wien (1999).

Leonhard, K. (1960). Über kindliche Katatonien. *Psychiatrie, Neurologie, medizinische Psychologie* **12,** 1–17.

Leonhard, K. (1986). Aufteilung der endogenen psychosen und ihre differenzierte Ätiologie 6th ed. Academic-Verlag, Berlin.

Lutz, J. (1937–1938). Über die Schizophrenie im Kindesalter. *Schweizer Archiv für Neurologie und Psychiatrie* **39,** 335–372; **40,** 141–163.

Macintosh, K. E., and Dissanayake, C. (2004). Annotation: The similarities and differences between autistic disorder and Asperger's disorder: A review of the empirical evidence. *J. Child Psychol. Psychiatry* **45,** 421–434.

Mahendra, B. (1981). Where have all the catatonics gone? *Psychol. Med.* **11,** 669–671.

Mechanisms and Presentations of Catatonia (2000). *CNS Spectrums* **5,** 10–11, 25–72.

Neumärker, K.-J. (1990). Leonhard and the classification of psychomotor psychoses in childhood and adolescence. *Psychopathology* **23,** 243–252.

Neumärker, K.-J., Dudeck, U., and Neumärker, U. (1994). Katatonien im Kindesalter. *Psycho.* **20,** 212–218.

Neumärker, K.-J. (1995a). Diagnostics, therapy and course of catatonic schizophrenias in childhood and adolescence. *In* "Endogenous Psychoses: Leonhard's Impact on Modern Psychiatry" (H. Beckmann and K.-J. Neumärker, Eds.), pp. 160–176. Ullstein Mosby, Berlin.

Neumärker, K.-J. (1995b). Periodic catatonia families diagnosed according to different operational criteria for schizophrenia: Preliminary results of linkage analysis. *In* "Endogenous Psychoses. Leonhard's Impact on Modern Psychiatry" (H. Beckmann and K.-J. Neumärker, Eds.), pp. 176–181. Ullstein Mosby, Berlin.

Neumärker, K.-J., Dudeck, U., Meyer, U., Neumärker, U., Schulz, E., and Schönheit, B. (1997). Anorexia nervosa and sudden death in childhood: Clinical data and results obtained from quantitative neurohistological investigations of cortical neurons. *Eur. Arch. Psychiatry Clin. Neurosci.* **247,** 16–22.

Neumärker, K.-J. (2001). Karl Bonhoeffer and the concept of symptomatic psychoses. *Hist. Psychiatry* **12,** 213–226.

Neumärker, K.-J. (2003). Leo Kanner: His years in Berlin, 1906–1924. The roots of autistic disorder. *Hist. Psychiatry* **14,** 205–218.

Neumärker, K.-J., and Bartsch, A. J. (2003). Karl Kleist (1879–1960): A pioneer of neuropsychiatry. *Hist. Psychiatry* **14,** 411–458.

Plaskuda, W. (1911). Über Stereotypien und sonstige katatonische Erscheinungen bei Idioten. *Zeitschrift für die gesamte Neurologie und Psychiatrie* **4,** 399–416.

Pönitz, K. (1913). Beitrag zur Kenntnis der Frühkatatonie. *Zeitschrift für die gesamte Neurologie und Psychiatrie* **20,** 343–357.

Raecke, J. (1909). Katatonie im Kindesalter. *Archiv für Psychiatrie und Nervenkrankheiten* **45,** 245–279.

Realmuto, G. M., and August, G. J. (1991). Catatonia in autistic disorder: A sign of comorbidity or variable expression. *J. Autism Dev. Disord.* **21,** 517–528.

Remschmidt, H., Martin, M., Schulz, E., Gutenbrunner, C., and Fleischhaker, C. (1992). The concept of "positive and negative schizophrenia." *In* "Child and Adolescent Psychiatry. Negative versus Positive Schizophrenia" (A. Marneros, N. C. Andreasen, and M. T. Tsuang, Eds.), pp. 219–242. Springer, Berlin.

Ritvo, E. R. (1976). Autism: From adjective to noun. In "Autism: Diagnosis, Current Research and Management" (E. R. Ritvo, Ed.), pp. 3–6. Spectrum Publications Inc., New York.

Rosebush, P. I., and Mazurek, M. F. (1999). Catatonia: Reawakening to a forgotten disorder. *Mov. Disord.* **14**, 395–397.

Rosoklija, G., Toomayan, G., Ellis, S. P., Keilp, J., Mann, J. J., Latov, N., Hays, A. P., and Dwork, A. J. (2000). Structural abnormalities of subicular dendrites in subjects with schizophrenia and mood disorders. *Arch. Gen. Psychiatry* **57**, 349–356.

Schaeffer, J. L., and Ross, R. G. (2002). Childhood-onset schizophrenia: Premorbid and prodromal diagnostic and treatment histories. *J. Am. Acad. Child Adolesc. Psychiatry* **41**, 538–545.

Schneider, K. (1914). Über Wesen und Bedeutung katatonischer Symptome. *Zeitschrift für die gesamte Neurologie und Psychiatrie* **22**, 486–505.

Schüle, H. (1898). Zur Katatonie-Frage. Eine klinische Studie. *Allgemeine Zeitschrift für Psychiatrie* **54**, 515–552.

Seeman, P., Bzowej, N. H., Guan, H. C., Bergeron, C., Becker, L. E., Reynolds, G. P., Bird, E. D., Riederer, P., Jellinger, K., Watanabe, S., and Tourtellotte, W. W. (1987). Human brain dopamine receptors in children and aging adults. *Synapse* **1**, 399–404.

Selemon, L. D., Rajkowska, G., and Goldman-Rakic, P. S. (1995). Abnormally high neuronal density in the schizophrenic cortex: A morphometric analysis of prefrontal area 9 and occipital area 17. *Arch. Gen. Psychiatry* **52**, 805–818.

Shah, P., and Kaplan, S. (1980). Catatonic symptoms in a child with epilepsy. *Am. J. Psychiatry* **137**, 378–379.

Spencer, T., Biedermann, J., Wozniak, J., and Wilens, T. (2000). Attention deficit hyperactivity disorder and affective disorders in childhood: Continuum, comorbidity or confusion. *Curr. Opin. Psychiatry* **13**, 73–79.

Spiel, W. (1961). "Die endogenen Psychosen des Kindes – und Jugendalters." S. Karger, Basel.

Stöber, G., and Ungvari, G. S. (Eds.), (2001). Catatonia: A new focus of research. *Eur. Arch. Psychiatry Clin. Neurosci.* **251** (Suppl. 1), 1–34.

Stöber, G., Seelow, D., and Ruschendorf, F. (2002). Periodic catatonia: Confirmation of linkage to chromosome 15 and further evidence for genetic heterogenity. *Hum. Genet.* **111**, 323–330.

Stone, W. L., Lee, E. B., Ashford, L., Brissie, J., Hepburn, S. L., Coonrod, E. E., and Weiss, B. H. (1999). Can autism be diagnosed accurately in children under 3 years? *J. Child Psychol. Psychiatry* **40**, 219–226.

Takano, A., Shiga, T., Kitagawa, N., Koyama, T., Katoh, C., Tsukamoto, E., and Tamaki, N. (2001). Abnormal neuronal network in anorexia nervosa studied with I-123–IMP SPECT. *Psychiatry Res.: Neuroimaging Section* **107**, 45–50.

Taylor, M. A. (1990). Catatonia. A review of a behavioral neurologic syndrome. *Neuropsychiatry, Neuropsychol. Behav. Neurol.* **3**, 48–72.

Taylor, M. A., and Fink, M. (2003). Catatonia in psychiatric classification: A home of its own. *Am. J. Psychiatry* **160**, 1233–1241.

Ungvari, G. S. (1993). The Wernicke-Kleist-Leonhard school of psychiatry. *Biol. Psychiatry* **34**, 749–752.

Vogt, H. (1909). Über Fälle von "Jugendirresein" im Kindesalter (Frühformen des Jugendirreseins). *Allgemeine Zeitschrift für Psychiatrie* **66**, 542–573.

Volkmar, F. R., and Cohen, D. J. (1991). Comorbid association of autism and schizophrenia. *Am. J. Psychiatry* **148**, 1705–1707.

Weygandt, W. (1907). Idiotie und Dementia praecox. *Zeitschrift für die Erforschung und Behandlung des jugendlichen Schwachsinns* **1**, 311–332.

Wieck, C. (1965). "Schizophrenie im Kindesalter." S. Hirzel, Leipzig.
Wing, L., and Shah, A. (2000). Catatonia in autistic spectrum disorders. *Br. J. Psychiatry* **176,** 357–362.
Wolff, S. (2004). The history of autism. *Eur. Child Adolesc. Psychiatry* **13,** 201–208.
Ziehen, T. (1915–1917). "Die Geisteskrankheiten des Kindesalters einschließlich des Schwachsinns und der psychopathischen Konstitutionen." (Pt. 1. 1915). Reuther & Reichard, Berlin.

A SYSTEMATIC EXAMINATION OF CATATONIA-LIKE CLINICAL PICTURES IN AUTISM SPECTRUM DISORDERS

Lorna Wing* and Amitta Shah[†]

*Centre for Social and Communication Disorders
Bromley, Kent BR2 9HT, United Kingdom
[†]Leading Edge Psychology, Purley CR8 2EA
United Kingdom

I. Introduction
II. Clinical Features of ASDs and Catatonia
   A. Autism Spectrum Disorders
   B. Catatonia
III. Manifestations of Catatonic Features in ASDs
IV. Three Studies of Catatonic Features in ASD
   A. Study A: Frequency of Catatonia-Like Features
   B. Study B: Comparative Frequency of Catatonia-Like Features
   C. Study C: Catatonia-Like Deterioration
V. Discussion
   A. Effect of Level of Ability and Age
   B. Relationship of Catatonia-Like Features to Catatonia-Like Deterioration
   C. Catatonia-Like Features in Learning Disability
   D. Prevalence of Catatonia-Like Deterioration
VI. Conclusions
   References

Three studies concerning catatonia-like clinical pictures in people with autism spectrum disorders (ASDs) referred to clinics are described. The first investigated the frequencies, in children and adults with autistic disorders, of 28 specific disorders of movement, speech, and behavior similar to those occurring in chronic catatonia spectrum conditions. The second compared the frequency of these items among groups of children with, ASDs, learning disabilities, specific language impairment, and a group with typical development, respectively. The third study examined the pattern of catatonia-like deterioration occurring in a minority of adolescents and adults with ASDs. The studies demonstrated the high frequency of catatonia-like features in people with autistic disorders. There was some tendency for improvement with increasing age, especially for those with IQ 70 or over. The items were also found in children with learning disabilities and specific language disorders but significantly less often. They occurred least often in the children with typical development. Severe catatonia-like deterioration occurred in 17% of those with autistic disorders, who were aged 15 years or over when assessed at a diagnostic center. A history of passivity in social interaction and impairment of expressive language were associated with the deterioration. No clear relationship was found between a history of catatonia-like features, singly or combined, and catatonia-like deterioration. The findings pose questions for future research.

I. Introduction

In the early 1940s, Leo Kanner was the first to adopt the term "autism" for a type of childhood developmental disorder (Kanner, 1943). Before Kanner's first paper, children with atypical development and unusual behavior had been described by other authors, but the differences in terminology make comparisons difficult. However, in the historical literature there are a few accounts of children and adolescents with behavior that is reminiscent of autism spectrum disorders (ASDs) combined with marked catatonic features. Henry Maudsley (1867) discussed what he referred to as "insanity" in children. One of the subgroups he suggested was "cataleptoid insanity," which probably included some children with autism and catatonia. De Sanctis (1908) and Earl (1934) also described clinical pictures that might well have included autism and catatonia-like conditions.

After Kanner's original paper, interest in autism and related conditions has grown exponentially, but few studies have considered the similarities between autism and catatonia spectrum disorders. The committee chaired by Mildred Creak (1964), set up to define the diagnostic criteria for "childhood schizophrenia," listed nine points. Point 8 was *Distortion in mobility patterns* and included "immobility as in katatonia." All the committee's criteria would now be recognized as relevant for autism. Wing and Attwood (1987) briefly described catatonia in autistic disorders. Other papers in the English language on this subject have since been published by Brasic *et al.* (2000), Dhossche (1998), Dhossche (2004), Gaziuddhin *et al.* (2005), Hare and Malone (2004), Leary and Hill (1996), Realmuto and August (1991), Wing and Shah (2000), and Zaw *et al.* (1999). Each of this small group of authors has pointed out the marked overlap between the clinical pictures of, on the one hand, acute and chronic catatonia and, on the other, the unusual patterns of movement, speech, and behavior found in ASDs.

This chapter explores the various manifestations of features of catatonia found in autistic disorders.

II. Clinical Features of ASDs and Catatonia

A. AUTISM SPECTRUM DISORDERS

In this chapter, the terms "ASDs" and "autistic disorders" will be used to cover the developmental conditions that have in common a triad of impairments affecting social interaction, social communication, and social imagination. This triad is associated with a narrow, repetitive pattern of activities and/or interests (Wing and Gould, 1979). The spectrum is similar to but wider than the conditions referred to as "pervasive developmental disorders" in Diagnostic and Statistical Manual, 4th ed. (DSM-IV) (APA, 1994), and International Classification of Diseases, 10th revision (ICD-10) (WHO, 1993). Virtually all the published sets of criteria for ASDs include the above impairments (although the impairment of imagination is often placed under social interaction or social communication) together with the repetitive pattern of activities. However, the clinical picture also often includes unusual reactions to sensory input, a variety of motor problems, and disordered patterns of behavior. Although commonly present, these features are not invariable and they do occur in other developmental disorders. Nevertheless, they are an important aspect of autistic disorders for both research into the nature of these conditions and their clinical implications.

B. CATATONIA

In its most acute, severe form, catatonia is characterized by absence of speech (mutism), absence of movement (akinesia), maintenance of postures (catalepsy), and waxy flexibility, which may alternate with episodes of excitement and bizarre behavior. In addition, a wide range of other chronic, less severe, and sometimes very subtle disorders of posture, movement, speech, and behavior are considered by various authors to be catatonic in nature (Bush *et al.*, 1996; Fink and Taylor, 2003; Joseph, 1992; Rogers, 1992). Fink and Taylor (2003) used the term "catatonia spectrum" to cover the whole range of manifestations. While some features appear in all the published lists, there is much variation in those that are included. The features involved are disparate and it is not clear precisely if and how they are related to each other.

III. Manifestations of Catatonic Features in ASDs

The motor problems and disorders of speech and behavior found in ASDs include features like those described in the catatonia spectrum (Leary and Hill,

1996; Wing and Shah, 2000) together with some that are reminiscent of postencephalitic parkinsonism (Damasio and Maurer, 1978). There are two ways in which these features can be manifested. First, many children and adults in the autism spectrum have one or more such features in early childhood. There is some tendency for these to become less marked with increasing age, especially, in more able adolescents and adults. Another factor is the increasing awareness of social pressures in some high functioning people in adolescence or adult life, so that they are careful not to show odd movements and behaviors in public, though they still engage in them when on their own. These features will be referred to as "catatonia-like features" to avoid a premature conclusion that they are related to typical catatonia.

Second, in a small minority, some catatonia-like and some parkinsonism-like features become very marked or appear for the first time and are severe enough to interfere with the activities of everyday life. This usually happens in adolescence or adult life, though can rarely be seen in childhood. In most cases the onset is gradual and the presence of all the features of classic stupor appears to be rare (Wing and Shah, 2000). This is referred to here as "catatonia-like deterioration," again to avoid the premature conclusion that it is the same as catatonia. A few individual case studies of adolescents and young adults with what appears to be typical catatonic stupor have also been published (Dhossche, 1998, 2004; Ghaziuddin et al., 2005; Hare and Malone, 2004; Realmuto and August, 1991; Zaw et al., 1999).

The data presented here are based on samples of children and adults referred to clinics. Data from three different studies, which illustrate aspects of the manifestations of catatonia-like clinical pictures, are presented. The first examines the occurrence of catatonia-like features in 200 children and adults with ASDs. The second compares the frequency of these features in children having autistic disorders with a group with learning disorders, another group with specific language disorders, and a group of children with typical development. The third study describes catatonia-like deterioration in adolescents and adults and compares two groups of people with autistic disorders, one with this type of deterioration and one without. The aim was to identify clinical features in a subgroup of people with ASDs that might predict the later onset of catatonia-like deterioration. The existence of such a vulnerable subgroup has been suggested by Dhossche (2004) and Ghaziuddin et al. (2005).

IV. Three Studies of Catatonic Features in ASD

In each of these studies, permission to use findings anonymously in research was obtained at the time of assessment from parents or caregivers, and the individuals with ASDs if they were able to understand.

A. Study A: Frequency of Catatonia-Like Features

1. Participants

The participants comprised 200 children and adults with ASDs, aged 32 months–38 years (mean 12 years 7 months, SD 8 years 1 month), of which 167 were males and 33 were females. They were seen consecutively at the National Autistic Society's Centre for Social and Communication Disorders, a specialist tertiary referral center for diagnosis and assessment, during the period 1994–1997.

2. Method

Information was obtained by interviewing parents or other caregivers using the 10th edition of the Diagnostic Interview for Social and Communication Disorders (DISCO). This is a schedule for the diagnosis of autism spectrum and related disorders, and assessment of individual needs. It enables information concerning history from birth and present state to be recorded systematically for a wide range of behaviors and developmental skills and is suitable for use with all ages and levels of ability (Leekam *et al.*, 2002; Wing *et al.*, 2002). For this study, all the interviews were conducted by Elliot House staff.

For 153 participants, formal psychological assessments appropriate for their level of ability were carried out. Forty-seven participants did not have IQ tests because they were too low in ability or, in a very few cases, were too uncooperative to be tested. For these individuals, the psychologist observed their behavior in structured and unstructured situations. This information, together with the developmental information from the DISCO, was used to make an estimate of the level of ability. Among the 200 participants, 18% had IQ under 50; 17%, 50–69; 29%, 70–89; 31.5%, 90–119; 4.5%, 120 or above. Although 72 individuals had IQs in the average or superior range, none was very high functioning or living independently of caregivers when assessed.

The study was designed to examine variations in clinical pictures and their relationship to conventional diagnostic criteria (Leekam *et al.*, 2000). It was not primarily intended to investigate catatonia-like behavior. However, the DISCO included 28 items similar to those found in lists of catatonia-like features (though not all such features were included in the DISCO). In the DISCO interview, there were a series of questions concerning all kinds of unusual behavior, including the 28 items discussed in this chapter. The parents or other caregivers were asked to describe each type of behavior, giving examples. They were asked what the behavior was like at its most severe, at what age this had occurred, and also how it was manifested at the time of the assessment. On the basis of information from the DISCO interview, direct observation and information from any other available source, the interviewer made two ratings for each item of behavior, one for its most severe manifestation ever, and one for its current form. In each case, the choice of ratings was marked, moderate, or absent. A "marked" rating was made if the behavior was seen every day whenever the opportunity arose. A "moderate" rating was used if it occurred frequently enough to be unusual and inappropriate but not as persistently as needed for a marked rating. To avoid confusion, the "most severe" manifestation could be rated "moderate" or "absent," if it had never occurred in severe form.

All the DISCO items were coded for computer entry and were analyzed using Statistical Package for the Social Sciences (SPSS).

3. Results

The 28 items selected from the DISCO are shown in Table I. These are classified into the following subgroups: (1) disorders of movements of body and face (9 items), (2) disorders of speech and vocalization (5 items), (3) disorders of behavior: eye contact (4 items); fascination with visual stimuli (3 items); and unsocial behavior (7 items). For each item the percentages of individuals given marked or moderate ratings, respectively, for the most severe manifestations are shown. Eighteen of the 28 items had been present in a marked form at some time in the life of the person concerned in over one-third of all participants. Adding the percentages of marked and moderate ratings showed that 17 items had been present in either marked or moderate form in over one-half of the participants. The items occurring in more than three-quarters of participants were: lack of facial expression, delayed echolalia, odd intonation, poor eye contact, and lack of cooperation. The lowest frequencies were found among those concerning fascination with visual stimuli. Table I also shows relevant findings from Study B. The percentages of 36 children with autistic disorders showing each item in marked or moderate form in their most severe manifestations are given. The proportions of individuals affected in the two studies were similar.

Table I shows the significant Spearman correlations ($p < 0.01$) between the IQ estimates for the participants in Study A and their ratings of marked, moderate, or absent for the most severe manifestations for each item. The correlations for all the items were negative—that is higher IQ tended to be associated with less marked manifestations of the item concerned. However, all correlations were small and only 10 of the 28 were significant above the 0.01 level. The correlations with IQ for items concerning speech and vocalizations were all significant.

The effects of age are also shown in Table I. The participants were divided into 2 subgroups—70 individuals with IQ under 70, and 130 with IQ of 70 or above. For each subgroup, chronological age in months was correlated with the ratings for the current manifestation of each item. For the majority of items in both subgroups, the Spearman correlations were negative, indicating a tendency for less marked manifestations with greater age. Most of these correlations were

TABLE I
MOST SEVERE MANIFESTATIONS OF CATATONIA-LIKE FEATURES IN PARTICIPANTS WITH ASD IN STUDY A AND STUDY B

	Study A				Correlation[d] by age and IQ		Study B
	Percentage ++[a]	Percentage +[b]	Total percentage[c]	Correlation[d] by IQ	IQ <70	IQ 70+	Total percentage[c]
Movement							
Odd gait	38.5	22.0	60.5		+0.41		52.8
Poor coordination	34.5	32.5	67.0	−0.19	−0.55	−0.51	63.0
Odd hand postures	48.5	11.5	60.0				61.1
Runs in circles	26.0	17.5	43.6		−0.31	−0.37	47.2
Rocks while sitting	22.0	11.0	33.0		+0.44		19.4
Complex body movements	21.0	4.5	25.5				25.0
Walks on tiptoe	26.0	11.0	37.0				58.3
Grimaces	34.5	19.5	54.0				55.6
Lacks facial expression	39.5	38.5	78.0			+0.23	58.3
Speech and vocalizations							
Immediate echolalia[e]	32.2	15.3	47.5	−0.20		−0.30	61.3
Delayed echolalia[e]	52.6	24.7	77.3	−0.21		−0.49	83.3
Odd intonation[e]	61.3	15.3	76.6	−0.24			60.0
Shrieks for no reason	37.0	14.5	51.5	−0.23			44.4
Laughs for no reason	39.5	23.5	63.0	−0.26		−0.26	69.4
Behavior:							
Eye contact							
Poor eye contact	61.0	17.5	78.5				75.0

(*Continued*)

TABLE I (*Continued*)

	Study A				Study B
	Percentage ++[a]	Percentage +[b]	Total percentage[c]	Correlation[d] by IQ	Correlation[d] by age and IQ	Total percentage[c]
Sideways glances	51.5	16.0	67.5	−0.21		47.2
Blank look in eyes	53.5	14.5	68.0			88.9
Stares	38.5	22.0	60.5			36.1
Visual fascinations						
Spins objects	28.0	9.0	37.0			33.3
Twists hands near eyes	13.5	10.0	23.5	−0.27		36.1
Closely inspects objects from different angles	28.5	11.0	39.5		−0.28	44.4
Unsocial behavior						
Shouts for no reason	35.5	12.0	47.5			58.3
Aggressive for no reason	41.5	20.5	62.0			66.7
Lack of cooperation	65.5	19.5	85.0		−0.31	69.4
Destructive	37.5	12.5	50.0			47.2
Strips in public	53.5	8.5	62.0	−0.21	−0.57	63.9
Inappropriate personal habits (e.g., plays with saliva)	27.5	16.5	44.0			55.6
Hyperactive	16.5	27.5	44.0	−0.20	−0.36	55.6

[a]++: Percentages of participants with items rated as marked.
[b]+: Percentages with items rated as moderate.
[c]Total: Sum of percentages of marked and moderate ratings.
[d]Spearman's correlation $p < 0.01$.
[e]Based on numbers of participants with sufficient speech to be rated.

small and only nine in the higher and two in the lower IQ subgroups reached significance. However, for another nine items, small positive correlations were found that indicated a tendency for *older* individuals to have problems on certain items. Only three of these correlations reached above the 0.01 level of significance. These were for odd gait and rocks while sitting for the lower IQ group and lacks facial expression for the higher IQ group.

The numbers of marked and moderate ratings for the most severe manifestations of all items were added for each participant. The same total was calculated for the current manifestations. From Table II, it can be seen that the means of the totals for those with IQ under 70 were higher than those with IQ of 70 or above, though the ranges of the totals were similar for both the most severe and the current manifestations. The differences were not significant. For both IQ groups, the means for the current were lower than those for the most severe manifestations but did not reach the 0.01 level of significance.

The total scores for the current manifestations of the 28 items were correlated with chronological age in months for the two IQ groups. Spearman's correlation was very low and not significant for those with IQ under 70. It was -0.27 ($p < 0.01$) for those with IQ 70 or above—that is, older participants with higher ability were somewhat more likely to have fewer of the features in marked or moderate form.

Level of anxiety was also rated, since this has been reported to be high in typical catatonia (Northoff, 2002) and in catatonia-like deterioration in autistic disorders (Wing and Shah, 2000). In its most severe manifestations, 39% had marked and 15.5% had moderate anxiety. There was a significant tendency for older people to be rated as having anxiety on the current ratings. This was true for those with IQ under 70 (Spearman's correlation $+0.57$, $p < 0.001$) and those with IQ of 70 or above (Spearman's correlation $+0.36$, $p < 0.001$).

TABLE II
TOTAL OF MARKED PLUS MODERATE RATINGS FOR PARTICIPANTS IN STUDY A

	Total ratings			
	Most severe manifestations		Current manifestations	
	IQ < 70	IQ 70+	IQ < 70	IQ 70+
Number of participants	70	130	70	130
Range of total ratings	6–26	1–27	5–25	1–25
Mean	16.9	14.2	14.7	11.7
SD	4.2	5.0	4.3	4.8

B. Study B: Comparative Frequency of Catatonia-Like Features

1. Participants

The parents of 82 children, 50 school-age children aged 80–140 months and 32 preschool children aged 34–67 months, participated in this study. There were four participant groups; children diagnosed as having ASD (18 low-functioning, 16 of whom were boys and 18 high-functioning, 15 boys); mild, moderate, or severe learning disability, excluding Down's syndrome (17 participants, 10 boys); specific language disorder without generalized learning disability (14 participants, 9 boys); and a group with typical development (15 participants, 9 boys). The children with disabilities were recruited from clinics and special schools in London, Kent, and Sussex. The typically developing children attended schools and nurseries in the local area.

2. Method

The study was designed primarily to test the reliability of DISCO ratings (Leekam *et al.*, 2002; Wing *et al.*, 2002). The parents of all the children taking part were interviewed using the DISCO. The interviewers for this study were two research psychologists trained by Elliot House staff.

3. Results

The percentages of the children with autistic disorders showing each of the 28 items in marked or moderate form in the most severe manifestations are given in Table I.

For each item, for each of the diagnostic groups and for the children with typical development, the numbers of children with marked, moderate, or no problems were counted for the most severe manifestations. For 26 items, more children with autistic disorders were rated as having marked or moderate problems than the other 3 groups. The exceptions were odd gait and rocking while sitting, for which the proportions of those with learning disability were higher than for the rest. More children with learning disability had marked or moderate problems than those with specific language impairment, except for delayed echolalia, noisiness, and aggressiveness, which occurred more often in those with specific language impairments. For 27 items, at least 2 children with learning disability had a marked or moderate rating. The exception was complex bodily movements. For 23 of the items, at least 1 child with specific language impairment had a marked or moderate rating. The exception for this group were complex bodily movements, walks on tiptoe, twists hands near eyes, shouts for no reason, and laughs for no reason.

The typically developing children were much less likely to show any of the items. Only three children in this group had a marked rating on one item—rocks while sitting, walks on tiptoe, and noisiness, respectively. Some had moderate ratings—four children on lack of cooperation; three on aggression; two on odd gait, walks on tiptoe, rocks while sitting, and immediate echolalia; one on odd intonation, poor eye contact, sideways glances, blank expression, hyperactivity, and shouts for no reason.

The total number of marked plus moderate ratings in the most severe manifestations for the 28 items was calculated for each child in each diagnostic group. Table III shows that the children with autistic disorders in the two IQ groups had a similar range for total number of marked plus moderate ratings. Inspection of tabulations for individual items showed the same picture for most of them. However, for laughs for no reason and inspection of objects from different angles, more than twice as many children in the lower IQ group had marked ratings.

Table III underlines the differences between the groups. Children with autistic disorders tended to have more problems than the rest, those with learning disability having the next highest totals, then, slightly lower, were those with specific language impairments. The difference between the children with ASDs (combined totals) and those with the other developmental disabilities (combined totals) was significant (Chi-square = 48.9, df = 21, p = 0.001). The difference between the higher IQ group (autism with IQ 70+ plus specific language disorder) and the lower IQ group (autism with IQ < 70 plus learning disabilities) was not significant. The children with typical development had many fewer problems than those with developmental disorders (combined totals) (Chi-square = 39.7, df = 21, p < 0.005).

TABLE III
TOTAL OF MARKED PLUS MODERATE RATINGS FOR THE MOST SEVERE MANIFESTATIONS FOR CHILDREN IN STUDY B

	Diagnosis				
	Autistic IQ < 70	Autistic IQ 70+	Learning disabled	Language impaired	Typically developing
Number of children in each group	18	18	17	14	15
Range of numbers of total ratings among the children	5–25	8–20	0–15	0–15	0–4
Mean	14.0	13.7	6.9	5.4	1.5
SD	4.6	3.7	4.5	4.2	1.4
Number with no marked or moderate ratings	0	0	1	1	5

TABLE IV
RATINGS OF THE MOST SEVERE MANIFESTATIONS OF ANXIETY FOR CHILDREN IN STUDY B

Anxiety ratings	Diagnosis					Total
	Autistic IQ < 70	Autistic IQ 70+	Learning disabled	Language impaired	Typically developing	
Marked	5	10	3	3	0	21
Moderate	4	6	6	4	4	24
None	9	2	8	7	11	37
Total children	18	18	17	14	15	82

The numbers of children with marked or moderate anxiety in the most severe manifestations are shown in Table IV. The largest number rated with anxiety occurred in those with autistic disorders and the fewest in the typically developing children. All but 2 of the 18 children with autism and IQ 70 or above were rated as having marked or moderate anxiety, compared with 10 of the 18 children with IQs under 70. None of the differences between pairs of subgroups or combined subgroups reached significance. The clearest difference was between the children with typical development and those with developmental disorders combined together, but this did not reach the $p = 0.01$ level.

C. STUDY C: CATATONIA-LIKE DETERIORATION

Full details of this study are given in the paper published in the British Journal of Psychiatry (Wing and Shah, 2000).

1. Participants

Five hundred and six individuals with ASDs, aged from 2 years 8 months to 50 years, were seen by Elliot House staff from 1991–1997. From this population, 30 individuals with catatonia-like deterioration were identified.

A comparison group of 115 people of a similar age range was selected from among the same population. The age range of those with catatonia-like deterioration when assessed was 17–50 years (mean 24.6), and for the comparison group it was 15–60 years (mean 25.1). There were 28 males (93%) in the group with catatonia-like deterioration and 91 males (79%) in the comparison group. The differences of age and sex were not significant.

2. Method

Parents or other long-term caregivers were interviewed using the DISCO. For eight people with catatonia-like deterioration and six in the comparison group, no parent was available for interview. As much information as possible was collected from past case records and care staff gave details of the current clinical picture.

3. *Results*

   a. *Clinical characteristics.* Catatonia-like deterioration was diagnosed when the features were severe enough to produce an obvious and marked deterioration in movement, the pattern of activities, and self-care and practical skills, compared with previous levels. The level of severity varied from those who could take part in some activities with much prompting by staff, to those, four in number, who had to be physically supported or sometimes carried by staff in order to move around.

   The clinical pictures comprised catatonia-like and parkinsonism-like features. The most frequently seen problems affecting posture and movement were: slowness and difficulty in initiating movements unless prompted (30 individuals), odd gait (27), odd stiff posture (19), freezing during actions (17), difficulty crossing lines (e.g., pavement cracks) (16), and inability to cease actions (7). All 30 individuals showed marked reduction in the amount of speech or complete mutism. The most frequent problems affecting aspects of behavior were: impulsive acts (16), bizarre behavior (12), sleeping during the day but awake at night (10), incontinence (10), and excited phases (7). None had waxy flexibility.

   b. *Proportion Affected.* The 30 adolescents and adults with catatonia-like deterioration represented 6% of the 506 referrals of all ages seen by Elliot House staff over 7 years. They comprised 17% of those who were aged 15 or over when referred to Elliot House.

   c. *Factors Associated with Catatonia-like Deterioration.* The study of those with catatonia-like deterioration and the comparison group showed that the following features were *not* significantly associated with the occurrence of the deterioration; age when assessed; gender; IQ; diagnostic subgroup (autism or Asperger syndrome); and history of epilepsy.

   Impairment of social interaction is the most fundamental of the triad of impairments that characterize ASDs. It can be manifested as aloofness and indifference to others, passive acceptance of social approaches, and active but odd, socially inappropriate approaches to others (Wing and Gould, 1979). In this study, half of those with catatonia-like deterioration were passive in social interaction, compared with only 17% of the comparison group (Chi-square = 14.98, df = 2, $p < 0.001$). Those with catatonia-like deterioration were also statistically more likely than the comparison group to be impaired in expressive language before the onset of the deterioration (Chi-square = 6.95, df = 1, $p < 0.01$).

   The adolescents and adults with catatonia-like deterioration were not able to say anything about experiences of their problems, with one exception. This young man had episodes of virtual immobility but at other times the problems were much less severe. He was able to explain that, when immobile, he knew he wanted to move but his body would not respond to his wishes, however hard he tried. None of the 30 was able to suggest any explanation for their deterioration, but parents suggested a variety of precipitating causes such as the stress of school exams, leaving the routine of school with nothing to replace it, losing a job, bereavement, or the difficulties of adolescence. However, many of the comparison groups had similar experiences.

  d. *Information From Study A.* Only 7 of the 30 with catatonia-like deterioration and 58 of the 115 in the comparison group were included in the 200 participants in Study A, whose parents were interviewed using the 10th edition of the DISCO. For 66 people not included in the 200 (15 with deterioration and 51 in the comparison group) most were seen earlier than the 200 in Study A and the DISCO data needed for research were, for various reasons, incomplete. As noted earlier, the parents of 14 participants (8 with deterioration and 6 in the comparison group) were not available for interview.

  The 7 people with catatonia-like deterioration included in the 200 in Study A were too few to detect statistically significant trends in the relationship between deterioration and the 28 features from the DISCO. The ratings for each feature were tabulated by the two subgroups selected from the 200 in Study A—7 with deterioration and 58 other people aged 15 years or over. Inspection showed that none of the 7 people with catatonia-like deterioration had either marked or moderate ratings of hyperactivity compared with 27 (47%) of the comparison group. Otherwise, there were no findings of note.

V. Discussion

  The DISCO was not specifically designed to cover all the items that various authors have listed as catatonic features. In two of the three studies described, adolescents and adults were included so that the data concerning childhood relied upon parents' memories, in some cases looking back over more than 30 years. However, 28 items related to catatonic features were included in the interview. Every one of the 200 participants in the study had had two or more of these items in marked or moderate form at some time. In their most severe manifestations, which in almost all cases occurred during childhood, the items were rated as marked or moderate for between 23% and 85% of the participants. The range was similar for the 36 participants in Study B with ASDs, who were all aged below 13 years.

A. Effect of Level of Ability and Age

  Overall, higher IQ (70 and above) and older age were, for most items, related to fewer and less marked manifestations but the correlations were low and most were not significant. Those in both IQ groups were significantly more likely to show anxiety with increasing age, probably because of increasing awareness of their differences from most other people. It is important to note that none of the three studies included any very high-functioning people with ASDs, who were able enough to live independently. If they had been included, the proportions showing the relevant features at some time in their lives were likely to have been lower. In the absence of complete population studies no estimates of the overall rates are possible. In Study B, the findings for the most severe manifestations of the 28 items were very similar for the higher and lower IQ groups of children with ASDs. This may have been a more accurate reflection of the true clinical picture because the parents did not have to remember so far back as for the adults in Study A. For those with higher IQs who tended to show improvement with age, the parents who were rating retrospectively may have forgotten the details and degree of severity of features shown in childhood.

Study B also found that almost all of the 28 items had occurred in some children with learning disabilities or specific language disorders, though the ranges of totals for individuals in these groups were significantly lower than for children with ASDs. These findings underlined the relationship of the motor, speech, and behavioral items to developmental disorders. Ten of the 15 typically developing children had shown between 1 and 4 items. Only three had shown one item in marked form.

Study C gave a frequency of catatonia-like deterioration among referrals to Elliot House of 17% for those aged over 15 years when seen. This rate is likely to be higher than would be found in a complete population study, because the Elliot House staff's special interest in catatonia-like clinical pictures had become known to those in the field of ASDs.

B. Relationship of Catatonia-Like Features to Catatonia-Like Deterioration

Because of the way the data were collected, it was not possible to examine in detail the relationship between the history of catatonia-like features and the development of catatonia-like deterioration. The full data on past history were available only for seven of the individuals with catatonia-like deterioration. Information on quality of social interaction and language development was available for all the participants. There was a significant tendency for those with catatonia-like deterioration to have previously been passive in social interaction and to be impaired in expressive language. These characteristics may be a forerunner of the dependence on prompting before starting actions and the mutism found in people with catatonia-like deterioration.

The three studies have confirmed the comparatively high rates of catatonia-like features and catatonia-like deterioration in people with ASDs. Apart from the tendency to passivity in social interaction and the impaired expressive language, no features that predicted the likelihood of catatonia-like deterioration were identified. The studies have not thrown any light on the relationship between these two clinical pictures apart from the fact that they both comprise unusual patterns of movement, speech, and behavior. It could be suggested that the presence of so many catatonia-like features makes people with ASDs especially vulnerable to catatonia-like deterioration. However, clinical experience has shown that such deterioration can occur in adolescents or adults, who are very high functioning and who, before the deterioration, had few if any catatonia-like features even in childhood. Furthermore, the majority of people with ASDs do not develop catatonia-like deterioration even if they have had many catatonia-like features.

C. Catatonia-Like Features in Learning Disability

Some children with learning disabilities and some with specific language disorders without autism have some of the catatonia-like features. Earl (1934) described typical catatonia-like features in some residents of an institution for people with learning disabilities, but it is possible that at least some of these people had ASDs. Rollin (1946) also used the term "primitive catatonic psychosis" in relation to a number of adolescents and adults with Down's syndrome who were also resident in an institution, some of whom developed typical catatonic stupor. Rollin noted that, among 73 individuals with Down's syndrome, 17 had marked "catatonic psychosis" and 5 were probably in the early phases. It is of interest that Rollin described these adolescents and adults as "introverts" and later mentions "schizoid personality" and repetitive stereotyped activities. The clinical pictures prior to the catatonic deterioration were not described in any detail. Haw *et al.* (1996) described stereotyped movement disorders in Down's syndrome. Howlin *et al.* (1995) calculated that about 10% of children with Down's syndrome also have an ASD. Apart from these papers, we do not know of any publications concerning any link with catatonia for any developmental conditions except the ASDs.

D. Prevalence of Catatonia-Like Deterioration

Among the 506 children and adults of all ages with ASDs seen by the Elliot House service from 1991–1997, 6% had catatonia-like deterioration. The 30 individuals affected were all aged 15 years or above when seen. A total of 175 people were in this age range when seen. The 30 individuals with catatonia-like deterioration represented 17% of this group (Wing and Shah, 2000). This finding cannot be extrapolated to the total population of people with autistic spectrum disorders since it is based on clinic referrals. There might have been an excess of people with catatonia-like problems because our interest in these conditions became more widely known during the period studied.

Billstedt *et al.* (2005) have published a population based follow-up study of 120 individuals aged 17–40 years with ASDs diagnosed in childhood. They found that 13 (12%) had severe motor initiation problems similar to the features used by the present authors as criteria for catatonia-like deterioration. A further four (3%) had the features in milder form, compared with eight individuals (5%) in the Wing and Shah (2000) study.

VI. Conclusions

The studies described in this chapter suggest that a small but significant proportion of adolescents and adults with ASDs develop catatonia-like deterioration. However, the studies pose more questions than answers. Clinical impression suggests that catatonia-like deterioration is related to events that are experienced as stressful. Why some people with ASDs react in this way while others respond in other ways remains a mystery. What is the relationship between catatonia-like features that are so common in autistic disorders and catatonia-like deterioration? Is there a subgroup of autism that is an early expression of catatonia, as suggested by Dhossche (2004)? If so, how can it be identified? Are people with other developmental disorders in which the features occur also prone to the deterioration? What is the real relationship of the deterioration in autistic disorders to typical catatonic stupor and to parkinsonism? These questions present challenges for future research.

References

American Psychiatric Association (APA) (1994). "Diagnostic and Statistical Manual of Mental Disorders," 4th ed. (DSM-IV). APA Press, Washington, DC.
Billstedt, E., Gillberg, C., and Gillberg, C. (2005). Autism after adolescence. Population based 13- to 22-year follow-up study of 120 individuals with autism diagnosed in childhood. *J. Autism Dev. Disord.* **35,** 351–360.

Brasic, J. R., Zagzag, D., Kowalik, S., Prichep, L., John, E. R., Barnett, J. Y., Bronson, B., Nadrich, R. H., Cancro, R., Buchsbaum, M., and Brathwaite, C. (2000). Clinical manifestations of progressive catatonia. *Ger. J. Psychiatry* **3,** 13–24.

Bush, G., Fink, M., Petrides, G., Dowling, F., and Francis, A. (1996). Catatonia. I. Rating scale and standardised examination. *Acta Psychiatr. Scand.* **93,** 129–136.

Creak, M. (1964). Schizophrenic syndrome in childhood: Further progress report of a working party. *Dev. Med. Child Neurol.* **4,** 530–535.

Damasio, A. R., and Maurer, R. G. (1978). A neurological model for childhood autism. *Arch. Neurol.* **35,** 777–786.

De Sanctis, S. (1908). Dementia praecocissima catatonica oder katatonie des fruheren kindersalters? *Folia Neurobiologica* **2,** 9–12.

Dhossche, D. (1998). Brief report: Catatonia in autistic disorders. *J. Autism Dev. Disord.* **28,** 329–331.

Dhossche, D. (2004). Autism as early expression of catatonia. *Med. Sci. Monit.* **10,** 31–39.

Earl, C. J. C. (1934). The primitive catatonic psychosis of idiocy. *Br. J. Med. Psychol.* **14,** 230–253.

Fink, M., and Taylor, M. A. (2003). "Catatonia: A Clinician's Guide to Diagnosis and Treatment." Cambridge University Press, Cambridge.

Ghaziuddin, M., Quinlan, P., and Ghaziuddin, N. (2005). Catatonia in autism: A distinct subtype? *J. Intellect. Disabil. Res.* **49,** 102–105.

Hare, D. J., and Malone, C. (2004). Catatonia and autistic spectrum disorders. *Autism* **8,** 183–195.

Haw, C. M., Barnes, T. R. E., Clark, K., Crichton, P., and Kohen, D. (1996). Movement disorder in Down's syndrome: A possible marker of the severity of mental handicap. *Mov. Disord.* **11,** 395–403.

Howlin, P., Wing, L., and Gould, J. (1995). The recognition of autism in children with Down syndrome–implcations for intervention and some speculations about pathology. *Dev. Med. Child Neurol.* **37,** 406–414.

Joseph, A. B. (1992). Catatonia. *In* "Movement Disorders in Neurology and Neuropsychiatry" (A. B. Joseph and R. R. Young, Eds.), pp. 335–342. Blackwell, Oxford.

Kanner, L. (1943). Autistic disturbances of affective contact. *Nerv. Child* **2,** 217–250.

Leary, M. R., and Hill, D. A. (1996). Moving on: Autism and movement disturbance. *Ment. Retard.* **34,** 39–53.

Leekam, S., Libby, S., Wing, L., Gould, J., and Gillberg, G. (2000). Comparison of ICD-10 and Gillberger's criteria for Asperger syndrome. *Autism* **4,** 11–28.

Leekam, S. R., Libby, S. J., Wing, L., Gould, J., and Taylor, C. (2002). The diagnostic interview for social and communication disorders: Algorithms for ICD-10 childhood Autism, and Wing and Gould autistic spectrum disorder. *J. Child Psychol. Pyciatry* **43,** 327–342.

Maudsley, H. (1867). Insanity of early life. *In* "The Physiology and Pathology of the Mind" (H. Maudsley, Ed.), 1st ed., pp. 259–293. Appleton, New York.

Northoff, G. (2002). What catatonia can tell us about "top-down modulation": A neuropsychiatric hypothesis. *Behav. Brain Sci.* **25,** 578–604.

Realmuto, G., and August, G. (1991). Catatonia in autistic disorder; a sign of co-morbidity or variable expression? *J. Autism Dev. Disord.* **21,** 517–528.

Rogers, D. (1992). "Motor Disorder in Psychiatry: Towards a Neurological Psychiatry." Wiley, Chichester.

Rollin, H. (1946). Personality in mongolism with special reference to the incidence of catatonic psychosis. *Am. J. Ment. Def.* **51,** 219–237.

Wing, L., and Attwood, A. (1987). Syndromes of autism and atypical development. *In* "Handbook of Autism & Pervasive Developmental Disorders" (J. Cohen, A. Donnellan, and R. Paul, Eds.), p. 15. Winston-Wiley, New York.

Wing, L., and Gould, J. (1979). Severe impairments of social interaction and associated abnormalities in children: Epidemiology and classification. *J. Autism Childh. Schizophr.* **9,** 11–29.

Wing, L., and Shah, A. (2000). Catatonia in autistic spectrum disorders. *Br. J. Psychiatry* **176,** 357–362.

Wing, L., Leekam, S. R., Libby, S. J., Gould, J., and Larcombe, M. (2002). The diagnostic interview for social and communication disorders: Background, inter-rater reliability and clinical use. *J. Child Psychol. Psychiatry* **43,** 307–325.

World Health Organization (WHO) (1993). "The ICD-10 Classification of Mental and Behavioural Disorders. Diagnostic Criteria for Research." World Health Organization, Geneva.

Zaw, F. K. M., Bates, G. D. I., Murali, V., and Bentham, P. (1999). Catatonia, autism and ECT. *Dev. Med. Child Neurol.* **41,** 843–845.

CATATONIA IN INDIVIDUALS WITH AUTISM SPECTRUM DISORDERS IN ADOLESCENCE AND EARLY ADULTHOOD: A LONG-TERM PROSPECTIVE STUDY

Masataka Ohta,* Yukiko Kano,[†] and Yoko Nagai[‡]

*Center for the Research and Support of Educational Practice
Tokyo Gakugei University, Koganei-Shi, Tokyo, Japan
[†]Graduate School of Medical Sciences
Kitasato University, Sagamihara, Kanagawa, Japan
[‡]School of Nursing, University of Shizuoka, Suruga-Ku, Shizuoka, Japan

I. Introduction
II. Subjects
III. Methods
   A. Criteria For Catatonia in this Study
   B. Severities of Catatonia
IV. Presentation of Cases
V. Results
   A. Symptoms and Severity of Catatonia
   B. Preceding Conditions
   C. Psychiatric Complication, Family History, and Medication
   D. Courses
VI. Discussion
VII. Suggestions on Treatment
VIII. Limitation of this Study
IX. Conclusions
   References

The objective is to cast light on diagnosis and catastasis, course, and comorbidity as concerned with catatonia in patients with autism spectrum disorders (ASDs) with respect to long-term prospective follow-up. Eleven patients (all male) were enrolled. The mean age and the mean follow-up duration were 27.6 years (standard deviation (SD) 5.5) and 18.7 years (SD 8.7), respectively. The mean IQ was 27 (SD 16.4). Information was garnered from medical case records; current examination and observation of patients, interview of parents, and questionnaires completed by parents or other caretakers. Informed consent was obtained from the parents. Criteria for catatonia in this study were: (1) abrupt stop of movements and maintenance of immobility or bizarre posture beginning in adolescence and early adult life, (2) such a cataleptic state had continued for at least several minutes and appeared many times a day to the point of interfering with daily activities. We described two typical catatonic cases of ASDs. The average onset age was 19 years (SD 6). In all cases, our diagnostic criteria of catatonia evaluating at worse are fully compatible with those of Diagnostic and Statistical Manual of Mental Disorders, 4th ed. (DSM-IV). In 8 out of 11, the onset of catatonia was clearly preceded by the appearance of slowness in movements accompanying the exacerbation of obsessive-compulsive symptoms. Catatonia was also found to have some connection with Tourette syndrome (3 cases), adjustment disorders ($N = 1$), and depressive mood disorders ($N = 1$). In one case, the manifestations of catatonia had to be distinguished from parkinsonism caused by antipsychotics.

Catatonia in ASDs seems to be a chronic condition in most cases. However, there were also a few cases in which catatonia repeatedly aggravated over short spans of time. Catatonia in ASDs may be considered an epiphenomenon of ASDs or a manifestation of comorbidity in adolescence or early adulthood.

I. Introduction

The concept of catatonia has broadened in recent years. It is no more restricted to schizophrenia but is also thought to occur in mood and organic disorders (Fink and Taylor, 2003). Realmuto and August (1991) reported three catatonic adolescents with autism and other psychiatric conditions. Among patients with autism in adolescence and early adulthood, motoric immobility is occasionally observed along with delayed action and repeated conduct. Wing (1996) defined "catatonia" as this sort of immobility. Although catatonia in autism spectrum disorders (ASDs) has been recognized among clinicians and researchers, the nature of catatonia in ASDs is yet to be clarified. The nature and treatment of this psychomotor syndrome as well as its scope and course remain virtually unidentified. Few published studies are available on the long-term course of catatonia in patients with ASDs.

We previously reported about eight ASD patients with catatonia, aged 20 or over, who had regularly visited the outpatient clinic at Department of Neuropsychiatry of Tokyo University Hospital (Ohta *et al.*, 1999). With an elapse of almost 6 years, "M," a child psychiatrist, seeing most of these ASD patients with catatonia, moved to the "Z" center. Many of the patients also did so later. In the meantime, there appeared new patients suspected of having catatonia.

This study is primarily designed to cast light on diagnosis and catastasis, onset and course, and association with complications as concerned with catatonia in patients with ASDs in terms of the long term prospective study. A few suggestions for treatment of catatonia in ASDs are made.

II. Subjects

The subjects came up to a total of 11, including 8 cases reported at the 40th Congress of the Japanese Society for Child and Adolescent Psychiatry (Ohta *et al.*, 1999). The three new cases were identified among ASDs patients, aged 20 or over, who regularly visited the "Z" Center from June to December in 2003, who fit into the criteria for catatonia, and who had symptoms that considerably hampered their everyday lives or had had them in the past. All the authors of this chapter were seeing at least one of those patients as the attending doctors or the clinical psychologist. The 11 patients fulfilled the criteria of autistic disorder in DSM-IV-Text Revision (TR) (APA, 2000). The average age at the time of initial diagnosis was 8.7 years (standard deviation (SD) 6.4) with the mean follow-up duration at 18.7 years (SD 8.7). All of them were male. The ages at the time of our investigation averaged 27.6 years (SD 5.5, age range: 21–40). Mean IQ was 17 on the Tanaka–Binet scale of intelligence (within the range of 13–70). As evaluated according to the Ohta Staging, which is an evaluation system of cognitive development in autistic children devised and standardized by Ohta *et al.* (Mutoh *et al.*, 2003; Ohta, 1987; Ohta *et al.*, 1989), three came on Stage II, four on III-1, two on III-2, and one on IV.

At the time of investigation, 8 of the 11 patients were visiting the doctor on a regular basis. The whereabouts of two cases reported in 1999 were unknown, and information about conditions of another case could be secured from the mother by telephone.

III. Methods

Regarding the course of the illness, data could be obtained from the statements made by the parents and the descriptions in the medical charts, as well as psychiatric interviews of each visit. The parents consented to our publishing the results of the study.

The term "catatonia," once solely attributed to schizophrenia, is broadly employed today as a behavioral syndrome for other disorders. Psychiatrically, we took catatonia as a failure to manifest spontaneous will and defined it as follows, while referring to descriptions about catatonic disorders in the Catatonic Disorder Due to a General Medical Condition and the Catatonic Features Specifier in the 4th edition of the DSM-IV-TR (APA, 2000), and those both in the Guideline (WHO, 1992) on Organic Catatonic Disorders in International Classification of Diseases, 10th revision (ICD-10) and Diagnostic Criteria for Research (DCR) (WHO, 1993), and those by Wing (Wing, 1996; Wing and Shah, 2000).

A. Criteria For Catatonia in this Study

Catatonia is a behavioral syndrome, and the severity changes during the span of a day and according to the mode of life, so that it would be difficult to come to grips with the loss of voluntary will. Therefore, we focused on movement that comes out and then stops halfway, in situations where the conditions may be accurately grasped by physicians at the outpatient clinic or the parents and other persons in their everyday lives. We picked up only the cases that fell under the category of this condition at the worst time.

1. In adolescence and early adult life they had abruptly stopped their movements and gotten locked into immobility or maintained bizarre posture.
2. Such a cataleptic state had continued for at least several minutes and appeared many times a day.
3. The disturbance caused clinically significant impairment in social, occupational, or other important areas of functioning, and continued for 3 months or more.
4. Clear drug-induced parkinsonism or cases in which the immobility could be explained by an inner state of absorption should be excluded.

B. Severities of Catatonia

The severities of catatonia were classified into "none," "mild," "moderate," and "severe," and the degree of severity was judged depending on the social impairment caused by compulsions referring to "interference due to compulsive behaviors" on the Yale-Brown obsessive-compulsive scale (Y-BOCS) (Goodman et al., 1989a,b). "Mild" represents slight impairment in social and vocational activities without hampering efficiency as a whole; "moderate," some degrees of impairment evidently existent in those activities; and "severe," the degree at which the patients and their families feel it measurably difficult to cope with.

IV. Presentation of Cases

Case 4: 27-year-old male; IQ 40, Ohta Stage III-2
This is a typical case of catatonia as described by Wing (1996), which continued for about 10 years.
There was nothing noteworthy about him in the prenatal, perinatal, and infantile periods. At 12 months of age he started toddling. At about 18 months, he appeared to lag far behind in language development and was markedly

TABLE I
DESCRIPTION OF 11 CASES WITH ASDs AND CATATONIA

Case	1	2	3	4	5	6
Sex	m	m	m	m	m	m
Outcome	Visiting	Visiting	Dropout	Visiting	Visiting	Dropout
Current age	31	30	40	27	25	31
Ohta Stage[a]	II	III-2	III-1	III-2	I-3	IV
Age of first visit (yy:mm)	9:06	15:03	7:11	3:05	5:06	23:05
Duration of follow-up (yy:mm) (Oct 2003)	18:00	14:08	32:07	23:10	19:11	5:06
IQ	17	30	22	40	13	70
Preceding slowness (age)	15	No	No	14	15	20
Preceding symptoms such as obsessive–compulsive symptoms (OCS)		Bad feeling, negativism, self-injurious, aggression		Excitement, OCS↑, manifestation of TS[b]	Aggression↑ ordering↑	
History of antipsychotics	Yes			Yes		
Age of manifestation of catatonia	23	15	19	19	18	21
Social situation at onset	Workshop	Special school	Workshop	Competitive job	Workshop	Workshop
Severity at worst	Mild	Moderate	Moderate	Severe	Moderate	Severe
Course and outcome of catatonia (yrs)	Suddenly developed and lasted for 7 months After that no catatonia	1st lasted few months 2nd occurred along with TS	Suddenly occurred with slowness and lasted for less than 1 year. After that time no catatonia	Has lasted in a mild form, but no TS symptoms	Difficulty in initiation has lasted	Subacutely occurred and lasted till the time of dropout
Second phase (yr)		19				
Epilepsy			Yes	Yes		
Psychiatric comorbidity		TS[b]		TS[b]		
Family history						

(*Continued*)

TABLE I (Continued)

Case	7	8	9	10	11
Sex	m	m	m	m	m
Outcome	Visiting	Dropout	Visiting	Visiting	Visiting
Current age	27	28	23	21	21
Ohta Stage[a]	III-1	III-1	III-2	III-1	II
Age of first visit (yy:mm)	15:11	3:11	3:05	3:08	4:03
Duration of follow-up (yy:mm) (Oct 2003)	11:07	23:03	20:01	17:10	18:01
IQ	19	14	32	27	13
Preceding slowness (age)	No	19	20	17	20
Preceding symptoms such as obsessive-compulsive symptoms (OCS)	Repetitive movement↑	Touching compulsion	Ordering	Sleep disturbance, ritual behavior	Touching compulsion, excitement
History of antipsychotics	Yes (2nd time)				Yes
Age of manifestation of catatonia	15	21	21	17	20
Social situation at onset	Special school	Workshop	Workshop	Workshop	Workshop
Severity at worst	Moderate	Severe	Moderate	Severe	Moderate
Course and outcome of catatonia (yrs)	1st suddenly occurred and lasted for 1 years. 2nd (21yr) occurred followed by eyes rolling and lasted for 8 months. After that time no catatonia.	Lasted for 2years, mitigated at the time of dropout	Separating from his sib, he entered a group home. Soon after disappeared	Lasted for less than 2 years	Lasted for more than 1year, but touching compulsion has lasted with the same intensity
Second phase (yr)	21				
Epilepsy				Yes	
Psychiatric comorbidity	Parkinsonism?		Adjustment disorder	Sleep disturbance	TS[b]
Family history			Sib: schizophrenia	Father: depression	

[a]Ohta Stage: Levels of cognitive development in autistic children devised and standardized by Ohta et al (1989).
[b]TS: Tourette syndrome.

hyperactive. He did not respond when his name was called out. When he turned 2-year old, he was diagnosed as having autism.

After the fourth birthday, he began having severe tantrums when he saw the very slightly disorganized tableware on the dinner table.

While in special classes at the elementary school (6-year old), his hyperactivity remained unabated. He threw in a fit of temper when he was not allowed to do familiar routines or when his schedule was upset. With the administration of haloperidol, his condition turned better. As a fifth grader, he had an epileptic seizure in autumn. Although EEG showed nothing abnormal, treatment with valproic acid commenced.

In January, of the sixth year at elementary school, Tourette syndrome (TS) appeared abruptly. The major symptoms he exhibited included facial grimacing, hiccupping, and rapid jerking of the body concurrent with the utterance of bizarre sounds. These well-defined forms of tics had disappeared when he moved up to the second year of junior high school. Several months later, however, self-injurious behavior emerged. There were sudden outbursts during which he banged his head against the table violently.

In October, he took to hanging his clothes on the hanger and taking them off again and again. He also repeatedly said "itadaki-masu," a short prayer of thanks before a meal in Japan. These repetitive actions lasted from 5 min to more than 10 min. Toward the end of the year, increased slowness affecting movements became conspicuous. In May, of the third year at junior high school, he again started loosing his temper easily. Daily doses of pimozide were increased to 4 mg and these temper tantrums diminished as a result.

After he entered senior high school for mentally handicapped children, he persisted to the characteristic pattern of behavior in everyday life without relapse of temper tantrums.

He finished senior high school and found a job at 18. It seemed that he somehow managed to do his own part, although it was said that he had poor ability of concentration. He rarely lost his temper. It seemed that all things were going well with him at home and work.

Repetitive actions increased in October, of the year, when he turned 19-year old. In January of the next year, he became slow moving. In April, he began exhibiting such symptoms as the repetition of bizarre behaviors, and freezing in postures during activity such as making tea. The manifestations of these abnormalities of behavior and posture lasted several minutes. In May, he became unusually concerned with keeping the tableware on the dinner table in perfect order or arranged exactly. Repetition of such words and phrases as "I'm home!", "Good night," and "itadaki-masu" increased in frequency. Also, repetition of a word or phrase just spoken by another person increased remarkably.

In April of the following year, he became fussier and manifested motoric immobility more frequently regardless of mood, which varied greatly from day to day. In September, he maintained a rigid posture while in standing position for several hours a day. Prompted to move, he wouldn't budge an inch. When forced to move, he returned to where he stood and continued to stand as stiff as a stature. This symptomatology caused disruption of his occupational functioning.

When the patient was additionally dosed with bromazepam (BZP) 4 mg, these symptoms tended to abate, but were not ameliorated completely. At the end of the third year, he resigned from his job and moved to a sheltered workshop in April. Through 1 year, after he became 21, motoric immobility occurred every day. Once it occurred, it lasted for 10 min to several hours.

When he was 22, the frequency of the psychomotor disturbance decreased to three times a week. Three years later, when he was 25, his condition improved considerably. The maintenance of a rigid posture lasted only several minutes. A few months later he moved to a residential welfare institution. As immobility had almost disappeared, the dosage of BZP 5 mg was tapered off.

Soon after that, he had an epileptic seizure after an absence of over 5 years. Once the seizure returned, it occurred once or twice a month. With an increase in the dosage of an antiepileptic drug, the occurrence rate was on the decline. Relapse of the seizure did not change the residual catatonic symptoms. Epileptic seizures occurred infrequently for the 2 years that followed and have not occurred for more than 1 year now. There are no MRI abnormalities. ECG examinations revealed only a slight degree of paroxysmal abnormalities.

Case 7: 27-year-old male; IQ 19, Ohta Stage III-1

This is the case in which catatonic symptoms appeared twice and disappeared quickly.

No abnormalities were observed during the prenatal and perinatal periods.

At age 1, he started walking by himself. It was not long before he exhibited signs of hyperkinesia, mutism, and apathy. From infancy, he had a strong inclination to adhere to a pattern of behavior in everyday life. Even now, he has been occupied in doing a ritualistic custom at mealtime and has a mania for collecting plastic models of monsters.

In July, when he was a first grade junior high school student, he began to act rudely or take a defiant attitude. When he got angry, he slapped or pinched his opponent's hand.

At the age of 15 months, he started walking up and down or extending his hands compulsively. It was around that time that motoric immobility emerged. He rejected any approach when he was prompted to move. However, it disappeared in a year or so.

When he was 16-year old, he visited us, as he could not shake off the self-injurious behaviors. With the use of an antipsychotic agent, self-injurious and aggressive behaviors were reduced notably.

His condition remained in remission till he reached 21 years and 6 months of age, when eye-rolling suddenly occurred. This symptom responded to treatment with anticholinergic medication.

Two months later, he began to show bizarre behavior. He stood motionless with one leg raised. He also began to threaten to scratch his family members with his nails. After his dose of antipsychotic medication was increased, these abnormalities gradually faded away.

At present, he is not aggressive and freezing has disappeared. He still follows his characteristic pattern of behavior in every day life. He entered the 2005 Special Olympics World Winter Games and won three Gold medals.

V. Results

Average age at onset and frequency in ASDs: The average age at onset of catatonia was 19 years (SD 6, age range: 15–23). Out of 69 cases, who were 20 years or over and visited outpatient clinic of "Z" center, 8 (11.6%) had current symptoms of catatonia or had a past history of catatonia. As for the remaining three cases, who had been followed-up till 3 years ago; the whereabouts of two were unknown, whereas the third one was under treatment in another hospital (See Table I).

A. Symptoms and Severity of Catatonia

In DSM-IV-TR (APA, 2000), five symptoms characterize catatonia or catatonic disorder due to a general medical condition—that is, the maintenance of imposed postures (catalepsy), including waxy flexibility, or the absence of movements

TABLE II
Distribution of DSM-IV Criteria for Catatonia in 13 Catatonic Episodes in Patients with ASD

Case	1	2	3	4	5	6	7	8	9	10	11	12
Times of catatonic phase	1	1	2	1	1	1	1	1	2	1	1	1
Motoric immobility	1	1	1	1	1	1	1	1	1	1	1	1
Excessive motor activity	0	1	1	0	0	1	0	0	0	0	0	1
Extreme negativism or mutism	0	1	1	0	1	1	1	1	1	1	1	1
Peculiarities of voluntary movement	1	1	1	1	1	1	1	1	1	1	1	1
Echolalia or echopraxia	0	1	1	1	1	0	1	0	0	0	0	0
Number of positive items	2	5	5	3	4	4	4	3	3	3	3	4

1: Present.
0: Not present.

(akinesia) with the manifestation of stupor; hyperactivity; extreme negativism (evidently nonmotivated resistance or the maintenance of a stiff posture against attempts to be moved), mutism; stereotypy, significant mannerism or the strangeness of volitional movements demonstrated by a significant grimace; and echolalia or echopraxia. Should there be two or more such symptoms, it can be argued that catatonia coexists. After "*M*" had selected probable catatonic cases on the basis of the worst state, "*K*" independently evaluated them, and confirmed the cases to have catatonia by the existence of more than two of the five symptoms. Four cases were considered severe, six cases moderate, and one mild (See Table II).

B. Preceding Conditions

Before the manifestation of a typical catatonic symptom in which movements come to a halt in a strange posture, eight cases had prodromal symptoms, typically a gradually emerging sluggishness with compulsive behaviors lasting for more than 1 year. In the other three cases, onset of catatonia was abrupt with no preceding prodromal phase.

C. Psychiatric Complication, Family History, and Medication

Three cases were diagnosed with TS. At present, one of them (Case 11) still has it (see Table I).

The obsessive-compulsive symptoms in Case 9 seemed to increase in periods of greater family turmoil and conflict with his schizophrenic sibling. Catatonia may have been precipitated by increased stress.

Case 10 had a family history of mood disorder. Complications with epilepsy were observed in three cases, and in one of them, epileptic seizures recurred twice or so a month after the alleviation of catatonia (Case 4). In Case 10, the onset of epilepsy came for the first time at the age of 20 after the alleviation of catatonia. Only three cases were on antipsychotics before onset of catatonia. Of the remaining 8 cases, Case 7 took antipsychotics when he had catatonia for the second time. For Case 7, above all, a discreet differentiation was required between catatonia and antipsychotic-induced parkinsonism (see the case presentation).

D. Courses

Catatonic symptoms showed considerable fluctuations during the span of a day, in all cases. Those changes could be observed even at the worst time. The alleviation of symptoms did not signify full improvements in the attitude of refusal or spontaneity. A review of the long-term course showed that catatonic symptoms came out twice in two cases. Of them, Case 2 was complicated with TS. With the exclusion of Case 6—whose whereabouts are unknown—and the inclusion of two cases with two catatonic episodes, the average duration was 27 months (SD 31.8, duration range 4–108 months). Out of nine cases interviewed on clinic visits or by telephone, five cases no longer had catatonia at the time of the current examination, whereas one case remained moderate and three cases were mild. When comparing the three cases with sudden onset to the eight cases with gradual onset, it was found that the rate of remission within 1 year was higher in the sudden onset cases (100%) than in the gradual onset cases (25%) with no statistical significance.

VI. Discussion

Wing and Shah (2000) operationally defined catatonia in individuals with ASDs. In their definition, four features were taken up—that is, (1) increased slowness affecting movements and verbal response, (2) difficulty in initiating and completing action, (3) increased reliance on physical or verbal prompting by others, and (4) increased passivity and apparent lack of motivation. As often-associated symptoms, they referred to (5) reversal of day and night, (6) parkinsonian features (tremor, eye-rolling, dystonia, odd stiff posture, freezing in postures, etc.), (7) excitement and agitation, and (8) an increase in repetitive and ritualistic behavior.

Unlike the criteria set forth by Wing and Shah, we adopted suspension in an odd posture as the core of the diagnostic criteria for catatonia in this study.

However, it was confirmed that our diagnostic criteria are fully compatible with those of DSM-IV-TR (APA, 2000), and their validity was ascertained.

Wing and Shah (2000) reported that the age-of-onset of catatonia ranged from 10 to 30 years of age, with a peak at 15–19 and the prevalence of catatonia was 6% in outpatients with ASDs. In our cases, the prevalence in ASDs was rather higher than that. But the age of onset roughly came within the ranges set by them. As the "Z" center is the tertiary facility for developmental disorders and a large portion of the patients visiting the center has various behavioral problems regardless of level of intelligence, the prevalence in our study would be higher than that in the previous study. Among ASD patients with remarkable social impairment, the prevalence of catatonia might be higher than that expected. It is said that catatonia can occur as intrinsic symptoms of ASDs or comorbid psychiatric condition of ASDs or aversive side effects related to antipsychotics (Chaplin, 2000; Dhossche, 1998; Leary and Hill, 1996; Realmuto and August, 1991).

First, we examined relationship between catatonia and ASDs in terms of comorbidity.

It is known that catatonia comes out regardless of levels of intelligence (Howlin, 2000). In this study, most of the subjects are individuals who had severe or moderate mental retardation. Therefore, it is difficult to diagnose complications with mood disorders and schizophrenia. On the other hand, TS is relatively easy to diagnose (Baron-Cohen et al., 1999; Kano et al., 1988) and was found in three subjects in this study.

They are the first cases of ASDs reported to have both catatonia and TS. It seems to be worthy to examine relationship between catatonia and TS, which is closely associated with ASDs.

Second, we examined the relationship between catatonia and ASDs from the viewpoint of course of catatonia.

We found that some cases developed catatonia with preceding gradual slowness and other cases had sudden onset of catatonic symptom, in accord with the findings of Wing and Shah (2000).

It may also be pointed out that there existed cases in which catatonia repeatedly aggravated over short spans of time.

In addition, it should be emphasized catatonia continued for more than 2 years on average, and there were cases with no significant change for nearly 9 years.

VII. Suggestions on Treatment

First of all, it should be emphasized that it is inappropriate to force ASD patients with catatonia to act on their own initiative.

And it should be considered that, for any clinical case, the severity of catatonia changes in a day. It is effective to approach catatonic ASD patients during minutes or hours when severity of catatonia diminishes within a day. The severity of catatonia often fluctuates throughout the day. It is most effective to approach catatonic ASD patients when catatonic symptoms are at their lowest point during the day.

Catatonic ASD patients assume a negative attitude toward approaches from other persons when the disease is at its worst, and it may well be argued that they offer strong resistance to treatment as suggested by Wing (1996). It seems to be impossible to approach such patients with oral instructions. However, the patients may be able to take an action, albeit at a slow pace, when their bodies are touched and moved toward the place to which he presumably wanted to move. As regards pharmacotherapy, it can be said from our experience that the use of both benzodiazepine and antipsychotics will be effective in the long run.

VIII. Limitation of this Study

There is the need to examine if the diagnostic criteria for catatonia set forth here are in harmony with those worked out by Wing and Shah. It is also necessary to prepare diagnostic criteria for the screening of catatonia and to systematically review the medical records of outpatients who are suspected of having catatonia.

It is convenient to screen patients for catatonia in ASDs at age 20, because, at that time, a comprehensive review and diagnostic assessment is done in order to file applications for pensions payable to physically or mentally handicapped in Japan. On the basis of outpatient services, there is the need to study patients at younger ages. As many ASD patients with severe or moderate mental retardation were taken up in this study, it was difficult to come to grips with mood disorders and schizophrenia. Though two catatonic ASD patients in this study carry a family history of mood disorders or schizophrenia, it cannot be hastily concluded which complications are closely related to catatonia. It is necessary to investigate relationship between catatonia and other complications in ASD cases without mental retardation.

IX. Conclusions

Our diagnostic criteria of catatonia are fully compatible with those of DSM-IV-TR (APA, 2000), and their validity was ascertained.

Catatonia in ASDs seems to be a chronic condition in most cases. However, there were also a few cases in which catatonia repeatedly aggravated over short spans of time. Catatonia in ASDs may be considered an epiphenomenon of ASD or a manifestation of comorbidity in adolescence or early adulthood.

Further studies in patients with ASDs are needed to compare different diagnostic criteria for catatonia and to examine the biological correlates of catatonia in ASDs.

References

American Psychiatric Association (APA) (2000). "Diagnostic and Statistical Manual of Mental Disorders," 4th ed. Text Revision (DSM-IV-TR). APA Press, Washington, DC.

Baron-Cohen, S., Scahill, V. L., Izaguirre, J., Hornsey, H., and Robertson, M. M. (1999). The prevalence of Gilles de la Tourette syndrome in children and adolescents with autism: A large scale study. *Psychol. Med.* **29,** 1151–1159.

Chaplin, R. (2000). Possible cases of catatonia in autistic spectrum disorders. *Br. J. Psychiatry* **177,** 180–181.
Dhossche, D. (1998). Brief report: Catatonia in autistic disorders. *J. Autism Dev. Disord.* **28,** 329–331.
Fink, M., and Taylor, M. (2003). "Catatonia: A Clinician's Guide to Diagnosis and Treatment." University Press, Cambridge.
Goodman, W. K., Price, L. H., Rasmussen, S. A., Mazure, C., Fleischmann, R. L., Hill, C. L., Heninger, G. R., and Charney, D. S. (1989a). The Yale-Brown obsessive-compulsive scale (Y-BOCS). Part I: Development, use, and reliability. *Arch. Gen. Psychiatry* **46,** 1006–1011.
Goodman, W. K., Price, L. H., Rasmussen, S. A., Mazure, C., Delgado, P., Heninger, G. R., and Charney, D. S. (1989b). The Yale-Brown obsessive-compulsive scale (Y-BOCS). Part II: Validity. *Arch. Gen. Psychiatry* **46,** 1012–1016.
Howlin, P. (2000). Outcome in adult life for more able individuals with autism or Asperger syndrome. *Autism* **4,** 63–83.
Kano, Y., Ohta, M., Nagai, Y., Yokota, K., and Shimizu, Y. (1988). Tourette's disorder coupled with infantile autism: A prospective study of two boys. *Jap. J. Psychiatry Neurol.* **42,** 49–57.
Leary, M., and Hill, D. A. (1996). Moving on: Autism and movement disturbance. *Ment. Retard.* **34,** 39–53.
Mutoh, N., Suzuki, H., Kano, Y., Nagai, Y., and Ohta, M. (2003). Ohta staging: Evaluation system of cognitive development for persons with autism spectrum disorder. *In* "16th Asian Conference on Mental Retardation Proceedings," pp. 353–361.
Ohta, M., Nagai, Y., and Kano, Y. (1999). Catatonia like symptoms in individuals with autism spectrum disorders in adolescence and early adulthood: Diagnosis and course. *Jap. J. Child and Adolescent Psychiatry* **40,** 50(in Japanese).
Ohta, M., Nagai, Y., and Kano, Y. (1989). On the cognitive developmental therapy for autistic children at the Day Care Center. *In* "An Interim Report for Mitsubishi Foundation: Studies on Treatment and Evaluation of Their Effectiveness of Autistic Children (directed by Ohta)," pp. 80–87.
Ohta, M. (1987). Cognitive disorders of infantile autism: A study employing the WISC, spatial relationship conceptualization, and gesture imitations. *J. Autism Dev. Disord.* **17,** 45–62.
Realmuto, G. M., and August, G. J. (1991). Catatonia in autistic disorder: A sign of comorbidity or variable expression? *J. Autism Dev. Disord.* **21,** 517–528.
Wing, L. (1996). "The Autism Spectrum," pp. 174–175. Constable, London.
Wing, L., and Shah, A. (2000). Catatonia in autistic spectrum disorders. *Br. J. Psychiatry* **176,** 357–362.
World Health Organization (WHO) (1992). "The ICD-10 Classification of Mental and Behavioral Disorders: Clinical descriptions and diagnostic guidelines." WHO, Geneva.
World Health Organization (WHO) (1993). "The ICD-10 Classification of Mental and Behavioral Disorders: Diagnostic Criteria for Research (DCR)." WHO, Geneva.

ARE AUTISTIC AND CATATONIC REGRESSION RELATED? A FEW WORKING HYPOTHESES INVOLVING GABA, PURKINJE CELL SURVIVAL, NEUROGENESIS, AND ECT

Dirk Marcel Dhossche* and Ujjwal Rout[†]

*Department of Psychiatry and Human Behavior, University of Mississippi Medical Center
Jackson, Mississippi 39216, USA
[†]Department of Surgery, University of Mississippi Medical Center
Jackson, Mississippi 39216, USA

  I. Introduction
 II. Early Autistic Regression
III. Late Autistic Regression
 IV. Catatonic Regression in Autism
  V. Are Autistic and Catatonic Regression Related?
 VI. Theoretical Treatment Implications
VII. History of ECT in Childhood Psychoses
VIII. GABA in Autistic Regression
 IX. GABA in Catatonic Regression
  X. GABA and the Mechanism of Action of ECT
 XI. ECS in GABA Models of Autism
XII. ECS and Purkinje Cell Survival
XIII. ECS and Neurogenesis
    References

> We have multitudes of facts, but we require, as they accumulate, organizations of them into higher knowledge; we require generalizations and working hypotheses.
> John Hughlings Jackson, M. D. (1835–1911)

Autistic regression seems to occur in about a quarter of children with autism. Its cause is unknown. Late-onset autistic regression, that is, after 2 years of age, shares some features with catatonic regression. A working hypothesis is developed that some children with autistic regression suffer from early-onset catatonic regression. This hypothesis cannot be answered from current data and is difficult to address in clinical studies in the absence of definite markers of autistic and catatonic regression. Treatment implications are theoretical and involve the potential use of anticatatonic treatments for autistic regression. Focus is on electroconvulsive therapy (ECT)—an established but controversial treatment that is viewed by many, but not all, as the most effective treatment for severe, life-threatening catatonic regression. Clinical trials of ECT in early- or late-onset autistic regression in children have not been done yet. The effects of electroconvulsive seizures—the experimental analogue of ECT—should also be tested in gamma-aminobutyric acid-ergic animal models of autistic regression, autism, catatonia, and other neurodevelopmental disorders. Purkinje cell survival and neurogenesis are putative outcome measures in these models.

I. Introduction

Autistic regression has become an important focus of inquiry. Parents have reported for a long time the sometimes-dramatic loss of communicative and social skills in children who are later diagnosed with autism. Fluctuations in the development of autistic children are not a new phenomenon. The original descriptions of Kanner (1943) contain several examples of uneventful early developmental courses of autistic children, until age 2. Kanner emphasized the presence of subtle abnormalities in the first few months of life, long before autistic symptoms became clear. For example, he reports: "It is therefore highly significant that almost all mothers of our patients recalled their astonishment at the children's failure to assume at any time an anticipatory posture preparatory to being picked up." These reports among others led him to hypothesize: "We must, then, assume that these children have come into the world with innate inability to form the usual, biologically provided affective contact with people, just as other people come into the world with innate physical or intellectual handicaps. He concluded: "For here we seem to have pure-culture examples of inborn autistic disturbances of affective contact."

The extent of any innate inability in affective or social repertoire of autistic children is still unknown. However, recent studies support the validity of autistic regression as a distinct event in about a quarter of cases with autistic disorder (AD) or pervasive developmental disorder not otherwise specified (PDD NOS). The occurrence of a regressive phase does not necessarily imply that prior development was completely normal. Massive and late-onset regression (i.e., after age 2 but before age 10) is the diagnostic feature of childhood disintegrative disorder (CDD). A regressive phase is not a feature of Asperger disorder (AsD), according to Diagnostic and Statistical Manual, 4th ed. (DSM-IV) (APA, 1994) descriptions. In this chapter, the terms autism or autism spectrum disorders (ASD) are used interchangeably for the group of AD, PDD NOS, and AsD (with the understanding that autistic regression is not a DSM-IV feature of AsD).

Catatonic regression is characterized by the onset of stereotypical movements, rigidity, mutism, and posturing. In its most severe form, patients become mute, immobile, and stuporous. The condition is then life threatening and needs emergency medical intervention. Catatonia has been reported infrequently in children

(Dhossche and Bouman, 1997a,b), although systematic studies in this age group have not been done. The oldest description of catatonic symptoms in a (3-year-old) child comes from de Sanctis (1908–1909) (see also Chapter 1 by Neumärker). Similarities between autistic and catatonic regression have been observed in individual cases (Dhossche, 2004). In this chapter, the literature on early and late autistic regression is reviewed. The symptoms of autistic and catatonic regression are compared. In the next sections, the hypothesis is developed that autistic regression may be an early form of catatonic regression, and theoretical treatment implications are discussed. Focus is on electroconvulsive therapy (ECT)—an established but controversial treatment that is viewed by many, but not all, as the most effective treatment for severe, life-threatening catatonic regression. Working hypotheses for clinical, biochemical, and animal research are submitted at the end.

II. Early Autistic Regression

The first signs are noticed sometime during the first two years of life. It would be difficult to pinpoint an exact time of onset.

Leo Kanner, M. D. (1973)

Autistic regression or the loss of communicative and social skills seems to be a significant feature in about 15–37% of children who will be diagnosed with ASD in later preschool or elementary school (Davidovitch *et al.*, 2000; Kurita, 1985). The occurrence of a regressive phase does not necessarily imply that prior development was completely normal. It is possible that subtle abnormalities were present but escaped detection. For example, at 12 months of age, children in a high-risk sample, who later developed autism, had atypicalities in eye contact, visual tracking, disengagement of visual attention, imitation, social smiling, marked passivity at 6 months, followed by extreme stress reactions, and a tendency to fixate on particular objects in the environment (Zwaigenbaum *et al.*, 2005). Although it is not known if some of these children experienced a regression before age 2, it is conceivable that the earlier signs may have been missed, especially in the face of a subsequent period of rapid decline in communication and socialization.

Two recent studies support the concept of autistic regression before age 2. In a prospective study (Lord *et al.*, 2004), early loss of words (before age 2) was examined in a cohort of children who were referred for assessment of possible autism at age 2 or younger. About 25% of children with ASD were described by their parents as having used words meaningfully and losing this skill in the second year of life. Word loss was specific for autism, as it was not observed in other developmental disorders such as Down syndrome or Fragile X. Almost all of the children with ASD, whose parents reported loss of words or nonspecific vocalizations, also seemed to experience loss of social skills before or at the time of the word loss.

In another study (Werner and Dawson, 2005), home videotapes of 56 children's first and second birthday parties were rated blindly. Fifteen children, later diagnosed with ASD, were thought to have had autistic regression. Twenty-one children, later diagnosed with ASD, were reported to have early abnormalities before 1 year of age without a regressive phase. Findings were compared with 20 normally developing children. Infants in the ASD group with regression showed similar communication skills as the control group at 12 months, but at age 24 months, they had similar defects as the ASD group with early-onset of symptoms and without regression. These findings provide support for regressive and nonregressive patterns of development in autism.

The cause of this decline is unknown. Lainhart *et al.* (2002) assessed the presence of the broader autism phenotype in families with probands with and without autistic regression and found no difference. These findings suggest that the familial genetic liability is of equal magnitude (but not necessarily by the same genes) for regressive and nonregressive autism and environmental events are unlikely to be the sole determinants of autistic regression in young children. The authors hope that the identification of specific genetic loci associated with autism will provide an explanation for the occurrence of autistic regression in some children with autism. Biological and environmental correlates of autistic regression are currently unknown as well as effective treatments.

In some children with loss of speech but without concurrent social and nonverbal deterioration, a form of epilepsy, that is, Landau-Kleffner syndrome (LKS) or acquired epileptiform aphasia, is diagnosed (Mantovani, 2000). Some consider autistic regression with epileptiform EEG as a form of focal epilepsy or atypical LKS and an indication for anticonvulsant medications, steroids, and, on an experimental basis, vagus nerve stimulation (Park, 2003) or surgical treatment (multiple subpial transection) (Nass *et al.*, 1999; Patil and Andrews, 1998). However, the relations between autistic regression, epilepsy and epileptiform EEG are not straightforward as a significant proportion of nonepileptic children without regression also has epileptiform EEG. Moreover, the correlation between autistic symptoms, language improvements, and medication-induced EEG normalization is inconsistent (Mantovani, 2000; Tuchman and Rapin, 1997). There are no controlled medical or surgical trials in autistic regression with or without EEG abnormalities.

III. Late Autistic Regression

Massive autistic regression after age 2 (but before age 10, as per DSM-IV) is the hallmark of CDD. Key elements for this diagnosis are (apparently) normal behavior for at least 2 years, acute and massive regression before age 10 followed by abnormalities in social interactions, communication, and behavior patterns as observed in autistic children. According to DSM-IV, autistic regression in CDD needs to occur before age 10. The disorder was first described by Heller (1909) in Austria and was later called Heller's dementia or CDD. DSM criteria for CDD have only been published in the 4th edition (APA, 1994).

A pooled estimate across four surveys showed 1.7 per 100,000 subjects (95% confidence interval 0.6–3.8 per 100,000) (Fombonne, 2002). Males predominate. The relationship between CDD and AD is unclear. Similar to AD, CDD is occasionally observed in association with medical conditions, for example, metachromatic leucodystrophy, Schilder's disease, neurolipidoses, tuberous sclerosis, subacute sclerosing panencephalitis, and seizures (Corbett et al., 1977; Creak, 1963; Malhotra and Singh, 1993; Rapin, 1995; Rivinus et al., 1975). No etiology is found in most cases. Deterioration occurs over the course of months. Residual symptoms are impaired social interaction, restricted language output, and repetitive behaviors.

The support that CDD is a late-onset-variant of AD comes from clinical studies showing phenomenological similarities between CDD and AD (Kurita et al., 1992) and from one report showing co-occurrence of (high functioning) AD and CDD in half brothers (Zwaigenbaum et al., 2000). Follow-up studies have suggested that older age of onset of autistic symptoms, as in CDD, may be associated with worse outcome (Kurita et al., 1992; Mouridsen et al., 1998; Volkmar and Cohen, 1989).

A range of EEG abnormalities has been reported in patients with CDD, along with high rates of clinical seizures at follow-up (Mouridsen et al., 1999). In a review of 77 cases with CDD, EEG was obtained in 45 cases. Twenty cases had a normal EEG, 4 cases had a borderline-normal EEG, and 21 cases had an abnormal EEG. No unique abnormality was reported. The pattern of EEG abnormalities and seizure disorder at follow-up was similar between cases with CDD and AD. Seizure disorders at follow-up may be more frequent in CDD compared to AD (Mouridsen et al., 1999).

IV. Catatonic Regression in Autism

Catatonic regression has been reported in adolescents with AD and AsD (Brasic et al., 1999; Dhossche, 1998; Realmuto and August, 1991; Wing and Shah, 2000; Zaw et al., 1999), but not in CDD. There is however considerable symptom overlap between autistic regression in CDD and catatonia. In Table I, the symptom overlap between autistic regression in CDD and catatonia is assessed by comparing symptoms of autistic regression as described in a series of 18 CDD patients (Kurita et al., 1992), and catatonic symptoms as described in 30 cases of childhood catatonia culled from the world literature (Dhossche and

TABLE I
Comparison of Symptoms of Autistic Regression in CDD and Catatonia Regression in Children and Adolescents

	Autistic regression % ($N = 18$)	Catatonia regression % ($N = 30$)
Symptoms of autistic regression in CDD		
Loss of expressive or receptive language	100	87
Loss of social skills or adaptive behavior	100	100
Loss of bowel or bladder control	22	45
Loss of play	+	+
Loss of motor skills	+	+
Symptoms of catatonic regression		
Stereotypies	94	24
Compulsive behavior	83	+
Hyperkinesis	83	15
Posturing/grimacing	+	52
Stupor (immobility)	−	80
Staring/avoidance of eye contact	+	49
Negativism	+	38
Rigidity	+	38
Waxy flexibility	−	62
Echolalia/echopraxia	+	14
Automatic obedience	−	10

Bouman, 1997a). Most catatonic symptoms (except stupor, waxy flexibility, and automatic obedience) have been observed in the regressive phase of CDD. All regressive symptoms have been observed in catatonia.

The psychological and biological correlates of catatonic regression in people with and without autism are unknown (Fink and Taylor, 2003; Wing and Shah, 2000). Acute deterioration should prompt investigations for underlying medical and neurological disorders, particularly seizure disorders. Sometimes, seizures cause catatonia (Primavera *et al.*, 1994). In his original description of catatonia, Kahlbaum (1874) also reported seizure-like symptoms in catatonia. The sparse literature on EEG findings in catatonia has been summarized by Fink and Taylor (2003), who find meager evidence for a direct connection between EEG measures and catatonic symptoms. Diffuse slowing has been reported in patients in catatonic stupor. Others have reported a dysrythmic EEG in catatonia consistent with nonconvulsive status epilepticus that resolved when the catatonia resolved. In a review of 30 published cases of pediatric catatonia (Dhossche and Bouman, 1997a), EEG was done in 23 cases. Seven of 23 EEGs were read as abnormal. All positive cases showed nonspecific findings of diffuse slowing, except clear epileptiform activity in one case. Abnormal EEGs were found both in cases with and without neurological conditions.

V. Are Autistic and Catatonic Regression Related?

The considerable symptom overlap between autistic and catatonic regression supports that both conditions are related. However, stupor, waxy flexibility, and automatic obedience have not been described in autistic regression. An important difference between autistic and catatonic regression is age-of-onset. Catatonic regression is typically observed at a later age than autistic regression. Is it possible that differences in age-of-onset alter symptom expression?

This principle is illustrated by inspecting the symptoms profiles of congenital versus adult-onset syphilis and neonatal (cretinism) versus adult-onset hypothyroidism. Congenital syphilis and syphilitic infection acquired in adulthood have different symptoms but are caused by the same spirochete. Both congenital and adult infections respond to the same treatment, that is, penicillin. In addition, congenital and adult disease forms often differ in degree of reversibility. For example, untreated congenital hypothyroidism (cretinism) is irreversible, but adult hypothyroidism is not. In summary, differences in age-of-onset of the same disorder may explain similar but slightly different symptoms profiles, as well as differences in degree of reversibility of the disorder.

Stupor/immobility, waxy flexibility, and automatic obedience have not been described in autistic regression in CDD. There are several possible explanations for this discrepancy. Waxy flexibility and automatic obedience may go unrecognized as their presence need to be elicited for accurate recording. This argument does not apply to immobility or stupor as their presence is obvious. One possible explanation is that cases with stupor may not be recognized as autistic regression in older children because a medical etiology is presumed even if all diagnostic studies are negative and/or the stupor is protracted in many cases. Mortality may be high if appropriate treatments for catatonic stupor are not started in a timely fashion. Fink and Taylor (2003) distinguish between benign and malignant forms of catatonia. The malignant form is acute in onset, systematically devastating, and associated with fever and autonomic instability. These children look as if they have a central nervous system infection (Slooter *et al.*, 2005). However, no specific infection is usually found. Some patients are considered to suffer from a coma of unknown etiology.

Another possibility is that younger age may be protective against the occurrence of stupor or immobility (Dhossche, 2004). This contention is supported by the study of Wing and Shah (2000) who reported that 17% of all autistic patients (age 15 and older), attending a tertiary referral center for autism in the UK, met full criteria for the catatonic syndrome. None of those under age 15 had the full syndrome although isolated catatonic symptoms were often observed. It is possible that prominent psychomotor retardation only arises in older autistic children and adolescents.

VI. Theoretical Treatment Implications

Treatment implications would be far reaching if autistic regression represents an early form of catatonic regression but should be regarded as theoretical. There is currently no scientific, ethical, or practical basis to support clinical trials of ECT in early- or late-onset autistic regression in children.

There are currently no controlled trials in autistic regression. There are many limitations that treatment trials of early autism are facing. First, early diagnosis of autism is difficult. Most children with severe autism can be diagnosed accurately by age 3 by experienced clinicians. However, early symptoms may be missed for various reasons but mostly because of the lack of a foolproof diagnostic test. There may be different types of autistic regression. If it is true that anticatatonic treatments only work in certain type of autistic regression, then there is currently no way to identify this type because subtypes of autistic regression have not been defined yet.

In the next sections, the focus is on ECT that is regarded by many, but not all, as the most effective treatment for acute catatonia. It is also the most controversial psychiatric treatment, especially in children and adolescents. There is widespread anti-ECT sentiment. The general public (Lauber et al., 2005) as well as many medical professionals (Ghaziuddin et al., 2001) continues to be aversive toward the use of electrically induced seizures for therapeutic aims. However, ECT is also viewed as a safe, effective, and life-saving treatment for affective disorder, acute psychosis, and especially catatonia in people of all ages (Abrams, 1992). Rey and Walter (1997) reviewed the literature on the use of ECT in children and adolescents. They found that overall improvement rates were 42% for schizophrenia, 63% for depression, and 80% for mania and catatonia. Indications, efficacy, safety, and side effects of ECT in children and adolescents were similar as those in adults. However, large controlled trials are lacking.

If it is true that autistic regression is early catatonic regression, ECT may improve autistic regression in some children (Dhossche and Stanfill, 2004). The reasoning is logical, but unpractical. The safety of ECT in children is unknown. Clinical trials of ECT in early- or late-onset autistic regression in children have not been done yet. The potential use of ECT for severe catatonia in adolescents and adults with autism will be discussed in later chapters (see Chapters 13 and 14 by Zaw and Fink, et al.).

In the following sections, the history of ECT in childhood psychoses will be reviewed. The scant evidence that gamma-aminobutyric acid (GABA) may play a role in autistic regression, acute catatonia, and mode of action of ECT is summarized. Finally, future studies of the effects of electroconvulsive seizures (ECS)—an accepted animal model of ECT—on animal models of autism (and catatonia), in particular those that involve GABAergic deficits, are described.

VII. History of ECT in Childhood Psychoses

> Everything has been thought of before, but the problem is to think of it again.
> J. W. Von Goethe (1749–1832)

Seminal papers of Kanner (1944) and Asperger (1944–1991) delineating autism were published during the 1940s. About that same time, findings in two large studies on ECT in children were added to the literature (Bender, 1947; Heuyer et al., 1943). Autism was not assessed in these studies because the autistic syndrome was just then being recognized as a separate entity. Findings from these studies add little to the hypothesis that ECT may be effective in autistic children. However, reviewing these early ECT studies is useful as a means of examining aspects of ECT use in young children.

Heuyer et al. (1943), in studies done in Paris, reported on the effects of ECT (then a new procedure), in a group of children and adolescents between ages 5½–19. Forty ECT courses were used in 37 children and adolescents who were diagnosed with a variety of disorders. Details of the ECT (number of treatments, frequency, treatment course duration, ECT device, and electrical parameters) were not reported. Most frequent diagnoses were "démence précoce" ($n = 6$), "troubles du caractère" ($N = 5$), and "tics" ($N = 5$). Electroconvulsive therapy was found effective in children with affective disorders (depression or mania) and in some cases with severe tic disorder. In other diagnoses, the effect was equivocal. Lang (1997) noted in Heuyer's biography that Heuyer did not find ECT, cardiazol, or any other biological treatment particularly useful in childhood psychoses.

Most importantly, none of the children had a worsening of symptoms. Children tolerated the procedure well and no adverse effects on cognitive development were found at 18 months' follow-up. The authors concluded: "*En neuropsychiatrie infantile, l'électrochoc est une méthode qui est sans danger et qui rend des services dans les limites que nous avons indiquées.*" (In child neuropsychiatry, ECT is a safe treatment that is useful in certain conditions—translation of Dirk M. Dhossche).

In the years 1942–1947, at New York City's Bellevue Hospital, Bender (1947) did the largest study ever done before or since on ECT and children. Ninety-eight children under the age of 12, 70 boys and 28 girls, were treated with ECT. The youngest patient was 4 years old. Typically, the children's treatment course consisted of 20 daily sessions of unmodified ECT (an early form of the treatment in which no muscle relaxants were used). Bender (1947) diagnosed all the children as schizophrenic based on "*pathology in behavior at every level and in every area of integration or patterning within the functioning of the central nervous system be it vegetative, motor, perceptual, intellectual, emotional, and social.*" About one-third of the children first showed symptoms within the first year of life and one-fourth between 3 and 5 years. In the rest, the disorder occurred later.

Almost all children improved after one course of ECT but in very few did their symptoms remit completely. Bender (1947) found the children: "*less disturbed, less excitable, less withdrawn, and less anxious. They were better controlled, seemed better integrated and more mature and were better able to meet social situations in a realistic fashion. They were composed, happier, and were better able to accept teaching or psychotherapy in groups or individually.*" Most of the children had minimal side effects. ECT did not harm their cognitive functioning and development.

The studies of Heuyer and Bender attest to the safety and feasibility of ECT in children. Their main limitation is the uncertainty about diagnostic criteria. Bender, for example, used criteria for schizophrenia that by modern standards are considered over broad and nonspecific. Therefore, it is unclear how these children would be diagnosed today. Now it is assumed that schizophrenic disorders as defined by DSM are very rare in prepubertal children and cannot be diagnosed reliably in children younger than 6 years. If any of the children in Bender's study met current criteria for autism, they should be found in the group with early-onset (before age 3). However, further speculation is futile because modern criteria for autism were not applied. The same diagnostic limitation applies to the French study.

Another limitation is the use of older ECT techniques. Since the time of Heuyer and Bender, ECT technique has greatly improved. It now consists of the following: pretreatment work-up, a schedule of 2 or 3 treatments per week, use of muscle relaxants, controlled stimulus selection, close seizure monitoring, and continuous electrocardiographic monitoring.

VIII. GABA in Autistic Regression

There have been no studies so far on the role of GABA dysfunction in autistic regression. In fact, there are no studies in the literature, at least to our knowledge, on any biological correlate of autistic regression. The issue warrants future research attention. However, GABA dysfunction has been implicated in the etiology and pathophysiology of autism (Dhossche *et al.*, 2002; Hussman, 2001). As it is still unclear how regressive and nonregressive autism are different, a review is given in the next sections on the possible role of GABA dysfunction in autism.

Hussman (2001) speculated that autism reflects dysfunction in a single factor, that is, decreased GABA inhibition, shared in common by many systems. In his model that recognizes the possibility of multiple etiological factors, suppression of GABAergic inhibition results in excessive stimulation of glutamate-specialized neurons and loss of sensory gating. Loss of inhibitory control may cause deterioration in the quality of sensory information due to the failure to suppress competing "noise," resulting in compensatory restrictions in sensory input to a narrow, repetitive, or controllable scope.

Formal criteria for a comprehensive theory of autism have also been formulated (Dhossche *et al.*, 2002). In brief, any viable theory must consider the protean nature of symptoms, course, and outcome in autistic people and should account for: (1) the early-onset of clinical abnormalities, (2) worsening of symptoms around puberty in a considerable number of patients, (3) the association between autism and epilepsy, and (4) genetic transmission of the disorder.

The evidence that central GABA dysfunction can account for these key features is briefly reviewed as follows:

1. The main inhibitory neurotransmitter in the mature brain is GABA as perhaps 25–40% of all terminals contain GABA. In early development, GABA has an excitatory trophic role affecting neuronal wiring, plasticity of neuronal network, and neural organization (Roberts, 2000). Interference with the trophic role of GABA may affect development of neuronal wiring, plasticity of neuronal network, and neural organization (Christie *et al.*, 2002; Herlenius and Lagercrantz, 2001; Varju *et al.*, 2001). For example, in mice, developmental changes in inhibitory synaptic currents in cerebellar neurons are determined primarily by developmental changes in $GABA_A$ receptor subunit expression. Overall, the effects of abnormal trophic GABA function could account for the brain abnormalities reported so far in autistic people (Courchesne *et al.*, 1988, 2001).

2. Decreased GABA inhibition in the hypothalamus is considered as an important trigger for onset of puberty (Genazzani *et al.*, 2000; Mitushima *et al.*, 1994). Adaptive changes in GABA function at the onset of, or during, puberty may worsen or induce disorders associated with underlying abnormalities of GABA function. Increased rates of seizure disorders (Deykin and MacMahon, 1979; Gillberg, 1991b), catatonia (Wing and Shah, 2000), and worsening of autistic symptoms or overall behavioral deterioration (Gillberg, 1991a; Gillberg and Schaumann, 1981) have been reported in autistic people around and after puberty.

3. In epilepsy, GABA has been strongly implicated (Petroff *et al.*, 1996). About 30% of autistic people develop some type of epilepsy (Gillberg, 1991b).

4. Genetic studies have implicated the proximal long arm of chromosome 15 in autism (Bakker *et al.*, 2003; Borgatti *et al.*, 2001; Cook *et al.*, 1997; Lauritsen *et al.*, 1999) and catatonia (Stöber *et al.*, 2000, 2002). Three $GABA_A$ receptor subunits genes (*GABRB3*, *GABRA5*, and *GABRG3*) are located at the chromosome location that includes the Prader–Willi syndrome (PWS)/AS region. Animal and human studies have suggested a role for these genes in the phenotypic expression of PWS (Ebert *et al.*, 1997) and Angelman syndrome (AS) (DeLorey *et al.*, 1998; Odano *et al.*, 1996). *GABRB3* have been associated with autism in several studies (Cook *et al.*, 1998; Menold *et al.*, 2001; Nurmi *et al.*, 2003), especially

in patients with increased levels of "insistence on sameness" (a composite score of difficulties with minor changes in personal routine or environment, resistance to trivial changes in environment, and compulsions/rituals) (Shao *et al.*, 2003) and savant skills (Nurmi *et al.*, 2003). A recent study found evidence for involvement in autism of *GABRA4*, possibly through interaction with *GABRB1*, another $GABA_A$ receptor subunit gene also located on chromosome 4p12 (Ma *et al.*, 2005).

Empirical support for GABA dysfunction in autism is limited. Elevated plasma GABA levels in autistic children were found in a case-report (Cohen, 1999) and case-series (Dhossche *et al.*, 2002). Reduced $^3[H]$-flunitrazepam labeled benzodiazepine bindings sites and $^3[H]$-muscimol labeled $GABA_A$ receptors have been reported in the hippocampus of autistic people (Blatt *et al.*, 2001), providing direct evidence of abnormal benzodiazepine-GABA receptor complexes in this brain region. Glutamic acid decarboxylase (GAD) is the enzyme responsible for normal conversion of glutamate to GABA in the brain. GAD exists in two isoenzymes, GAD_{65} and GAD_{67}, which are products of two independently regulated genes (Soghomonian and Martin, 1998). In a post-mortem study (Fatemi *et al.*, 2002), brain levels of both isoenzymes were reduced approximately by 50% in the parietal and cerebellar cortices of autistic patients. GAD deficiency in autism may be due to, or associated with, brain abnormalities in levels of glutamate/GABA, or transporter/receptor density.

IX. GABA in Catatonic Regression

The single most important observation leading to believe that GABA dysfunction plays a role in catatonic regression is the often dramatic response to treatment with benzodiazepines, that is, positive modulators of the benzodiazepine/$GABA_A$ receptor complex (Bush *et al.*, 1996; Fricchione *et al.*, 1983). Other effective treatments for catatonia, that is, barbiturates, zolpidem, carbamazepine, and ECT (Green, 1986; Sanacora *et al.*, 2003), also enhance GABA function. Efficacy of serotonergic agents and antipsychotics in catatonia has been less well documented but seems less consistent. In fact, many psychotropic medications, including antipsychotics and antidepressants, have been associated anecdotally with the emergence of catatonia (Fink and Taylor, 2003).

If one assumes a central role of GABA dysfunction in catatonia, the scope of GABA functions in the normal brain should allow the expression of catatonia when deficiencies in GABA function develop. Roberts (2000) has formulated a central role of GABA as neurotransmitter used by neurons that exert tonic inhibition of neural circuits for innate or learned behavioral sequences. In its extreme forms, catatonia is characterized by immobility alternating with purposeless agitation. Both behaviors can be viewed as opposite primitive reflexes in response to overwhelming stress or danger that are expressed when innate, genetically preprogrammed neuronal circuits are released from tonic inhibition. Following Jackson's (1958) hierarchical concept of dissolution, immobility or hyperactivity are "positive" symptoms caused by the removal of the influence of higher centers. The neuronal circuitry that is involved in catatonia is not well defined, but probably involves frontal cortex, parietal cortex, basal ganglia, and the cerebellum (see Chapter 9 by Dhossche *et al.*).

A few general criteria for any viable GABA theory of catatonia are as follows:

1. The theory has to accommodate findings from GABAergic theories of schizophrenia and affective disorders as catatonia occurs in both disorders.
2. GABA dysfunction in the hypothalamus may account for severe autonomic dysfunction in malignant catatonia.
3. Treatments that relieve catatonia should enhance GABA function, directly or indirectly.
4. Genetic studies in catatonia, schizophrenia, and affective disorders should provide some support for involvement for genes affecting GABA function.

Evidence that GABA dysfunction in catatonia satisfies these criteria is summarized as follows:

1. GABA theories have been formulated for schizophrenia (Roberts, 1972; van Kammen, 1977), psychosis (Kalkman and Loetscher, 2003; Keverne, 1999), and affective disorders (Brambilla *et al.*, 2003; Emrich *et al.*, 1980; Petty, 1995) including bipolar disorder (Petty *et al.*, 1993).

2. A prominent neurotransmitter in the hypothalamus is GABA (Decavel and van den Pol, 1990) suggesting the importance of inhibitory hypothalamic circuits for regulation of stress responses by the hypothalamic–pituitary–adrenal (HPA) (Engelmann *et al.*, 2004; Herman and Cullinan, 1997).

3. Most currently used psychotropic medications, including benzodiazepines, antipsychotics (Zink *et al.*, 2004), selective serotonin reuptake inhibitors (Bhagwagar *et al.*, 2004; Sanacora *et al.*, 2002; Tunnicliff *et al.*, 1999), phenelzine (Baker *et al.*, 1991; McManus *et al.*, 1992), and anticonvulsants seem to enhance GABA neurotransmission, albeit through different mechanisms. The possible role of GABA in the mechanism of action of ECT will be discussed in the next section.

4. There is some evidence that GABA-related genes are involved in affective disorder and schizophrenia. *GABRA5* has been associated with bipolar disorder in two studies (Otani *et al.*, 2005; Papadimitriou *et al.*, 1998). In a genome scan of catatonia, a linkage signal in the region 15q11.2-q21.1 was found (Stöber *et al.*, 2000). There are no (family-based) gene association studies in catatonia available in the literature. Findings from a family-based association study in a sample of

children and adolescents with childhood-onset schizophrenia (COS) ($n = 72$) suggested that the gene encoding GAD (67) may be a common risk factor for schizophrenia (Addington et al., 2005).

Empirical evidence for GABA dysfunction in catatonia comes from one single receptor-imaging study. Findings in a benzodiazepine ligand-binding study of catatonic patients have shown a decreased density of $GABA_A$ receptors in the left sensorimotor cortex in akinetic catatonia (Northoff et al., 1999). Other, more circumstantial, evidence comes from biochemical and genetic studies. Cerebrospinal fluid (CSF) levels of GABA were decreased in 11 patients with Neuroleptic Malignant syndrome, a condition that may be related to catatonia, compared with 8 controls (Nisijima and Ishiguro, 1995). In that study, levels of noradrenalin were increased, but levels of 5-hydroxyindoleacetic acid (5-HIAA), serotonin's main metabolite, were slightly, but not significantly, decreased. Further evidence of impaired GABA function in catatonia must await future studies.

X. GABA and the Mechanism of Action of ECT

> Like Man, ECT is at the end of an evolutionary line, but, also like Man, rather than facing imminent extinction it is flourishing. I do not see this millennium bringing any exciting new advances in ECT instrumentation or technique—indeed, it is hard to see how the treatment might be further improved at this point other than through refinements in patient selection, prediction of response, and more effective dissemination of knowledge.
> Richard Abrams, M. D. (2002)

The mechanism of action of ECT in affective disorders, psychosis, and catatonia is unknown. Abrams (2002) finds ECT awaiting its Lavoisier. Nonetheless, enhanced GABA function remains a viable neurochemical mechanism of action of ECT. ECT-induced enhancement of GABA function may explain the increase in seizure threshold that typically occurs across an ECT course, as supported by rodent studies (Green, 1986). Restoration of GABA inhibitory hypothalamic function may be pivotal. This would be in line with the diencephalic hypothesis of the mechanism of action of ECT (Abrams, 2002; Fink and Nemeroff, 1989).

A few studies suggest a direct role of GABA in ECT. In an iomazenil-SPECT study, increased benzodiazepine receptor uptake in cortical regions (except temporal cortices) was found 1 week after successful bitemporal ECT (Mervaala et al., 2001). In an MRS study, two-fold increased brain GABA levels were found in depressed patients after a course of ECT (Sanacora et al., 2003). The small number of patients precluded making correlations between clinical improvement and increased brain GABA level. Previously, it was reported that CSF GABA increased by 50% after ECT (Lipcsey et al., 1986). In another MRS study, cortical glutamate–glutamine levels in the left anterior cingulum of depressed patients normalized after ECT but only in responders (Pfleiderer et al., 2003). In nonresponders, levels remained low. Limitations in the study's MRS methodology did not allow obtaining separate measurements for GABA because of overlapping resonances of glutamate, glutamine, and GABA.

The reported increase in brain or CSF plasma after ECT is contrary to findings in a study of plasma GABA in depressed patients treated with ECT. Plasma GABA levels tended to decrease for about 1 hour after ECT (Devenand et al., 1995). Limitations of this study include variable storage times known to increase GABA levels in plasma and CSF and PRN administration of chloral hydrate that enhances GABA function similar as barbiturates, during the ECT course.

In summary, there are some direct evidences of a role of GABA in the mechanism of action of ECT, but larger studies need to demonstrate positive correlations between GABAergic measures and clinical changes.

XI. ECS in GABA Models of Autism

Animal models provide a way to examine the effects of ECS—an accepted animal model and experimental analogue of ECT—on neurodevelopmental disorders.

Arguably, the lack of definite molecular mechanisms associated with autism and subsequent uncertainty of the appropriateness of various animal models are the most important obstacles for finding rational targets for effective prevention and treatment in autism (Andres, 2002; Murcia et al., 2005; Seong et al., 2002). Although GABA dysfunction may not be present in all cases of autism, there is evidence that abnormalities in $GABA_A$ receptor may be present in at least a subgroup of people with autism (Blatt et al., 2001), as well as in other related disorders, for example, AS (DeLorey et al., 1998; Odano et al., 1996). The effect of ECS is yet to be tested in animal models of autism. Next, two putative GABAergic models of autism, that is, the valproic acid (VPA)-treated rodent and GABRB3 mutant, are proposed to be suitable models to assess the effects of ECS on GABA function and $GABA_A$ receptors.

First, maternal exposure to valproate during pregnancy is associated with higher (and dose-dependent) risks for malformation compared with other antiepileptics (Vajda and Eadie, 2005). Prenatal exposure to sodium valproate was the antiepileptic most commonly associated with autism in the study of Rasalam et al. (2005). None of the children with fetal anticonvulsant syndrome-associated autism (including those exposed to valproate) seemed to have a regressive phase or loss of skills.

Prenatal exposure to VPA is a promising animal model of nonregressive autism (Schneider and Przewlocki, 2005). Offspring of female rats injected with VPA on day 12.5 of gestation show brain abnormalities including smaller cerebella with fewer Purkinje cells, similar as in autistic patients (Ingram *et al.*, 2000; Rodier *et al.*, 1996). In addition, the rats exhibit autistic-like behaviors that appear before puberty: (1) lower sensitivity to pain and higher sensitivity to nonpainful stimuli, (2) diminished acoustic prepulse inhibition, (3) locomotor and repetitive/stereotypic-like hyperactivity combined with lower exploratory activity, and (4) decreased number of social behaviors and increased latency to social behaviors. The role of GABA in this model of autism has not yet been examined. However, many of the effects of VPA are mediated via GABA receptors (Owens and Nemeroff, 2003).

A second model consists of mutant mice with deficiency of the beta 3 subunit of the $GABA_A$ receptor. The beta 3 subunit gene has been associated with autism or a subgroup of autism in a few studies (Cook *et al.*, 1998; Martin *et al.*, 2000; Nurmi *et al.*, 2003; Shao *et al.*, 2003). The association between the beta 3 subunit gene and autistic regression has not been examined yet. Autistic-like behaviors are present in these mutants (DeLorey *et al.*, 1998). Targeted disruption of beta 3 subunit gene in mice results in 50% reduction in $GABA_A$ receptor binding in the neonatal and adult brains. These mice exhibit impaired GABA function and autistic like behaviors. This includes seizures, learning and memory deficits, repetitive behavior, hyperactivity, and a disturbed rest–activity cycle.

Heterozygotes display less behavioral abnormalities with phenotypic features indermediate between knockout and wild types. There is also one study supporting (partial) genomic imprinting of the *GABRB3* gene, at least in mice (Liljelund *et al.*, 2005). The issue is important because of the genomic imprinting that occurs in Angelman and PWS in humans. Findings suggest that both homozygous and heterozygous mice could be useful to model certain aspects of beta 3 subunit deficient expression.

XII. ECS and Purkinje Cell Survival

Purkinje cell survival is a potential outcome measure for treatment studies of autism. One of the most consistent findings in autism is the selective vulnerability of the cerebellar Purkinje cell (Kemper and Bauman, 1993; Ritvo *et al.*, 1986). Purkinje cells are large inhibitory neurons that receive input from both parallel fibers (from granule cells) and climbing fibers (from the inferior livery nucleus). GABA is the major inhibitory neurotransmitter in the mammalian cerebellum. Cerebellar granule, Purkinje, and deep nuclear neurons are known to receive GABAergic afferents. A high metabolic demand combined with constant input from the inferior olive and large amounts of calcium stores and influx makes the Purkinje neuron vulnerable to oxidative stress causing necrotic cell death. Functional deficiencies in $GABA_A$ receptors may interfere with cerebellar development through imbalances between excitatory and inhibitory pathways and changed expression of genes involved in apoptosis, calcium regulation, neurodegeneration, and neurotransmission.

The exact timing of abnormal processes in autism is uncertain. Absence of gliosis in an autopsy study (Bauman *et al.*, 1997) suggests that abnormalities in autism are not a result of insult after birth. However, other studies in autism (Ahlsen *et al.*, 1993; Bailey *et al.*, 1998) support that nerve cell damage and Purkinje cell loss occur postnatally. This would be consistent with the occurrence of autistic regression in a subgroup of children. GABAergic models provide an opportunity to assess the effects of ECS on Purkinje cell survival.

XIII. ECS and Neurogenesis

It has been known for several years that seizure activity increases neurogenesis of dentate granule cells (Bengzon *et al.*, 1997; Parent *et al.*, 1997) and in the hippocampus (Masden *et al.*, 2000; Scott *et al.*, 2000) as well as in the frontal cortex (Madsen *et al.*, 2005) in adult rodents. Hellsten *et al.* (in press) reported that ECS-treatment in adult rats increased the number of endothelial cells in the dentate gyrus of the hippocampus by 30%, resulting in a 16% increase in vessel length. Neurogenesis and angiogenesis may be important for the therapeutic effects of ECT and antidepressant medications in mood disorders.

The effect of seizures on the developing brain has been examined in previous studies, but many questions remain (Holmes *et al.*, 2002). Single prolonged seizures in immature rats do not seem to cause cell loss, sprouting, or impairment of learning, memory, and behavior. However, recurrent seizures during the first week of life result in impairment of learning and a lower seizure threshold when rats are studied as adults (Holmes, 2005; Holmes *et al.*, 1998).

Wasterlain found a reduction of cell number in developing rat brain after seizures (Wasterlain, 1976, 1978; Wasterlain and Plum, 1973), suggesting a reduction rather than increase of neurogenesis in the immature brain after seizures. These findings were supported in a recent study indicating that recurrent seizures in the neonatal rat are associated with reductions of newly born granule cells (McCabe *et al.*, 2001). The authors warn that recurrent seizures during early life are different from those occurring in the mature brain and can have detrimental effects on brain development.

The literature lacks reflection on possible therapeutic effects of seizure activity in the immature brain mainly because the concept is still foreign to neurological thinking and practice. Differential effects of spontaneous versus induced seizures have not been studied well. Macrocephaly (Bailey *et al.*, 1993; Dementieva *et al.*, 2005; Kanner, 1943) and abnormal brain growth (Courchesne *et al.*, 2001), particularly early overgrowth (Courchesne *et al.*, 2003), are probable indicators of abnormal neurogenesis or apoptosis in autism. These processes will be intensely studied in the future. At the same time, more data on seizure-induced changes in progenitor cells in the mature and immature brain will become available. Possible therapeutic effects of induced seizures on abnormal neurogenesis should not be dismissed prematurely.

Acknowledgment

Research support from the Thrasher Research Fund, Salt Lake City, Utah, is gratefully acknowledged.

References

Abrams, R. (1992). "Electroconvulsive Therapy," 2nd ed. Oxford University Press, New York.
Abrams, R. (2002). "Electroconvulsive Therapy," 4th ed. Oxford University Press, New York.
Addington, A. M., Gornick, M., Duckworth, J., Sporn, A., Gogtay, N., Bobb, A., Greenstein, D., Lenane, M., Gochman, P., Baker, N., Balkissoon, R., Vakkalanka, R. K., Addington, A. M., Gornick, M., Duckworth, J., Sporn, A., Gogtay, N., Bobb, A., Greenstein, D., Lenane, M., Gochman, P., Baker, N., Balkissoon, R., Vakkalanka, R. K., Weinberger, D. R., Rapoport, J. L., and Straub, R. E. (2005). GAD1 (2q31.1), which encodes glutamic acid decarboxylase (GAD67), is associated with childhood-onset schizophrenia and cortical gray matter volume loss. *Mol. Psychiatry* **10,** 581–588.
Ahlsen, G., Rosengren, L., Belfrage, M., Palm, A., Haglid, K., and Hamberger, A. (1993). Glial fibrillary acidic protein in the cerebrospinal fluid of children with autism and other neuropsychiatric disorders. *Biol. Psychiatry* **33,** 734–743.
American Psychiatric Association (APA) (1994). "Diagnostic and Statistical Manual of Mental Disorders," 4th ed. APA Press, Washington, DC.
Andres, C. (2002). Molecular genetics and animal models in autistic disorder. *Brain Res. Bull.* **57,** 109–119.
Asperger, H. (1944). Autistic psychopathy in childhood. *In* "Autism and Asperger's Syndrome" (U. Frith, Ed.). Cambridge University Press, Cambridge.
Bailey, A., Luthert, P., Bolton, P., Le Couteur, A., Rutter, M., and Harding, B. (1993). Autism and megalencephaly (letter). *Lancet* **341,** 1225–1226.
Bailey, A., Luthert, P., Dean, A., Harding, B., Janota, I., and Montgomery, M. (1998). A clinicopathological study of autism. *Brain* **121,** 889–905.
Baker, G., Wong, J., Yeung, J., and Coutts, R. (1991). Effects of the antidepressant phenelzine on brain levels of gamma-aminobutyric acid (GABA). *J. Affect. Disord.* **21,** 207–211.

Bakker, S., van der Meulen, E., Buitelaar, J., Sandkuijl, L., Pauls, D., Monsuur, A., van't Slot, R., Minderaa, R., Gunning, W., Pearson, P., and Sinke, R. (2003). A whole-genome scan in 164 Dutch sib pairs with attention-deficit/hyperactivity disorder: Suggestive evidence for linkage on chromosomes 7p and 15q. *Am. J. Hum. Genet.* **72,** 1251–1260.

Bauman, M., Filipek, P., and Kemper, T. (1997). Early infantile autism. *In* "The Cerebellum and Cognition" (J. Schmahmann, Ed.), pp. 367–386. Academic Press, New York.

Bender, L. (1947). One hundred cases of childhood schizophrenia treated with electric shock. *Trans. Am. Neurol. Soc.* **72,** 165–169.

Bengzon, J., Kokaia, Z., Elmer, E., Nanobashvili, A., Kokaia, M., and Lindvall, O. (1997). Apoptosis and proliferation of dentate gyrus neurons after single and intermittent limbic seizures. *Proc. Natl. Acad. Sci. USA* **94,** 10432–10437.

Bhagwagar, Z., Wylezinska, M., Taylor, M., Jezzard, P., Matthews, P., and Cowen, P. (2004). Increased brain GABA concentrations following acute administration of a selective serotonin reuptake inhibitor. *Am. J. Psychiatry* **161,** 368–370.

Blatt, G., Fitzgerald, C., Guptill, J., Booker, A., Kemper, T., and Bauman, M. (2001). Density and distribution of hippocampal neurotransmitter receptors in autism: An autoradiographic study. *J. Autism Dev. Disord.* **31,** 537–543.

Borgatti, R., Piccinelli, P., Passoni, D., Dalpra, L., Miozzo, M., Micheli, R., Gagliardi, C., and Balottin, U. (2001). Relationship between clinical and genetic features in "inverted duplicated chromosome 15" patients. *Pediatr. Neurol.* **24,** 111–116.

Brambilla, P., Perez, J., Barale, F., Schettini, G., and Soares, J. (2003). GABAergic dysfunction in mood disorders. *Mol. Psychiatry* **8,** 721–737.

Brasic, J., Zagzag, D., Kowalik, S., Prichep, L., John, E., Liang, H., Klurchko, B., and Cancro, R. (1999). Progressive catatonia. *Psychol. Rep.* **84,** 239–246.

Bush, G., Fink, M., Petrides, G., Dowling, F., and Francis, A. (1996). Catatonia. II. Treatment with lorazepam and electroconvulsive therapy. *Acta Psychiatr. Scand.* **93,** 137–143.

Christie, S., Miralles, C., and De Blas, A. (2002). GABAergic innervation organizes synaptic and extrasynaptic GABA$_A$ receptor clustering in cultured hippocampal neurons. *J. Neurosci.* **22,** 684–697.

Cohen, B. (1999). Elevated levels of plasma and urine gamma-aminobutyric acid: A case study of an autistic child (letter). *Autism* **3,** 437–440.

Cook, E., Lindgren, V., Leventhal, B., Courchesne, R., Lincoln, A., Shulman, C., Lord, C., and Courchesne, E. (1997). Autism or atypical autism in maternally but not paternally derived proximal 15q duplication. *Am. J. Hum. Genet.* **60,** 928–934.

Cook, E., Courchesne, R., Cox, N., Lord, C., Gonen, D., Guter, S., Lincoln, A., Nix, K., Haas, R., Leventhal, B., and Courchesne, E. (1998). Linkage-disequilibrium mapping of autistic disorders, with 15q11-13 markers. *Am. J. Hum. Genet.* **62,** 1077–1083.

Corbett, J., Harris, R., Taylor, E., and Trimble, M. (1977). Progressive disintegrative psychosis of childhood. *J. Child Psychol. Psychiatry* **18,** 211–219.

Courchesne, E., Yeung-Courchesne, R., Press, G., Hesselink, J., and Jernigan, T. (1988). Hypoplasia of cerebellar vermis lobules VI and VII in autism. *N Engl. J. Med.* **318,** 1349–1354.

Courchesne, E., Karns, C. M., Davis, H. R., Ziccardi, R., Carper, R. A., Tigue, Z. D., Chisum, H. J., Moses, P., Pierce, K., Lord, C., Lincoln, A. J., Pizzo, S., Schreibman, L., Haas, R. H., Akshoomoff, N. A., and Courchesne, R. Y. (2001). Unusual brain growth patterns in early life in patients with autistic disorder: An MRI study. *Neurology* **24,** 245–254.

Courchesne, E., Carper, R., and Akshoomoff, N. (2003). Evidence of brain overgrowth in the first year of life in autism. *JAMA* **290,** 337–344.

Creak, E. (1963). Childhood psychosis: A review of 100 cases. *Br. J. Psychiatry* **109,** 84–89.

Davidovitch, M., Glick, L., Holtzman, G., Tiroh, E., and Safir, M. (2000). Developmental regression in autism: Maternal perception. *J. Autism Dev. Disord.* **30,** 113–119.

Decavel, C., and van den Pol, A. (1990). GABA: A dominant neurotransmitter in the hypothalamus. *J. Comp. Neurol.* **302,** 1019–1037.

DeLorey, T., Handforth, A., Anagnostaras, S., Homanics, G., Minsassian, B., Asatourian, A., Fanselow, M., Delgado-Escueta, A., Ellison, G., and Olsen, R. (1998). Mice lacking the beta3 subunit of the GABAA receptor have the epilepsy phenotype and many of the behavioral characteristics of Angelman syndrome. *J. Neurosci.* **18,** 8505–8514.

Dementieva, Y., Vance, D., Donnelly, S., Elston, L., Wolpert, C., Ravan, S., DeLong, G., Abramson, R., Wright, H., and Cuccaro, M. (2005). Accelerated head growth in early development of individuals with autism. *Pediatr. Neurol.* **32,** 102–108.

Devenand, D., Shapira, B., Petty, F., Kramer, G., Fitzsimons, L., Lerer, B., and Sackheim, H. (1995). Effects of electroconvulsive therapy on plasma GABA. *Convulsive Ther.* **11,** 3–13.

Deykin, E., and MacMahon, B. (1979). The incidence of seizures among children with autistic symptoms. *Am. J. Psychiatry* **136,** 1310–1312.

de Sanctis, S. (1908). Dementia praecocissima catatonica oder Katatonie des frühen Kindesalters? *Folia Neuro-biologica* **2,** 9–12.

Dhossche, D., and Bouman, N. (1997a). Catatonia in an adolescent with Prader-Willi Syndrome. *Ann. Clin. Psychiatry* **4,** 247–253.

Dhossche, D., and Bouman, N. (1997b). Catatonia in children and adolescents. *J. Am. Acad. Child Adolesc. Psychiatry* **36,** 870–871.

Dhossche, D. (1998). Catatonia in autistic disorders (brief report). *J. Autism Dev. Disord.* **28,** 329–331.

Dhossche, D., Applegate, H., Abraham, A., Maertens, P., Bencsath, A., Bland, L., and Martinez, J. (2002). Elevated plasma GABA levels in autistic youngsters: Stimulus for a GABA hypothesis of autism. *Med. Sci. Monit.* 1–6.

Dhossche, D. (2004). Autism as early expression of catatonia. *Med. Sci. Monit.* **10,** 31–39.

Dhossche, D., and Stanfill, S. (2004). Could ECT be effective in autism? *Med. Hypotheses* **63,** 371–376.

Ebert, M., Schmidt, D., Thompson, T., and Butler, M. (1997). Elevated gamma-aminobutyric acid (GABA) levels in individuals with either Prader-Willi syndrome or Angelman syndrome. *J. Neuropsychiatry Clin. Neurosci.* **9,** 75–80.

Emrich, H., von Zerssen, D., Kissling, W., Moller, H., and Windorfer, A. (1980). Effect of sodium valproate on mania. The GABA-hypothesis of affective disorders. *Archiv fur Psychiatrie und Nervenkrankheiten* **229,** 1–16.

Engelmann, M., Landgraf, R., and Wotjak, C. (2004). The hypothalamic-neurohypophysical system regulates the hypothalamic-pituitary-adrenal axis under stress: An old concept revisited. *Front. Neuroendocrinol.* **25,** 132–149.

Fatemi, S., Halt, A., Stary, J., Kanodia, R., Schulz, S., and Realmuto, G. (2002). Glutamic acid decarboxylase 65 and 67 kDa proteins are reduced in autistic parietal and cerebellar cortices. *Biol. Psychiatry* **52,** 805–810.

Fink, M., and Nemeroff, C. (1989). A neuroendocrine view of ECT. *Convulsive Ther.* **5,** 296–304.

Fink, M., and Taylor, M. (2003). "Catatonia: A Clinician's Guide to Diagnosis and Treatment." Cambridge University Press, Cambridge.

Fombonne, E. (2002). Prevalence of childhood disintegrative disorder. *Autism* **6,** 149–157.

Fricchione, G., Cassem, N., Hooberman, D., and Hobson, D. (1983). Intravenous lorazepam in neuroleptic-induced catatonia. *J. Clin. Psychopharmacol.* **3,** 338–342.

Genazzani, A., Bernardi, F., Monteleone, P., Luisi, S., and Luisi, M. (2000). Neuropeptides, neurotransmitters, neurosteroids, and the onset of puberty. *Ann. NY Acad. Sci.* **900,** 1–9.

Ghaziuddin, N., Kaza, M., Ghazi, N., King, C., Walter, G., and Rey, J. (2001). Electroconvulsive therapy for minors: Experiences and attitudes of child psychiatrists and psychologists. *J. ECT* **17,** 109–117.

Gillberg, C., and Schaumann, H. (1981). Infantile autism and puberty. *J. Autism Dev. Disord.* **11,** 365–371.

Gillberg, C. (1991a). Outcome in autism and autistic-like symptoms. *J. Am. Acad. Child Adolesc. Psychiatry* **30,** 375–382.

Gillberg, C. (1991b). The treatment of epilepsy in autism. *J. Autism Dev. Disord.* **21,** 61–77.

Green, A. (1986). Changes in gamma-aminobutyric acid biochemistry and seizure threshold. *Ann. NY Acad. Sci.* **462,** 105–119.

Heller, T. (1909). Über Dementia infantilis (Verblödungsprozeß im Kindesalter). *Zeitschrift für die Erforschung und Behandlung des jugendlichen Schwachsinns auf wissenschaftlicher Grundlage* **2,** 17–28.

Hellsten, J., West, M. J., Arvidsson, A., Ekstrand, J., Jansson, L., Wennström, M., and Tingström, A. (in press). Electroconvulsive seizures induce angiogenesis in adult rat hippocampus. *Biol. Psychiatry*.

Herlenius, E., and Lagercrantz, H. (2001). Neurotransmitters and neuromodulators during early human development. *Early Hum. Dev.* **65,** 21–37.

Herman, J., and Cullinan, W. (1997). Neurocircuitry of stress: Central control of the hypothalamo-pituitary-adrenocortical axis. *Trends Neurosci.* **20,** 78–84.

Heuyer, G., Bour, and Leroy, R. (1943). L'électrochoc chez les enfants. *Ann. Med. Psychol. (Paris)* **2,** 402–407.

Holmes, G., Gaiarsa, J.-L., Chevassus-Au-Louis, N., and Ben-Ari, Y. (1998). Consequences of neonatal seizures in the rat: Morphological and behavioral effects. *Ann. Neurol.* **44,** 845–857.

Holmes, G., Kahzipov, R., and Ben-Ari, Y. (2002). Seizure-induced damage in the developing human: Relevance of experimental models. *Prog. Brain Res.* **135,** 321–334.

Holmes, G. (2005). Effects of seizures on brain development: Lessons from the laboratory. *Pediatr. Neurol.* **33,** 1–11.

Hussman, J. (2001). Suppressed GABAergic inhibition as a common factor in suspected etiologies of autism (letter). *J. Autism Dev. Disord.* **31,** 247–248.

Ingram, J., Peckham, S., Tisdale, B., and Rodier, P. (2000). Prenatal exposure of rats to valproic acid reproduces the cerebellar anomalies associated with autism. *Neurotoxicol. Teratol.* **22,** 319–324.

Jackson, J. (1958). Evolution and dissolution of the nervous system. In "Selected Writings of John Hughlings Jackson" (J. Taylor, Ed.), pp. 45–118. Stapes, London.

Kahlbaum, K. (1874). "Die Katatonie oder das Spannungsirresein." Verlag August Hirshwald, Berlin.

Kalkman, H., and Loetscher, E. (2003). GAD67: The link between the GABA-deficit hypothesis and the dopaminergic and glutamatergic theories of psychosis. *J. Neural Transm.* **110,** 803–812.

Kanner, L. (1943). Autistic disturbances of affective contact. *Nerv. Child* **2,** 217–250.

Kanner, L. (1944). Early infantile autism. *J. Pediatr.* **25,** 211–220.

Kanner, L. (1973). "Childhood Psychosis: Initial Studies and New Insights." Wiley, Washington, DC.

Kemper, T., and Bauman, M. (1993). The contribution of neuropathologic studies to the understanding of autism. *Neurol. Clin.* **11,** 175–187.

Keverne, E. (1999). GABAergic neurons and the neurobiology of schizophrenia and other psychoses. *Brain Res. Bull.* **48,** 467–473.

Kurita, H. (1985). Infantile autism with speech loss before the age of thirty months. *J. Am. Acad. Child Adolesc. Psychiatry* **24,** 191–196.

Kurita, H., Kita, M., and Miyake, Y. (1992). A comparative study of development and symptoms among disintegrative psychosis and infantile autism with and without speech loss. *J. Autism Dev. Disord.* **22,** 175–188.

Lainhart, J., Ozonoff, S., Coon, H., Krasny, L., Dinh, L., Dinh, E., Nice, J., and McMahon, W. (2002). Autism, regression, and the broader autism phenotype. *Am. J. Med. Genet.* **113,** 231–237.

Lang, J.-L. (1997). "Georges Heuyer, Fondateur de la Pedopsychiatrie. Un Humaniste du XX[e] Siecle." Expansion Scientifique Publications, Paris.

Lauber, C., Nordt, C., Falcato, L., and Rossler, W. (2005). Can a seizure help? The public's attitude toward electroconvulsive therapy. *Psychiatry Res.* **134,** 205–209.

Lauritsen, M., Mors, O., Mortensen, P., and Ewald, H. (1999). Infantile autism and associated autosomal chromosome abnormalities: A register-based study and a literature survey. *J. Child Psychol. Psychiatry* **40,** 335–345.

Liljelund, P., Handforth, A., Homanics, G., and Olsen, R. (2005). $GABA_A$ receptor beta 3 subunit gene-deficient heterozygous mice show parent-of-origin and gender-related differences in beta 3 subunit levels, EEG, and behavior. *Dev. Brain Res.* **157,** 150–161.

Lipcsey, A., Kardos, J., Prinz, G., and Simonyi, M. (1986). Der Effect der electrokonvulsiven Therapie auf den GABA-Spiegel des Liquor cerebrospinalis. *Psychiatrie, Neurologie, und medizinische Psycholologie Leipzig* **38,** 554–555.

Lord, C., Shulman, C., and DiLavore, P. (2004). Regression and word loss in autistic spectrum disorders. *J. Child Psychol. Psychiatry* **45,** 936–955.

Ma, D. Q., Whitehead, P. L., Menold, M. M., Martin, E. R., Ashley-Koch, A. E., Mei, H., Ritchie, M. D., DeLong, G. R., Abramson, R. K., Wright, H. H., Cuccaro, M. L., Hussman, J. P., Gilbert, J. R., and Pericak-Vance, M. A. (2005). Identification of significant association and gene–gene interaction of GABA receptor subunit genes in autism. *Am. J. Hum. Genet.* **77,** 377–388.

Madsen, T., Yeh, D., Valentine, G., and Duman, R. (2005). Electroconvulsive seizure treatment increases cell proliferation in rat frontal cortex. *Neuropsychopharmacology* **30,** 27–34.

Malhotra, S., and Singh, S. (1993). Disintegrative psychosis of childhood: An appraisal and case study. *Acta Paedopsychiatrica* **56,** 37–40.

Mantovani, J. (2000). Autistic regression and Landau-Kleffner syndrome: Progress or confusion? *Dev. Med. Child Neurol.* **42,** 349–353.

Martin, E. R., Menold, M. M., Wolpert, C. M., Bass, M. P., Donnelly, S. L., Ravan, S. A., Zimmerman, A., Gilbert, J. R., Vance, J. M., Maddox, L. O., Wright, H. H., Abramson, R. K., DeLong, G. R., Cuccaro, M. L., and Pericak-Vance, M. A. (2000). Analysis of linkage disequilibrium in gamma-aminobutyric acid receptors subunit genes in autistic disorder. *Am. J. Med. Genet.* **96,** 43–48.

Masden, T., Treschow, A., Bengzon, J., Bolwig, T., Lindvall, O., and Tingstrom, A. (2000). Increased neurogenesis in a model of electroconvulsive therapy. *Biol. Psychiatry* **47,** 1043–1049.

McCabe, B., Silveira, D., Cilio, M., Cha, B., Liu, X., Sogawa, Y., and Holmes, G. (2001). Reduced neurogenesis after neonatal seizures. *J. Neurosci.* **21,** 2094–2103.

McManus, D., Baker, G., Martin, I., Greenshaw, A., and McKenna, K. (1992). Effects of the antidepressant/antipanic drug phenelzine on GABA concentrations and GABA-transaminase activity in rat brain. *Biochem. Pharmacol.* **43,** 2486–2489.

Menold, M., Shao, Y., Wolpert, C., Donnelly, S., Raiford, K., Martin, E., *et al.* (2001). Association analysis of chromosome 15 $GABA_A$ receptor subunit genes in autistic disorder. *J. Neurogenet.* **15,** 245–259.

Mervaala, E., Kononen, M., Fohr, J., Saastamoinen, M., Korhonen, M., Kuikka, J., Viinanmaki, H., Tammi, A., Tiihonen, J., Partanen, J., and Lehtonen, J. (2001). SPECT and neuropsychological performance in severe depression treated with ECT. *J. Affect. Disord.* **66,** 47–58.

Mitushima, D., Hei, D., and Terasawa, E. (1994). Gamma-aminobutyric acid is an inhibitory neurotransmitter restricting the release of luteinizing hormone-releasing hormone before the onset of puberty. *Proc. Natl. Acad. Sci. USA* **91,** 395–399.

Mouridsen, S., Rich, B., and Isager, T. (1998). Validity of childhood disintegrative psychosis: General findings of a long-term follow-up study. *Br. J. Psychiatry* **172,** 263–267.

Mouridsen, S., Rich, B., and Isager, T. (1999). Epilepsy in disintegrative psychosis and infantile autism: A long-term validation study. *Dev. Med. Child Neurol.* **41,** 110–114.

Murcia, C., Gulden, F., and Herrup, K. (2005). A question of balance: A proposal for new mouse models of autism. *Int. J. Dev. Neurosci.* **23,** 265–275.

Nass, R., Gross, A., Wisoff, J., and Devinsky, O. (1999). Outcome of multiple subpial transections for autistic epileptiform regression. *Pediatr. Neurol.* **21,** 464–470.

Nisijima, K., and Ishiguro, T. (1995). Cerebrospinal fluid levels of monoamine metabolites and gamma-aminobutyric acid in neuroleptic malignant syndrome. *J. Psychiatr. Res.* **3**, 233–244.
Northoff, G., Steinke, R., Czcervenka, C., Krause, R., Ulrich, S., Danos, P., Kroph, D., Otto, H., and Bogerts, B. (1999). Decreased density of $GABA_A$ receptors in the left sensorimotor cortex in akinetic catatonia: Investigation of in vivo benzodiazepine receptor binding. *J. Neurol., Neurosurg. Psychiatry* **67**, 445–450.
Nurmi, E., Dowd, M., Tadevosyan-Leyfer, O., Haines, J., Folstein, S., and Sutcliffe, J. (2003). Exploratory subsetting of autism families based on savant skills improves evidence of genetic linkage to 15q11–13. *J. Am. Acad. Child Adolesc. Psychiatry* **42**, 856–863.
Odano, I., Anezaki, T., Ohkubo, M., Yonekura, Y., Onishi, Y., Inuzuka, T., Takahashi, M., and Tsuji, S. (1996). Decrease in benzodiazepine receptor binding in a patient with Angelman syndrome detected by iodine-123 iomazenil and single-photon emission tomography. *Eur. J. Nucl. Med.* **23**, 598–604.
Otani, K., Ujike, H., Tanaka, Y., Morita, Y., Katsu, T., Nomura, A., Uchida, N., Hamamura, T., Fujiwara, Y., and Kuroda, S. (2005). The GABA type A receptor alfa 5 subunit gene is associated with bipolar I disorder. *Neurosci. Lett.* **381**, 108–113.
Owens, M., and Nemeroff, C. (2003). Pharmacology of valproate. *Psychopharmacol. Bull.* **37**(Suppl.), 17–24.
Papadimitriou, G., Dikeos, D., Karadima, G., Avramopoulos, D., Daskalopoulou, E., Vassilopoulos, D., and Stefanis, C. (1998). Association between the $GABA_A$ receptor alpha 5 subunit gene locus (GABRA5) and bipolar affective disorder. *Am. J. Med. Genet.* **81**, 73–80.
Parent, J., Yu, T., Leibowitz, R., Geschwind, D., Sloviter, R., and Lowenstein, D. (1997). Dentate granule cell neurogenesis is increased by seizures and contributes to aberrant network plasticity in the adult hippocampus. *J. Neurosci.* **17**, 3727–3738.
Park, Y. (2003). The effects of vagus nerve stimulation therapy on patients with intractable seizures and either Landau-Kleffner syndrome or autism. *Epilepsy Behav.* **4**, 286–290.
Patil, A., and Andrews, R. (1998). Surgical treatment of autistic epileptiform regression. *J. Epilepsy* **11**, 1–6.
Petroff, O., Rothman, D., Behar, K., and Mattson, R. (1996). Low brain GABA level is associated with poor seizure control. *Ann. Neurol.* **40**, 908–911.
Petty, F., Kramer, G., Fulton, M., Moeller, F., and Rush, A. (1993). Low plasma GABA is a trait-like marker for bipolar illness. *Neuropsychopharmacology* **9**, 125–132.
Petty, F. (1995). GABA and mood disorders: A brief review and hypothesis. *J. Affect. Disord.* **34**, 275–281.
Pfleiderer, B., Michael, N., Erfurth, A., Ohrmann, P., Hohmann, U., Wolgast, M., Fiebich, M., Arolt, V., and Heindel, W. (2003). Effective electroconvulsive therapy reverses glutamate/glutamine deficit in the left anterior cingulum of unipolar depressed patients. *Psychiatric Research: Neuroimaging* **122**, 185–192.
Primavera, A., Fonti, A., Novello, P., Roccatagliata, G., and Cocito, L. (1994). Epileptic seizures in patients with acute catatonic syndrome. *J. Neurol., Neurosurg. Psychiatry* **57**, 1419–1422.
Rapin, I. (1995). Autistic regression and disintegrative disorder: How important the role of epilepsy? *Semin. Pediatr. Neurol.* **2**, 278–285.
Rasalam, A., Hailey, H., Williams, J., Moore, S., Turnpenny, P., Lloyd, D., and Dean, J. (2005). Characteristics of fetal anticonvulsant syndrome associated autistic disorder. *Dev. Med. Child Neurol.* **47**, 551–555.
Realmuto, G., and August, G. (1991). Catatonia in autistic disorder: A sign of comorbidity or variable expression. *J. Autism Dev. Disord.* **21**, 517–528.
Rey, J., and Walter, G. (1997). Half a century of ECT use in young people. *Am. J. Psychiatry* **154**, 595–602.

Ritvo, E., Freeman, B., Scheibel, A., Duong, T., Robinson, H., Guthrie, D., and Ritvo, A. (1986). Lower Purkinje cell counts in the cerebella of four autistic subjects: Initial findings of the UCLA-NSAC autopsy research report. *Am. J. Psychiatry* **143,** 862–866.

Rivinus, T., Jamison, D., and Graham, P. (1975). Childhood organic neurological disease presenting as a psychiatric disorder. *Arch. Dis. Child.* **50,** 115–119.

Roberts, E. (1972). Prospects for research on schizophrenia: A hypothesis suggesting that there is a defect in the GABA system in schizophrenia. *Neurosci. Res. Program Bull.* **10,** 468–482.

Roberts, E. (2000). Adventures with GABA. Fifty years on. *In* "GABA in the Nervous System. The View at Fifty Years" (D. Martin and R. Olsen, Eds.), pp. 1–24. Lippincott Williams and Wilkins, Philadephia, PA.

Rodier, P., Ingram, J., Tisdale, B., Nelson, S., and Romano, J. (1996). Embryological origin for autism: Developmental anomalies of the cranial nerve motor nuclei. *J. Comp. Neurol.* **370,** 247–261.

Sanacora, G., Mason, G., Rothman, D., and Krystal, J. (2002). Increased occipital cortex GABA concentrations in depressed patients after therapy with selective serotonin reuptake inhibitors. *Am. J. Psychiatry* **159,** 663–665.

Sanacora, G., Mason, G., Rothman, D., Hyder, F., Ciarcia, J., Ostroff, R., Berman, R., and Krystal, J. (2003). Increased cortical GABA concentrations in depressed patients receiving ECT. *Am. J. Psychiatry* **160,** 577–579.

Schneider, T., and Przewlocki, R. (2005). Behavioral alterations in rats prenatally exposed to valproic acid: Animal model of autism. *Neuropsychopharmacology* **30,** 80–89.

Scott, B., Wojitowicz, J., and Burnham, W. (2000). Neurogenesis in the dentate gyrus of the rat following electroconvulsive shock seizures. *Exp. Neurol.* **165,** 231–236.

Seong, E., Seasholtz, A., and Burmeister, M. (2002). Mouse models for psychiatric disorders. *Trends Genet.* **18,** 643–650.

Shao, Y., Cuccaro, M., Hauser, E., Raiford, K., Menold, M., Wolpert, C., Ravan, S., Elston, L., Decena, K., Donnelly, S., Abramson, R., Wright, H., DeLong, G., Gilbert, J., and Pericak-Vance, M. (2003). Fine mapping of autistic disorder to chromosome 15q11-q13 by use of phenotypic subtypes. *Am. J. Hum. Genet.* **72,** 539–548.

Slooter, A., Braun, K., Balk, F., van Nieuwenhuizen, O., and van der Hoeven, J. (2005). Electroconvulsive therapy of malignant catatonia in childhood. *Pediatr. Neurol.* **32,** 190–192.

Soghomonian, J., and Martin, D. (1998). Two isoforms of glutamate decarboxylase: Why? *Trends Pharmacol. Sci.* **19,** 500–505.

Stöber, G., Saar, K., Ruschendorf, F., Meyer, J., Nurnberg, G., Jatzke, S., Franzek, E., Reis, A., Lesch, K., Wienker, T., and Beckmann, H. (2000). Splitting schizophrenia: Periodic catatonia susceptibility locus on chromosome 15q15. *Am. J. Hum. Genet.* **67,** 1201–1207.

Stöber, G., Seelow, D., Ruschendorf, F., Ekici, A., Beckmann, H., and Reis, A. (2002). Periodic catatonia: Confirmation of linkage to chromosome 15 and further evidence for genetic heterogeneity. *Hum. Genet.* **111,** 323–330.

Tuchman, R., and Rapin, I. (1997). Regression in pervasive developmental disorders: Seizures and epileptiform electrocephalogram correlates. *Pediatrics* **99,** 560–566.

Tunnicliff, G., Schindler, N., Crites, G., Goldenberg, R., Yohum, A., and Malatynska, E. (1999). The $GABA_A$ receptor complex as a target for fluoxetine action. *Neurochem. Res.* **24,** 1271–1276.

Vajda, F., and Eadie, M. (2005). Maternal valproate dosage and foetal malformations. *Acta Neurol. Scand.* **112,** 137–143.

van Kammen, D. P. (1977). Gamma-aminobutyric acid (GABA) and the dopamine hypothesis of schizophrenia. *Am. J. Psychiatry* **134,** 138–143.

Varju, P., Katarova, Z., Madarasz, E., and Szabo, G. (2001). GABA signaling during development: New data and old questions. *Cell Tissue Res.* **305,** 239–246.

Volkmar, F., and Cohen, D. (1989). Disintegrative disorder or "late onset autism". *J. Child Psychol. Psychiatry* **30,** 717–724.

Wasterlain, C., and Plum, F. (1973). Vulnerability of developing rat brain to electroconvulsive seizures. *Arch. Neurol.* **29,** 38–45.

Wasterlain, C. (1976). Effects of neonatal status epilepticus on rat brain development. *Neurology* **26,** 975–986.

Wasterlain, C. (1978). Neonatal seizures and brain growth. *Neuropaediatrie* **9,** 213–228.

Werner, E., and Dawson, G. (2005). Validation of the phenomenon of autistic regression using home videotapes. *Arch. Gen. Psychiatry* **62,** 889–895.

Wing, L., and Shah, A. (2000). Catatonia in autistic spectrum disorders. *Br. J. Psychiatry* **176,** 357–362.

Zaw, F., Bates, G., Murali, V., and Bentham, P. (1999). Catatonia, autism, and ECT. *Dev. Med. Child Neurol.* **41,** 843–845.

Zink, M., Schmitt, A., May, B., Muller, B., Braus, D., and Henn, F. (2004). Differential effects of long-term treatment with clozapine or haloperidol on GABA transporter expression. *Pharmacopsychiatry* **37,** 171–174.

Zwaigenbaum, L., Szatmari, P., Mahoney, W., Bryson, S., Bartolucci, G., and MacLean, J. (2000). High functioning autism and childhood disintegrative disorder in half brothers. *J. Autism Dev. Disord.* **30,** 121–126.

Zwaigenbaum, L., Bryson, S., Rogers, T., Roberts, W., Brian, J., and Szatmari, P. (2005). Behavioral manifestations of autism in the first year of life. *Int. J. Dev. Neurosci.* **23,** 143–152.

SECTION II
ASSESSMENT

PSYCHOMOTOR DEVELOPMENT AND PSYCHOPATHOLOGY IN CHILDHOOD

Dirk M. J. de Raeymaecker

Department of Child Psychology and Psychiatry
"Kores" GGZ Group-Europoort
Rotterdam, The Netherlands

I. Three Vital Steps in Motor Development
   A. Smooth, Spontaneous Movements: "The Graceful, Flexible, and Elegant Motor Behavior of Babies" (Prechtl, 1981, p. 208)
   B. Acts and Effects: Focused, Precise Movements—Intentional and Goal-Oriented
   C. Motor Acts with Symbolic and Technical Significance
II. Human Motility: A View From Psychoanalysis
   A. Mobility and Psychoanalytic Therapy
   B. Children in the Hospital: Reactions to Motor Restraint
   C. Motor Achievements and Their Psychodynamic Significance
III. Developmental Psychopathology and Motor Impairment
   A. Pre- and Paranatal Complications and Disturbed Motor and Behavior Patterns
   B. ADHD of the Combined Type
   C. Dissociated Motor Development
   D. Low Birth Weight Children and Their Motor Development
   E. Developmental Coordination Disorder
   F. PDD and Body Ego Problems
   G. Stereotypic Movement Disorder
   References

The sensorimotor developmental phase, leading to a gradual acquisition of skilled actions, is of crucial importance for the young child and its growing sense of competence.

Three vital steps in motor development are mentioned: first, the smooth and spontaneous movements of the "graceful and elegant" baby, expression of his well-being and vitality, with their profound effect on the mother–infant relationship; second, the emergence of intentional and goal-oriented acts leading to Funktionslust and playful repetitions; and finally, the development of symbolic acts and increasing technical capacity to use playthings in imaginative play.

The psychodynamic significance of the most important motor milestones for the child's ego development is set out. Motility is one of the most important avenues for exercising such functions as mastery, integration, reality testing (self-preservation), and control of impulses. One may consider this early childhood period of rapid motor development as the motor phase of ego and libido development. Hence, many forms of developmental psychopathology are attended with motor impairment or insufficient motor mastery and integration. From that clinical perspective pass in review: perinatal complications and motor disturbance, attention deficit/hyperactivity disorder, dissociated motor development, low birth weight children and their developmental difficulties, developmental coordination disorder, aspects of pervasive developmental disorder, and stereotypic movement disorder.

I. Three Vital Steps in Motor Development

A. Smooth, Spontaneous Movements: "The Graceful, Flexible, and Elegant Motor Behavior of Babies" (Prechtl, 1981, p. 208)

Smooth general movements, apparently without intention, are the expression of the baby's well-being and vitality: in the first 2 months post term, these general movements have a "writhing quality"; at the age of 6–9 weeks post term, the form and character of these general movements change from the writhing type into a "fidgety pattern" (Prechtl, 1997).

These movements presuppose a normal muscle tone and contribute to the proprioceptive and kinesthetic sense of the child (bodily awareness); this proprioceptive sense is of enormous importance for posture and balance in the future. These graceful movements (without rigidity) are, according to Prechtl (1997), a sign of neurobiological health *in utero* and in the first year of life. Such a cute, trampling infant has a catching effect on the mother–infant relationship. From the beginning of their relationship, the baby's body posture and muscle tonus seem to have a profound effect on the mother's movements and way of holding him (Wolff, 1969).

B. Acts and Effects: Focused, Precise Movements—Intentional and Goal-Oriented

Slowly but surely, the young child becomes aware of the fact and has the growing insight (thought is internalized action, according to Freud) that there is a link between his act and the effect (Wallon, 1968). The child will repeat successful movements. Experiments have shown that "young infants will suck to produce movement in a mobile or to control the presentation of colorful slides" (Zelazo, 1976, pp. 90, 91).

Via trial and error and with much effort, the young child will adjust his tentative acts until firm sensorimotor schemes are the expression of the growth of his executive ego.

This is called by Piaget (1973), the sensorimotor intelligence. The child is getting to know his body and the world of immediate experience by the use of his senses and limbs. It is "practical" intelligence, according to Piaget. It results in Funktionslust and the first steps of "a sense of competence" called by Piaget, "le plaisir d'être cause" (the pleasure of being causal).

The child's curiosity is elicited. The human infant not only discovers the strength and vigor of his legs, it "goes through a phase of closely watching his playful hand and finger movements" (Prechtl, 1981, p. 207). His own pleasurable sounds surprise the baby and call forth mother's attention. Step-by-step the young child masters the use of tongue and mouth, called by René Spitz, the "primal cavity," seat of oral pleasure (food intake as well as nonnutritive sucking, so important for discharge of tension and affective self-regulation) and language.

Milestones of this phase are as follows:

1. The quiet alert state ("alert inactivity" or "quiet waking state"). The baby is in a quiet state of consciousness. There are only a few gentle movements, and he seems to be in a state of pleasurable vigilance, oriented both to his own body and especially to external perception. According to Stern (1985, p. 232), this quiet alert state favors learning in the child "eyes and ears trained on the external world."
2. Eye–hand coordination and visually guided reaching: from a primitive palmer grasping reflex to manual dexterity and precision grip (neat pincer) posture control and firm equilibrium

   - Head balance
   - Rolling over
   - Prone progression: leading to elegant going on all fours
   - Sitting firmly
   - Upright posture/standing
   - Free walking

All those steps are to be seen not only as maturational products, as progressive "unfolding" (Connolly, 1981, p. 217) of innate possibilities, but also as intrinsically linked to good mothering and holding according to Winnicott and parental availability according to Robert Emde, or to say it with Heinz Prechtl, developmental neurologist. "The infant's extensive neuromotor repertoire must be viewed in the context of the child–mother interaction" (Prechtl, 1971, pp. 137, 138; Zelazo, 1976, p. 97).

C. Motor Acts with Symbolic and Technical Significance

During the second year of life (the practicing period, according to Margareth Mahler), there is increasing motor mastery in the child. He learns to stand on his feet, explore the room, and experience distance and separation with respect to mother; during the rapprochement phase (18 months), the child will go back to mother for emotional refueling. Meanwhile, the child takes the important step "from action to language." The child not only learns that all everyday things have names but also learns that complex and laboriously acquired acts are defined by simple verbs, such as to eat, to sit, to walk, to sleep, to laugh; he learns the signal function of mother's forefinger, used by her when she forbids him, saying no in words and by shaking her head in the same horizontal direction as her index finger. This same index finger can be used for pointing to something and focus attention on it (important precursor of language skill) and to add, "I want it, give it to me." One gesture has a huge significance and enormous affective power (unfortunately, mother will teach you to be patient, look at her while asking and use the word: "please"), Michelangelo makes use of this symbolic act (between God and Adam) to depict the creation of man in the Sixteen Chapel. The "give and take" interactions between child and mother, according to Jerome Bruner, lay the foundation of "communicative give and take" by words, gestures, affective exclamations, and imperative body acts such as temper tantrums. When you talk about yourself you learn to say: "I, mine, me"; when you talk about mother you learn to say: "you." This precise referencing (the verb implicates an intrapsychic form of pointing) is a formidable task especially for children with pervasive developmental disorder (PDD). In his "Concluding Remarks" at the first International Congress of Infant Psychiatry, Erikson (1983, pp. 427, 428) spoke of the enigma of the development of "a sense of I."

In his tentative words, Erikson linked the first emergence of a sense of I with the experience of growing up, of realizing that one can and does "stand on one's feet," and of newly "grasping" something (basic sensorimotor acts of human infancy).

Given the affective significance of the first other (mother), the baby will perform many activities with mother as central point of reference. The child will imitate mother's movements and actions. Much later, the child will develop the capacity for "deferred imitation." Piaget gives the example of a 2-year-old little girl. After the visit of a young adult woman with a beautiful hat, the little girl imitates the lady with jaunty steps, pretending to wear a beautiful hat. Such a playful act not only evokes an absent person, but it is also the expression of your ego ideal to be grown up like the beautiful madam.

This capacity for playful identification with masculine or feminine traits will reach a peak in the Oedipal phase.

According to Anna Freud (1992, pp. 89, 90), especially the boy wishes to impress the mother with his strength, skill, and bodily perfection. "If you watch the interplay of mother and child in the phallic phase, you will be surprised how often one word returns in the exclamations of the child, and that word is look!" "It's a continuous demand on the mother: look how well I can do this, how I can do that, look at what I have just done." Acts, movements, and body control not only give birth to an intrapsychic sense of competence, they also enhance the child's self-esteem by love and praise (by the key figures, the parents).

1. *Clinical Significance*

The reader should bear this in mind when the clinical child is considered with his developmental lags in bodily skill: his poor motor coordination, clumsy acts, inability to sit still, given his hyperactive drivenness; his inability to stop his actions, to think first and consider the whole situation (i.e., being reflective), given his impulsivity. In the worst case, the child's body ego is unstable.

The more we move in the continuum of developmental psychopathology toward severe forms of PDDs, the less self-evident motor skills are integrated in the child's personal equipment.

2. *Gestalt Perception and Technical Insight*

An intriguing aspect in the child's development is the child's growing insight in the technical aspects of things and movements. A 3- or 4-year-old boy knows "at first sight" that this car in the playroom is a police car, and the other one is an ambulance or a fire engine (the Gestalt enigma).

The boy will act and play accordingly with the cars, based on their technical aspects. The little girl of the same age immediately recognizes the puppet house, the furniture, and the different dolls and starts her imaginative play. Observing the charm and the skill of those children during their play, we learn to comprehend Winnicott's words: playing is the psychical elaboration of physical function. This technical and constructive play at age 4 or 5 is the child's first step to feel at home in a technical world.

As a school child (between 7 and 12), the child has the task to develop "a sense of industry," according to Erikson (1971, pp. 249–251). His orientation on things and tools, his increasing pleasure in task completion ("the finished product of the activity" according to Anna Freud, 1965, p. 81) gives him status in the classroom and in the youth movement on the sports ground.

The schoolboy's favorite position on the football field betrays something of his intimate relations to his contemporaries in the symbolic language of attack, defense, ability or inability to compete, to succeed, to adopt the active masculine role, etc. (Anna Freud, 1965, p. 20).

II. Human Motility: A View From Psychoanalysis

A. Mobility and Psychoanalytic Therapy

Restriction of mobility (lying down on the coach) is meant by Freud to inhibit the patient's aggressive acts during the session and open a window on his wishes and fantasies and his internal motivations and preoccupations via free verbal associations.

In contrast, the child's freedom of movement in psychotherapy confronts the therapist with many problems. The child's unconscious presses the child to act; when the child endangers his own or the therapist's physical safety, when he tends to damage property, or when his acts are too provocative in the sexual sense, the therapist cannot help interfere.

According to Anna Freud (1965, p. 30), free association seems to liberate in the first instance the sexual fantasies of the adult patient; free action, even if only comparatively free, acts in a parallel way on the aggressive trends. What children overwhelmingly act out in the transference are therefore aggressions.

1. *Clinical Consequence*

Children have the developmental task to control and inhibit the aggressive side of their motility. The inhibited neurotic child is frequently characterized by excessive repression of his aggression, shyness, hesitation, and restricted motility. The attention deficit/hyperactivity disorder (ADHD) child with his motor disinhibition is frequently characterized by impulsivity and drivenness, aggressive outbursts, and an expansive behavior style. The therapeutic goals (and setting) are in accordance with those characteristics: more internal and external freedom for the neurotic child, more behavioral control for the ADHD child.

B. Children in the Hospital: Reactions to Motor Restraint

Bergmann and Anna Freud (1965) gave vivid descriptions of the impact of immobilization on, for instance, orthopedic young patients.

Besides acceptance and revolt regarding the confining treatments, in many cases a distinct inhibition of verbal expression was parallel with immobilization, as if the restraint enforced on children's limbs spread further and affected more highly differentiated motor functions. Young children often regressed in verbalization, or when just having learned to speak, lost the achievement altogether, other children occasionally became taciturn. Such losses were merely transitory.

With the cast or other restrictive measures removed, verbalization reinstated itself.

Katie, 4-year old, was in a plaster cast because of clubfeet. During the period of immobilization, it was noticed that although she was very intelligent and used educational toys very well, she would only point to the ones she wished to have but would not ask for them. Her vocabulary became extremely limited and consisted of only a few words such as mommy, dada, doggie, and by-by.

Coinciding with the removal of the cast, she suddenly reverted to the former self, speaking adequately in short sentences.

That removal of the cast and recovery of motor function can result in developmental spurts was also shown by the case of Craig, a 6-year-old boy, who was considered to be mentally retarded. After he had spent a year on a Bradford frame, he was gradually returning to ambulation.

At that time he was prone to temper tantrums and behaved in babyish ways. For weeks he kept walking with no other interest apparent. But once ambulation was established satisfactorily, to everybody's surprise he settled down in school, worked at his age level, and gave no further evidence of mental backwardness (Bergmann and Freud, 1965, pp. 59–68).

C. Motor Achievements and Their Psychodynamic Significance

This section is inspired, besides the authors mentioned, by the seminal papers of Mittelman (1954) and Kaplan (1965).

Following moving objects with the eyes: One of the first signs of the developing ego.

The baby finds his hands and learns that they stay with him, unlike other objects which come and go: Distinction self–outside world.

Bringing things to the mouth: The baby's major zone of gratification.

The baby continuously identifies himself (according to Freud) with the movements and things in the outside world; by his senses and motility he behaves automatically like his mother: we say he has incorporated her and this results in identification with her. Identification with the beloved *active mother* is a tremendous advance in ego development.

The child gradually links his sounds and utterances (prespeech acts) with the approach of mother and her comforting care; this maternal response leads to the infant's recognition of language as a communicative device. The link between body movements, affective expression, and mother's care is according to Spitz the basis for (the later) philosophical and epistemological stance: *post hoc ergo propter hoc*!!

Object permanence: The child pulls of whatever hides the reached for object. He drops and throws things and delights in retrieving them. With the establishment of permanence of persons and things, the child is able to form a lasting attachment to mother.

According to Mahler (1971), "the achievement of free upright locomotion (coupled with the beginning of representational intelligence) constitutes the most decisive contribution to, and organizer of, the psychological birth of the human being."

Within a year the baby has made this important transition: From becoming earth-bound to becoming upright (Kestenberg and Buelte, 1983).

In her Center for Parents and Children (Roslyn, New York), Judith Kestenberg and associates studied carefully the development of the young child as seen through bodily movement.

In the first half of the first year, the upper "prehensile part" of the baby is predominant: his arms and hands are his major instruments and they are "the mediators" between himself and the world around him. The hands bring the things into the baby's "near space," and his outer space is limited within "reach distance." His activities take place in the "horizontal" plane while the baby is lying on the floor or in bed. When he is sitting (with support) on mother's lap the horizontal plane is the surface of the table in front of them, or the palm of their hands, or mother's lap. The baby cannot separate himself from mother by locomotion. There is symbiotic closeness between baby and mother.

In the second part of the first year there is a strong shift in the ego interests of the baby. The lower stabilizing part of his body is in focus now. The baby discovers his legs. They support his weight. To achieve this, he uses his antigravity muscles to attain the upright, erect posture. So he becomes part of the adult world (*Homo erectus*!). He is ready to "confront" the world and gets a panoramic view of the world. In the vertical plane, the baby can change from up to down in space and corresponding feelings (elation–depression) emerge. Practical distinctions are learnt. "Little" things lie on the floor, big objects stand up. Slowly but surely, the baby acquires equilibrium and learns to stand firmly on his feet—this leads to an internal sense of "stability."

The toddler still holds on the mother's legs but wishes to abandon her in order to explore the room. Here ambitendency starts, forerunner of later ambivalence.

In her book, *The Magic Years*, Fraiberg (1972) depicts the child who conquers the erect posture. The child persists, despite falls and hurts, until the final triumph and until he can stand firmly on his feet. When the child dares to stand alone, he has mastered the anxiety to be without support. (Karl Abraham used to link a phobia in his adult patients with locomotion anxiety in their previous development) When the child sets his first footstep, he is alone and separated. He experiences the unique core of his person. He discovers the I.

His ability to move freely is "the great event in the child's life" according to Anna Freud and Burlingham (1944). "Some children behave as if they where drunk with the idea of space and even of speed; they crawl, walk, march, and run, and revert from one method of locomotion to the other with the greatest of pleasure."

In his book, *The First Year of Life*, Spitz (1965) pays attention to the impact of locomotion on the dyadic relationship. When locomotion starts, there is a radical transformation in the relationship between the mother and the infant. Now the child strives for autonomy, and he can move out of mother's reach. Now the mother is forced to curb and prevent the child's initiatives. There is a shift in the child from passivity to activity in this period—a turning period in their relationship. As a consequence, the exchange between mother and child now centers around bursts of infantile activity and maternal commands and prohibitions. Maternal intervention will have to rely increasingly on gesture and word because locomotion puts space between mother and child. The surge for activity is so strong that the child will not tolerate being forced back to passivity without resistance.

Sphincter control: A new milestone furthering the socialization of the child—body awareness (in the lower part of the body (hurry-up/where is the pot), anticipatory thinking leading to adequate actions (squatting and relieving), resulting in that final word: done/ready!

A triumph for "the potting couple." The child accepts and takes over the mother's attitude to cleanliness. Oppositional trends are canalized into body mastery.

The importance of body strength and masculine c.q. feminine charm of bodily movements in the phallic phase has already been mentioned. The body virtuosity of this phase has been entitled by Ernst Kris as "the melody of movement."

"Trial action in thought" (Freud, A., 1965, p. 172) makes it possible for the child to insert reasoning between the arising of an instinctual wish and the behavior aimed at its fulfillment (the opposite of impulsivity in ADHD children).

When the child's "muscular actions" come under the control of the sensible ego instead of serving the impulses of the id, it is another important step toward socialization (Freud, A., 1965, p. 173).

The school age child makes the transition from "unrepressed" freer affectomotor expressions to the "repressed" purposeful activities, such as the sedentary actions of writing and reading, which utilize the new capacity of the fine muscle organization. There is increasing pleasure in games requiring equipment and in the rules of the game, they are thus a complex achievement dependent on contributions from many areas of the child's personality such as the endowment and intactness of the motor apparatus (Freud, A., 1965, p. 81).

For example, cadenced, pleasurable sensorimotor games, accompanied by rhythmic chants, such as girls skipping rope to chants of:

Teddy Bear Teddy Bear
turn around
Teddy Bear Teddy Bear
touch the ground

At about the age of 10, the readiness of the child to carry impulses into action shows a crucial change. Up to that year, if toys are available, children automatically engage in play activities in psychotherapeutic sessions. After that age, they prefer verbal communications with the therapist.

III. Developmental Psychopathology and Motor Impairment

A. Pre- and Paranatal Complications and Disturbed Motor and Behavior Patterns

Children who are excessively mobile and restless (or their opposite) with uncontrolled behaviour and poor school ability, is a common problem according to Prechtl and associates (Prechtl, 1961; Prechtl and Stemmer, 1962), and were the object of their careful investigations. Decades ago, they laid emphasis on the consequences of mother–child interaction and upbringing. They describe three types of abnormal motor behavior in babies.

1. Babies born in breech presentation: show changes of flexion and extension reflexes of the leg
2. Hyperexcitability syndrome: characteristics of the baby:

    - Hyperactive
    - Easily startled
    - Has a very low threshold of the Moro-response
    - Most impressive symptom: a tremor of low frequency and high amplitude that was superimposed on the spontaneous movements and the elicited responses
    - Mothers who do not know that their baby is neurologically abnormal think he is very easily frightened
    - A few years later, most of these babies show a choreiform dyskinesia (see choreiform syndrome below)
    - Other symptoms in these infants and children are:
    - A very short concentration span
    - Hypermotilility
    - Instability of mood
    - Some degree of learning difficulty, especially in reading, writing, and arithmetic
    - Hyperexcitability occurs significantly more frequently among boys.

3. Apathy syndrome: the baby's characteristics:

    - Hypokinetic
    - Somnolent

- Hypotonic
- If a response is elicited, the movements are very slow and weak

Follow-up: At the age of 3 or 4, 70% of the children with these postnatal abnormal findings have behavior problems; in the control group of the same age (consisting of neurologically nonpathological babies), Prechtl and associates found 20% behavior problems.

In the discussion after the presentation of these findings, the hypothesis was raised that besides the brain damage effects, the baby's abnormal motor patterns disturb the interaction with the mother and this initial disturbance (may) continue to exert its effects in the next developmental phases.

The choreiform syndrome was described in hyperactive school children, "referred to the neurologist mainly because of poor school performance" (experimental group: 50 children, aged 9–12 years). The choreiform movements (in the extremities and the head) are slight and jerky, are of short duration and occur suddenly.

They can be detected electromyographically in fully relaxed muscles but are *most clearly observable in stress situations*.

The muscles of the tongue, face, neck, and trunk were involved in 100% of the cases; in 92% of the children, the eye muscles were affected resulting in disturbances of conjugate movement and difficulty in fixation and reading (!!).

In discussing the behavior problems of many of the children with the choreiform movements, Prechtl and associates point *first* to the obvious fact that when "the capacity for delicate movements of the eyes or hands" is affected, the child's psychic development may be disturbed. In 90% of the children there were more or less severe reading difficulties.

Second, the majority of the children's parents declared that even at an early age their children had displayed particularly wild, unrestrained behavior, clumsiness, inability to concentrate on any plaything for long, and very labile mood fluctuating between timidity and outburst of aggression.

Third, at school age the children do not possess required abilities, such as the capacity for sitting still, prolonged concentration, and differentiated fine motor control, required for reading and writing.

This results in the child's understandable failure that produces new burdens such as constant admonishments from teachers. Such children are caught in a vicious circle and become progressively more and more neurotic, either to the point of a breakdown or as sometimes happens, until an understanding adult demands help for them.

Finally, these behavior abnormalities are in most cases associated with a rapid functional development and a basically lively intelligence; the children stand and walk early; they promise at first much more than they are later able to fulfill, and this disguises the existence of cerebral injury and causes parents to be disappointed at their children's subsequent failures.

B. ADHD of the Combined Type

Many aspects of the ADHD child's motor disturbances (restlessness, hyperactivity, impulsive drivenness, diminished capacity for prolonged, sustained attention, poor school performance, diminished capacity for quiet imaginative play) and the resulting difficulties at home and school are very well dealt with by Prechtl and associates in their vision on the child with the hyperexcitability syndrome, and alluded to in the previous chapters on motor development and ego development, especially the capacity to postpone actions (the transition from an impulsive to a more reflective attitude).

C. Dissociated Motor Development

A very puzzling syndrome of delayed motor development, characterized by Hagberg and Lundberg (1969) as a condition where the traditional milestones for fine motor development (grasping, transferring from one hand to the other, pencil grasp) appear at the expected age, while gross motor skills (sitting up, crawling, standing, walking) are markedly delayed without findings of other obvious neurological signs. Lesigang (1977) collected a group of 71 children, suffering from what he called "Reifungsdissoziation" (dissociation of maturation). Clinical picture: the children refuse to stand and walk well into their second year of life, they show generalized reduction of muscle tone, with well-preserved strength and lively spontaneous movements, and are irritable, with temper tantrums and stereotyped movements.

Most important clinical signs (Haidvogel, 1979; Lundberg, 1979):

- Muscular hypotonia
- Bottom shuffling: propulsion in sitting position
- Late patterns of learning to sit actively, without help
- Joint laxity
- "Sitting on air": when held in vertical suspension the children keep their legs "firmly in the air" with flexion, moderate abduction, and outward rotation of the hips (refusal to stand)
- Girls are overrepresented in the syndrome

The retardation in gross motor milestones results, of course, in a delay of the growing sense of competence and independence during the toddler phase. The parent's sense of competence is also undermined; even with the aid of adequate psychomotor therapy; parents are left with questions such as: Will the child catch up and when? Is it organic? Is patience or pressure the best approach? If there are no obvious neurological signs, and given the fact that the child has many temper tantrums, shows refusal to stand, and clings tenaciously to bottom shuffling, do doctors doubt about the parents "firmness in their upbringing"? (De Raeymaecker, 1988).

D. Low Birth Weight Children and Their Motor Development

Clinical research has shown that the low birth weight (LBW) child and his mother face great problems during the sensorimotor phase of development, in which the supine baby evolves to an active explorer (De Raeymaecker, 1985, 1988; De Raeymaecker and Mettau, 1984; Sanders-Woudstra *et al.*, 1983).

From the extensive literature we cite some important research findings from experts in the field, documenting the LBW child's motor problems.

1. *Premature is More Sensory than he is Motor*

Shirley (1939) of Harvard School of Public Health observed a behavior syndrome characterizing prematurely born children. She points to their special difficulties with manual–motor and postural and locomotion control. In the conclusion of her research paper she states that much of the behavior of prematures becomes understandable if we assume that he is more sensory than he is motor. He is less capable than the term child of making an adequate motor adjustment to the stimulus. His reach exceeds his grasp.

2. *Syndrome of Previous Prematurity*

Berges *et al.* (1969) from Port-Royal Hospital in Paris describe with the utmost care and precision a "syndrome of previous prematurity." In the first year of development, the prematures show a tendency to posturomotor delay and general instability. Later, when they have caught up, as far as developmental milestones and IQ are concerned, they still show "instrumental" disturbances. The authors have elaborated a test of "imitation of gestures" to detect the problems which prematures can have with practognosia and body image in comparison with a term born controls. Needless to say that the origins of those sequels are to be found in the early sensorimotor phase of development and "holding environment."

3. *Syndrome of Transient Dystonia Associated with LBW*

Drillien (1972), from Edinburgh, identified a syndrome of transient dystonia associated with LBW. In a 5-year follow-up study, Drillien finds that many very-LBW infants are dystonic and have problems with postural development. As Drillien says, the history given by the mother is typical, and suggestive of the diagnosis, even before the examination commences.

When questioned the mother reports the following symptoms:

1. Marked irritability: constant crying, feeding difficulties
2. The child is "very jittery" and "jumpy"

3. The child is "easily startled" by noises and changes of posture
4. The child is "rather stiff to handle" the following:
   - When being dressed: getting arms into sleeves
   - Nappy changing: difficulty in abducting the groins for cleaning
   - Washing and drying hands: the infant's hands are kept tightly fisted
   - Bathing: the baby displays dislike of the bath, arching his back as soon as his feet touch the ground
5. The baby is "very good at standing up" contrasting with the difficulty the mother has with getting it to sit. The mother often says, "he stood on his toes like a ballet dancer"

In the above-mentioned French study, it was also noted that the delay of sitting independently (station assise sans appui) was of particular importance for the later development of the premature.

4. *Motor Development and Respiratory Distress Syndrome*

In *Infants Born at Risk*, volume I, Field *et al.* (1979) find for infants surviving the respiratory distress syndrome (RDS) a marked discrepancy between the mental and motor development. She suggests the importance of early intervention because of the strikingly low-motor performance and the delay in language production skills. In *Infants Born at Risk* volume II, Field *et al.* (1983) brings again a 5-year follow-up study of infants who suffered perinatal problems. She is surprised to find that the Bayley Motor Scale score at 8 months has predictive value for the next years. She says explicitly that the most critical intervention needs identified by her follow-up include "the motor delays" of preterm RDS infants.

The consequences for the LBW child's ego strength at the age of 3, was the subject of our thesis (De Raeymaecker, 1981).

E. DEVELOPMENTAL COORDINATION DISORDER

The Diagnostic and Statistical Manual, 4th ed. (DSM-IV) (APA, 1994) definition speaks for itself:

Performance in daily activities that require motor coordination is substantially below that expected given the persons chronological age and measured intelligence. The manifestations of this disorder vary with age and development: younger children may display delays in motor milestones, dropping things, clumsiness, difficulties with tying shoelaces, buttoning shirts, and zipping pants. Older children may display difficulties with the motor aspects of assembling puzzles, building models, playing balls, and handwriting.

Prevalence: according to DSM-IV the prevalence of developmental coordinated disorder has been estimated to be as high as 6% for children in the age range of 5–11 years (!)

The reader should bear in mind that this disorder is not due, by definition, to a general medical condition (no specific neurological disorder) and does not meet criteria for a PDD. Again, such a disorder undermines the child's sense of competence, an important part of the person's ego strength; it gives rise to difficult interactions between child and parents; at school age the disorder will contribute to "a sense of inferiority," the opposite of Erikson's "a sense of industry."

F. PDD and Body Ego Problems

Several diagnostic DSM-IV criteria of the PDDs, as discussed in the following sections, are closely linked to the integration of motor skills in a coherent, purposeful personality.

1. *Qualitative Impairment in Social Interaction*

A lack of spontaneous seeking to share enjoyment, interests, or achievements with other people, for example by a lack of *showing*, *bringing*, or *pointing out* objects of interest (these are exactly the normal give and take actions between the child and the mother, leading to communication and later to phallic pride when the child demonstrates his skills to mother).

Marked impairment in the use of multiple nonverbal behaviors such as eye-to-eye gaze, facial expression, body postures, and gestures to regulate social interaction.

2. *Qualitative Impairments in Communication*

Delay in spoken language, not accompanied by an attempt to compensate through alternative mode such as gesture or mime (expressive body acts).

Lack of varied make-believe play or social initiative play (during play, the child's body is his main instrument).

3. *Repetitive and Stereotyped Patterns of Behaviors*

Stereotyped and repetitive motor mannerisms (e.g., hand or finger flapping or twisting, or complex whole body movements).

The movements lack symbolic significance and technical skill and are not goal oriented.

4. *Loss of Purposeful Hand Skills in Rett's Disorder*

Followed by characteristic stereotyped hand movements (resembling hand-wringing or hand-washing); appearance of poorly coordinated gait or trunk movements (loss of eye–hand coordination and visually guided reaching resulting normally in manipulative dexterity).

5. *Subtle Forms of PDD not otherwise Specified*

This type of children is sometimes called odd, or atypical, or described as having multiplex developmental disorder. They display a puzzling combination/alternation of normal and sometimes extremely pathological behavior.

Besides other forms of developmental pathology, their weak and fragile capacity for higher achievements of motor control, especially difficulties with the use of their bodily skills in the confrontation with their contemporaries, are at the center of their disturbance.

School boys of that type with normal intelligence and corresponding educational level may lack in football play; either they isolate themselves in a narcissistic way from peers and coach, with (hidden) fantasies of grandeur, or they behave in a childish way on the playground, or they do not engage in competitive play at all. Girls of that age may be fixated in a world of imaginative play, avoiding the painful contact with their peers. Emotional breakdown and school refusal lie in wait after some dramatic episodes of being teased. Frequently skilled behavior and infantile dependence on the mother went together in previous phases of their development. The drawings, the fantasies, and identifications of these children may reveal their fragile body ego. Sometimes careful psychomotor assessment by a skilled psychomotor therapist is necessary to detect the child's bodily insecurities.

G. Stereotypic Movement Disorder

The DSM-IV features are: motor behavior that is repetitive, often seemingly driven and nonfunctional. It interferes markedly with normal activities or results in self-inflicted bodily injury that is significant enough to require medical treatment.

- The clinician knows that such behavior can be part of mental retardation, but if that is not the case and if the child is not overtly autistic, he may look for hidden symptoms of PDD of the nonspecified type. (Given the risk that the central core of the person and a good body image are poorly developed)
- Repetitive (rhythmic) body movements may be within broad normality during infancy: c.q. rocking and head banging in well-defined settings, when the child is alone in his bed or in his play-pen.
- Repetitive movements of the hands may be more ominous when the child is in a state of withdrawal and when the hand movements and the facial expressions of the child are vague and out of context; on the other hand, when the child accompanies his enthusiasm, for example, when seeing interesting playthings with fluttering arm movements, these can be considered as affectomotor expressions. Immature children, such as LBW

children, often display such atypical behavior without being autistic in their relationships.
- Stereotypic movement disorder may also occur in association with severe sensory defects (blindness and deafness) and may be more common in institutional environments in which the individual receives insufficient stimulation.

The dramatic consequences of the loss of "appropriate maternal care" were described by Ribble (1943, pp. 40, 41) in her beautiful book: *The Rights of Infants.*

We quote: after the loss of maternal care, the average child will develop various forms of automatic activity. Head rolling, head banging, and various other forms of hyperactivity, including masturbation, frequently develop. In this way normal behavior activities become dissociated into troublesome habits. An actual case will show the logical sequence. *Baby Sally*, who was breast-fed and cared for only by her mother, developed well for the first $4\frac{1}{2}$ months. At this time the mother was suddenly called away from home and the baby had to be weaned abruptly. The child was left in charge of an aunt who gave her the most conscientious care, exactly as it had been prescribed by a pediatrician. For fear that the infant would be handled too much, the aunt had been instructed not to pick her up; accordingly, she did not hold the baby for feedings or fondle her. During the first week of the mother's absence no noteworthy reaction was observed. The child did not cry and apparently slept soundly at night. However, a marked pallor developed in spite of scrupulous hygienic attention to fresh air, sunlight, and diet. By the end of the second week, the aunt was horrified to discover that when going to sleep the child rolled her head violently, at time knocking it on the side of the crib. Not long after this, she began to bang her head with her fists, at times picking at her hair and actually pulling it out. This behavior continued off and on for 2 months. Finally, in desperation, the aunt decided that a psychiatrist must see the child because she had the idea that some serious disease had developed. In the meantime, the mother had returned, and it was not difficult to outline a course of treatment, which she carried out in an excellent and even exaggerated fashion, so that the child rapidly recovered, soon being restored to an entirely normal condition. Breast feedings were, of course, not resumed, but the excellent mothering enabled this infant to overcome the results of the combination of sudden weaning and loss of stimulus satisfaction.

References

American Psychiatric Association (APA) (1994). "Diagnostic and Statistical Manual of Mental Disorder," 4th ed. (DSM-IV). APA Press, Washington, DC.

Berges, J., Lezine, I., Harrison, A., and Boisselier, F. (1969). Le syndrome de l'ancien prématuré. Recherche sur sa signification. *Revue de Neuropsychiatrie Infantile* **17,** 719–779.

Bergmann, T., and Freud, A. (1965). "Children in the Hospital," Chapter 8, pp. 59–67. International Universities Press, New York.

Connolly, K. J. (1981). Maturation and ontogeny of motor skills. *In* "Maturation and Development: Biological and Psychological Perspectives" (K. J. Connoly and H. F. R. Prechtl, Eds.), pp. 216–230. Heinemann Medical Books, London.

De Raeymaecker, D. M. J. (1981). The ego under observation. Childpsychiatric study of 40 three-year-old low birth weigth children and 40 three-year-old full term normal birth weight children. Thesis: Erasmus University Rotterdam, Davids Decor, Alblasserdam.

De Raeymaecker, D. M. J., and Mettau, J. W. (1984). Follow-up of VLBW children: A contribution to the profile of the vulnerable child. *In* "Kongressberichte 1984 Kinderchirugie (Surgery in Infancy and Childhood)," pp. 197–203. Hippokrates Verlag, Stuttgart.

De Raeymaecker, D. M. J. (1988). Very low birth weight and "dissociation of maturation" the hazards of the sensori-motor development. *Acta Pædiatr. Scand.* **77**(344), 71–80.

Drillien, C. M. (1972). Abnormal neurologic signs in the first year of life in low-birhtweight infants: Possible prognostic significance. *Devl. Med. Child Neurol.* **14,** 576–584.

Erikson, E. H. (1971). Het kind en de samenleving. "Childhood and Society." Het Spectrum N.V./ Aula Boeken 181, Utrecht/Antwerpen.

Erikson, E. H. (1983). Concluding remarks: Infancy and the rest of life. *In* "Frontiers of Infant Psychiatry" (J. D. Call, E. Galenson, and R. L. Tyson, Eds.), pp. 425–428. Basic Books, Inc., New York.

Field, T., Dempsey, J. R., and Shuman, H. H. (1979). Developmental assessments of infants surviving the respiratory distress syndrome. *In* "Infants Born at Risk" (T. M. Field, Ed.), pp. 261–280. SP Medical & Scientific Books, New York.

Field, T., Dempsey, J. R., and Shuman, H. H. (1983). Five year follow-up of preterm respiratory distress syndrome and post-term postmaturity syndrome infants. *In* "Infants Born at Risk" (T. Field and A. Sostek, Eds.), pp. 317–335. Grune & Stratton, New York.

Fraiberg, S. H. (1972). De eerste stappen naar de zelfstandigheid. *In* "De Magische Wereld Van Het Kind (The Magic Years)," pp. 59–65. Paul Brand, Bussum.

Freud, A., and Burlingham, D. (1944). Infants Without Families. *In:* "The writings of Anna Freud," Vol. III. International Universities Press, Inc., New York.

Freud, A. (1965). Normality and pathology in childhood: Assessments of development. *In* "The Writing of Anna Freud," Vol. VI. International Universities Press, New York.

Freud, A. (1992). The Harvard lectures. *In* "Lecture 6: Love Identification and Superego" (J. Sandler, Ed.), pp. 79–90. The Institute of Psychoanalysis, Karnac Books, London.

Hagberg, B., and Lundberg, A. (1969). Dissociated motor development simulating cerebral palsy. *In* "Neuropädiatrie," **1,** 187–199.

Haidvogel, M. (1979). Dissociation of maturation: A distinct syndrome of delayed motor development. *Dev. Med. Child Neurol.* **21,** 52–57.

Kaplan, E. B. (1965). Reflections regarding psychomotor activities during the latency period. *Psychoanal. Study Child* **20,** 220–238.

Kestenberg, J., and Buelte, A. (1983). Prevention, infant therapy, and the treatment of adults: Periods of vulnerability in transition from stability to mobility and vice versa. *In* "Frontiers of Infant Psychiatry" (J. D. Call, E. Galenson, and R. L. Tyson, Eds.), pp. 200–216. Basic Books, New York.

Lesigang, C. (1977). Reifungsdissoziation: Eine typische Variante der statomotorischen Entwicklung. *In* "Wiener Klinische Wochenschrift," **98,** 53–58.

Lundberg, A. (1979). Dissociated motor development. *Neuropediatrie* **10,** 161–182.

Mahler, M. (1971). Excerpts from a discussion by Margaret Mahler. New York Psychoanalytic Society Meeting. *In* "Children and Parents" (J. Kestenberg, Ed.), pp. 210–214. Jason Aronson, New York.
Mittelman, B. (1954). Motility in infants, children and adults. Patterning and psychodynamics. *Psychoanal. Study Child.* **9,** 142–177.
Piaget, J., and Inhelder, B. (1973). La psychologie de l'enfant 5th ed. Presses Universitaires de France, Paris (Collection: Que Sais-Je? No. 369).
Prechtl, H. F. R. (1961). Neurological sequelae of prenatal and paranatal complications. *In* "Determinants of Infant Behaviour" (B. M. Foss, Ed.), pp. 45–48. Methuen, London.
Prechtl, H. F. R., and Stemmer, C. J. (1962). The choreiform syndrome in children. *Dev. Med. Child Neurol.* **4,** 119–127.
Prechtl, H. F. R. (1971). Motor behaviour in relation to brain structure. *In* "Normal and Abnormal Development of Brain and Behaviour," pp. 133–145. Leiden University Press, Leiden.
Prechtl, H. F. R. (1981). The study of neural development as a perspective of clinical problems. *In* "Maturation and Development: Biological and Psychological Perspectives" (K. J. Connolly and H. F. R. Prechtl, Eds.). Clinics in developmental medicine, No. 77/78, pp. 198–215. Heinemann Medical Books, London.
Prechtl, H. F. R. (1997). Spontaneaous motor activity as a diagnostic tool. A Scientific Illustration of Prechtl's Method. Video Guide Published by the GM Trust Medical Guide, Karl Franzen University, Department of Physiology, Graz.
Ribble, M. A. (1943). "The Rights of Infants: Early Psychological Needs and their Satisfaction." Columbia University Press, New York.
Sanders-Woudstra, J. A. R., Visser, H. K. A., De Raeymaecker, D. M. J., Skoda, I., and Uleman-Vleeschdrager, M. (1983). Low-birthweight children in their first two years of life and their difficulties. *In* "Frontiers of Infant Psychiatry" (J. D. Call, E. Galenson, and R. L. Tyson, Eds.), pp. 282–290. Basic Books, New York.
Shirley, M. (1939). A behaviour syndrome characterizing prematurely-born children. *Child Dev.* **10,** 115–127.
Spitz, R. (1965). The impact of locomotion on dyadic relations. *In* "The First Year of Life," pp. 180–185. International University Press, New York.
Stern, D. N. (1985). "The Interpersonal World of the Infant. A View from Psychoanalysis and Developmental Psychology." Basic Books, New York.
Wallon, H. (1968). L'acte et l'effect (chapitre 4)/L'acte moteur (chapitre 6). *In* "L'Evolution Psychologique de L'enfant." Librairie Armand Colin, Paris.
Wolff, P. H. (1969). Mother–infant relations at birth. *In* "Modern Perspectives in International Child Psychiatry," pp. 80–97. Oliver & Boyd, Edinburgh.
Zelazo, Ph. R. (1976). From reflexes to instrumental behavior. *In* "Developmental Psychobiology: The Significance of Infancy" (L. P. Lipsitt, Ed.), pp. 87–104. Laurence Erlbaum Associates, Publishers, Hillsdale, NJ.

THE IMPORTANCE OF CATATONIA AND STEREOTYPIES IN AUTISTIC SPECTRUM DISORDERS

Laura Stoppelbein,* Leilani Greening,* and Angelina Kakooza[†]

*Department of Psychiatry and Human Behavior, Center for Psychiatric Neuroscience
University of Mississippi Medical Center, Jackson, Mississippi 39216
[†]Makerere University, Kampala, Uganda

I. Introduction
II. Catatonia and ASD
   A. Epidemiology
   B. Characteristic Symptoms
   C. Risk Factors
   D. Etiology
   E. Assessment
   F. Treatment
III. Stereotypic Movement Disorder
   A. Epidemiology
   B. Etiology
   C. Assessment
   D. Treatment
   References

Motor disturbances are often observed in individuals with autistic spectrum disorders (ASDs) and recognized as diagnostic features of these disorders. The movement disorders characteristically associated with autism include stereotypies and self-injurious behavior. Yet, individuals with ASD may also be at the risk for catatonia. Although not as frequent as stereotypies, up to 17% of older adolescents and adults with autistic disorder may have severe catatonic-like symptoms. Catatonia may be a comorbid risk factor of autism that warrants further empirical and clinical evaluations. Clinicians may need to be attentive to more subtle signs of catatonic-like symptoms in individuals diagnosed with ASDs, especially as they enter adolescence and young adulthood. Stress has been implicated as a possible precursor for symptoms; however, its role has not been empirically proven as a potential risk factor. Clinicians might also need to assess for signs of significant declines in motor movements, as this appears to be a useful diagnostic indicator of catatonic-like symptoms. The literature on stereotypies and autism is more extensive than for catatonia and ASDs, probably because of the higher rate of stereotypies with autism. Explanations for the occurrence of stereotypies range from genetic to behavioral contingencies, with evidence for a multifactor explanation. Assessment measures often include items that assess for stereotypies to aid with diagnosing these symptoms in individuals with autism. Treatment for stereotypies is largely behavioral at the present time and requires consistent reinforcement of treatment gains to manage the symptoms successfully. An important area of future research in autism is the relation among different types of motor abnormalities, including stereotypies and catatonia.

I. Introduction

Motor disturbances are commonly linked to physical conditions, such as Parkinson's disease and Huntington's chorea, but can also occur with psychiatric conditions, including schizophrenia and autistic spectrum disorders (ASDs). Schizophrenia, for example, can be accompanied by catatonia and waxy flexibility that are diagnostic features of the subtype, catatonic schizophrenia. Catatonia also has been observed among individuals with autistic disorder. Research suggests that when present with ASDs, catatonia may represent a different clinical phenomenon that is referred to as "autistic catatonia." Although still in its infancy, research on this concept offers important implications for conceptualizing and treating individuals with autistic disorder. Hence, the literature on catatonia and autistic disorder is reviewed in this chapter. While the prevailing clinical data are mainly in the form of case reports, these can provide useful information for conceptualizing and treating individuals with autistic disorder who present with catatonia. Other movement disorders, specifically stereotypies, are also discussed because they are diagnostic features of ASDs and overlap in symptomatology with catatonia. Reviews of the clinical presentations for each type of motor disturbance will reveal similarities and differences that may aid with differential diagnosis.

II. Catatonia and ASD

Catatonia is a life-threatening condition marked by a paucity of motor movements, negativism, and mutism. In its severe form, catatonic symptoms can be expressed as episodes of excitement and restlessness superimposed on pervasive mutism, akinesia, catalepsy, and waxy flexibility. Wing and Shah (2000) described a minority of cases with autistic disorder and "catatonia-like deterioration" that includes marked catatonia-like and parkinsonian-like features. These symptoms usually appear gradually and tend to eventually lead to significant deterioration in movement and self-care behaviors but rarely to a classic catatonic stupor. Wing and Shah (2000) observed that most of the 30 individuals with ASDs and catatonia exhibited slowness and difficulty initiating movements unless prompted, odd gait, odd stiff posture, freezing during actions, difficulty in crossing lines or cracks on the pavement, and an inability to cease actions. All of the individuals also showed a significant reduction in speech production or else mutism. Less pronounced and subtle symptoms included aberrations of posture, movement, speech, and behavior. These less severe symptoms are described as "catatonic in nature" because they do not necessarily match the classic symptoms of catatonia and, therefore, may represent a different condition. The diverse manifestations of the catatonic-like symptoms likely contribute to the general lack of agreement on the definition of this condition and its cardinal symptoms (Ungvari and Carroll, 2004). Although not classified as a separate diagnostic category in the Diagnostic and Statistical Manual of Mental Disorders, 4th ed., Text Revision (DSM-IV) (APA, 2000), catatonia does present with some unique features that have led to five features that are considered representative of catatonic disorder. These five symptoms include—motoric immobility, excessive motor activity, extreme negativism or mutism, peculiarities of voluntary movement, and echolalia or echopraxia. The features are based on clinical observations of adults and not of younger age groups. Although there are no standardized criteria for children or adolescents, Wing and Shah (2000) described features of catatonia that are characteristic of children and adolescents who may present with this disorder. These symptoms include—increased slowness affecting movement and verbal responses, difficulty initiating and completing actions, increased reliance on physical or verbal prompting by others, increased passivity and apparent lack of motivation, reversal of day and night, parkinsonian features, excitement and agitation, and increased repetitive, ritualistic behavior.

A. EPIDEMIOLOGY

Historically, catatonia had been observed in patients with schizophrenia. Consequently, most of the research conducted to date has been with populations diagnosed with psychotic conditions. Although epidemiological data reveal a wide range of prevalence rates for catatonic symptoms in individuals diagnosed with schizophrenia (4–67%), lower rates (7–17%) are reported for individuals hospitalized with more acute onsets of psychotic symptoms (Caroff et al., 2004). Higher rates (up to 67%) have been observed in individuals diagnosed with other psychiatric conditions including depression, bipolar disorder, personality disorders, or organic brain disorder (Caroff et al., 2004; Fein and McGrath, 1990).

The prevalence rate for severe catatonic-like deterioration appears to be rare, with 6% of 506 adolescents and adults with ASDs reportedly found to exhibit such symptoms across 7 years (Wing and Shah, 2000). These individuals comprised 17% of the patients over the age of 15 who participated in the study. The 6% prevalence rate for deteriorating symptoms is similar to the 7–10% prevalence rate reported for acutely ill psychiatric patients (Ghaziuddin et al., 2005; van der Heijden et al., 2005) but higher than the 1.3% rate reported for a cohort of psychiatric patients tested between 1990 and 2001 (van der Heijden et al., 2005). Overall, it does not appear that individuals with autism are necessarily at a greater risk for catatonia than other psychiatric populations. However, catatonia may be observed more often in children with ASDs than in children with other developmental delays such as learning and language-disordered children (see Chapter 2 by Wing and Shah). Given that catatonia occurs in other psychiatric conditions and reaction to medications (Northoff, 2002; Takaoka and Takata, 2003), biological underpinnings that are not necessarily specific to autism may help explain symptom expression.

Observations of catatonic-like symptoms among individuals with ASDs are in sharp contrast to the excessive stereotypic behaviors that are considered diagnostic features of autism. However, some of the items reflecting catatonic-like features in Wing and Shah's Chapter 2 of pre- and school-aged children included stereotypic behaviors (e.g., runs in circles, body rocking, toe-walking, and spins objects), raising some questions about the overlap of catatonic and stereotypic symptoms in assessments. There tends to be a general bias to diagnose catatonia in its severe form rather than in more subtle manifestations, thereby increasing the risk of overlooking less salient symptoms. Finally, catatonia has historically been conceptualized as a type of schizophrenia in psychiatry and not investigated as systematically in individuals with ASDs as with other psychiatric disorders (Taylor, 2004; Ungvari et al., 2005; van der Heijden et al., 2005).

B. Characteristic Symptoms

To date, research on "catatonic autism" has largely consisted of case reports. These case descriptions share in common similar symptomatology including manifestations of immobility, extreme negativism, mutism, and peculiarities of voluntary movements co-occurring with episodes of echolalia and bursts of hyperactivity (Dhossche, 2004; Ghaziuddin et al., 2005). In Wing and Shah's (2000) assessment of 30 individuals having more severe features of catatonia, a combination of catatonic-like and parkinsonian type of symptoms were revealed. All showed significant paucity of speech or complete mutism, and most exhibited slowness and difficulty initiating movements unless prompted, as well as odd gait and stiff posture. Other symptoms included freezing during actions, difficulty crossing lines on the pavement, inability to cease actions, and impulsive and bizarre behavior. Less than half showed sleeping during the day, while staying awake at night, incontinence, and excited phases, and none of the individuals exhibited waxy flexibility. Many of these symptoms match Wing and Shah's list of essential features described earlier for diagnosing catatonia in children and adolescents. Hare and Malone (2004) have proposed three essential symptoms for "autistic catatonia" that are similar to Wing and Shah's (2000) description of symptoms, but that are restricted to a paucity of movement. These symptoms include freezing when carrying out actions and being resistant to prompting, very slow voluntary movements, and stopping in the course of movement and requiring prompting to complete actions.

C. RISK FACTORS

Risk factors for the development of catatonia-like symptoms in ASDs include a significant decline in motor movement (Ghaziuddin et al., 2005). Age also appears to be a risk factor. All of the individuals with ASDs in Wing and Shah's (2000) study who manifested catatonia-like deterioration were either adolescents or adults; none were children. Case descriptions of individuals with autism who developed catatonia also manifested symptoms during adolescence or young adulthood (Dhossche, 1998; Ghaziuddin et al., 2005; Realmuto and August, 1991; Zaw et al., 1999).

The role of stressful events prior to the onset of catatonic symptoms is still open to debate. Parents of the individuals in Wing and Shah's (2000) study who exhibited catatonic-like deterioration reported a variety of precipitating events, ranging from school examinations to bereavement. All of these events were considered stressful for the individual; however, as noted by the researchers, a comparison group of adolescents and adults with autism had experienced similar events without developing catatonic-like symptoms. Wing and Shah (2000) also noted that there were no significant differences in the age, gender, IQ, history of seizure disorder, or in the diagnostic subgroup for autism noted between the individuals showing symptoms and the comparison control group. The most common risk factor identified in this study was passivity in social situations and impaired expressive language skills that predated the onset of catatonic symptoms. Case reports of adolescents with autism and catatonia also describe passivity with symptoms of depression, obsessive compulsive, and nonspecific psychotic symptoms preceding the onset of catatonia (Dhossche, 1998; Zaw et al., 1999). Realmuto and August (1991) have suggested in their case description that these psychiatric conditions might account for the increased risk of catatonia among individuals with autism.

Depression is often noted as a cause of catatonic symptoms in general, yet it is unclear if this association can be generalized to individuals with ASDs. In their case report of an adolescent male with autism and catatonia, Ghaziuddin et al. (2005) suggest that obsessive slowing may be a precursor to catatonia in people with autism and that regressive behavior may result from the emergence of catatonic symptoms. While informative, further research is warranted before drawing definitive conclusions about possible risk factors for catatonia in individuals with autism.

D. Etiology

Etiological explanations for catatonia include a genetic predisposition, especially for periodic catatonia. At the molecular level, chromosome 15q15 has been strongly linked to this specific subtype. Research appears to support a single gene model and an autosomal dominant mode of inheritance with reduced penetrance (Stöber, 2004). However, family constellations as well as other psychosocial and environmental factors can modify the penetrance of the disorder.

Other hypothesized causes include structures of the brain including the frontal lobe (Taylor, 1990) and the basal ganglia (Rogers, 1991). Fricchione (2004) has implicated a network of systems that includes the basal ganglia–thalamo (limbic)–cortical circuits. Fricchione (2004) elaborated further on his hypothesis and suggested that a disruption in the gamma-aminobutyric acid (GABA) and dopamine (DA) balance likely contributes to the development of catatonia. Dhossche et al. (2002) also has implicated a dysregulation of GABA as a shared risk factor for both autism and catatonia. Hare and Malone (2004) argue, however, that catatonic symptoms are expressions of autism and should not necessarily be conceptualized as comorbid symptoms. Hence, they suggest that sensory, perceptual, and neurocognitive systems underlying the cause of autism might explain catatonia in ASDs. Any hypothesis is mere speculation at this point, pending further empirical evaluations.

E. Assessment

Although not specific to catatonia, measures of autism including the Autism Diagnostic Interview-Revised (ADI-R) (Lord et al., 1994) and the Childhood Autism Rating Scale (CARS) (Schopler et al., 1988) include questions about psychomotor movements that could be diagnostic of catatonia. It is important to note that these items were neither developed specifically for this purpose, nor were they intended to be used to diagnose catatonia. Some of the items, characteristic of catatonia, are also characteristic of autism (e.g., posturing), thereby increasing the risk for misdiagnoses. To help with differential diagnosis,

Ghaziuddin *et al.* (2005) suggest that the key issue in diagnosing catatonia in autism is the emergence of "new" symptoms or a "change" in the type and pattern of premorbid functioning.

One measure of ASDs, the Diagnostic Interview for Social and Communication Disorders (DISCO), includes 28 items that specifically address catatonic-like features. Caregivers describe each type of behavior listed and give examples to aid the interviewer with the assessment of symptoms. Information from the interview, direct observation, and data from other available sources are considered to rate the presence and severity of the behavior. Wing and Shah (see Chapter 2 by Wing and Shah in this volume) found that children with autistic disorders were rated with more marked and moderate problems on the 28 items than children with learning disabilities or a specific language disorder and a normally developing group of children. There was no significant difference between the groups of children with autistic disorder where IQ scores were either above or below 70.

Over the past two decades there has been a growing interest in developing standardized measures specifically for diagnosing catatonia (Mortimer, 2004). While many of these measures have been used to assess for catatonia in patients with a diagnosis of schizophrenia or a mood disorder, the psychometric properties of these measures have not been adequately assessed with other psychiatric populations. The Modified Rogers Scale (Rogers *et al.*, 1991) is one such measure that has been found to distinguish reliably between depressed individuals with and without catatonia, as well as between depressed individuals with catatonia and individuals with Parkinson's disease. Items with the best specificity from the scale have been extracted and were renamed the Rogers Catatonia Scale (Starkstein *et al.*, 1996). Factor analysis of this scale has revealed two primary factors, hypokinetic and hyperkinetic, that account for approximately 64% of the variance (McKenna *et al.*, 1991).

A second measure of catatonia, the Bush–Francis Catatonia Rating Scale (BFCRS) (Bush *et al.*, 1996) also appears to be a reliable and moderately valid measure of catatonia. While some measures of catatonia assess the presence or absence of symptoms, the BFCRS measures the severity of 23 signs of catatonia (e.g., mutism, grimacing, echolalia, impulsivity, and combativeness). A truncated 14-item version is available as a quick screener. However, this measure has not been evaluated for use with children or with individuals with autism, thereby limiting its clinical utility with certain populations.

F. Treatment

Research on treatment is largely limited to case reports and there have been no empirically based treatment outcome studies that would aid in the development of evidence-based practices for the treatment of catatonia in autism.

Consequently, pharmacological agents are often utilized to treat accompanying depressive, regressive, and psychotic symptoms, but these are not always successful for addressing the catatonic symptoms, especially in severe cases (Dhossche, 1998; Ghaziuddin *et al.*, 2005; Realmuto and August, 1991). Electroconvulsive therapy (ECT) has been used with some success, but this option is reserved for the most severe and life-threatening cases (Ghaziuddin *et al.*, 2005; Zaw *et al.*, 1999). For an in-depth discussion of treatment issues, refer to the Treatment Section (Section III) in this book.

III. Stereotypic Movement Disorder

Stereotypic movement disorder (SMD) is characterized by repetitive nonfunctional motor or vocal responses that are severe enough to interfere with psychosocial functioning or to cause physical injury (APA, 2000; LaGrow and Repp, 1984). Although recognized as a separate diagnostic category, stereotypic movements can occur with pervasive developmental disorders (PDDs), obsessive-compulsive disorder (OCD), hair pulling, and tic disorders. If occurring with these disorders, then SMD is not diagnosed.

A. EPIDEMIOLOGY

Stereotypies are most often seen in individuals with mental retardation (25%) and autism (85%) (Volkmar *et al.*, 1986); however 2–3% of the general population of children and adolescents (APA, 2000; Rojahn *et al.*, 1998) and up to 15–20% of pediatric populations (Matthews *et al.*, 2001) exhibit stereotypic behaviors. Some of the more common stereotypies observed among individuals with autism include rocking (65%), toe-walking (57%), arm, hand, or finger flapping (52%), and twirling (50%) (Volkmar *et al.*, 1986). These behaviors are often the focus of clinical attention because they can interfere with learning and the application of learned skills (Morrison and Rosales-Ruiz, 1997). Early intervention is strongly recommended as they tend to be prognostic of more severe self-injurious behaviors (Guess and Carr, 1991; Schroeder *et al.*, 1990).

The age of onset for stereotypic behaviors is typically toddlerhood. Symptoms generally remit by age 5 in normally developing children but persist in children with developmental delays, reaching a peak in adolescence. Stereotypies tend to decline gradually thereafter with the exception of adults with severe or profound mental retardation who can show symptoms for years. In its most severe form, SMD can result in physical injuries, as in the case of self-biting, head banging, scratching, and hair pulling, if untreated.

B. ETIOLOGY

Since stereotypic behaviors are closely intertwined with autism, biochemical abnormalities implicated in autism may be pertinent for understanding the biological basis for stereotypical behaviors. As with many other psychiatric conditions, serotonin (5-HT) and the 5-HT system have been linked to the development of autism. Serotonin is involved in many aspects of human behavior including sleep, pain, motor function, appetite, and others (Volkmar and Anderson, 1989). Research has consistently revealed that up to 50% of individuals with autism are hyperserotonemic (Geller *et al.*, 1982). The mechanism through which 5-HT influences the symptoms of autism, however, is still unknown. Findings specifically related to repetitive and stereotypic behaviors suggest that decreases in 5-HT levels are related to *decreased* stereotypies. Consistent with this hypothesis, Curzon (1990) found that rats administered with agonists for 5-HT activity engaged in behaviors that closely resemble stereotypies. However, contrary to this finding, others have reported a decrease in stereotypies when 5-HT is inhibited from being reabsorbed by the presynaptic neuron (Powell *et al.*, 1997). Similarly, McDougle *et al.* (1996) observed an exacerbation of stereotypic behavior among a sample of adults with autism after 5-HT was reduced through the depletion of its precursor, tryptophan. This may be explained by the effects of central versus peripheral 5-HT that may differentially influence the manifestation of stereotypic behaviors.

Opioids have also been implicated in the development of self-injurious and stereotyped behaviors. According to the opioid hypothesis, when individuals engage in self-injurious behavior (SIB), the brain releases neurochemicals, such as endorphins, that block pain and produce mild euphoria. Although this may seem paradoxical, it is believed that continued self-injury actually blocks the painful stimulation that it would ordinarily produce, contributing to the maintenance of this behavior. Individuals who engage in SIB and other stereotypies may be motivated by sensations of euphoria and escape or avoidance from painful stimulation. Using animal models, researchers have observed higher rates of SIB and stereotypies after administrations of opiate agonists and lower rates after administrations of opiate antagonists (Dantzer, 1986; Iwamoto and Way, 1977). Both Campbell *et al.* (1993) and Rojahn *et al.* (1998) reported similar observations with humans, however, the declines in stereotypies observed by Campbell *et al.* (1993) after administering naltrexone, an opiate antagonist, did not reach statistical significance.

Research has examined the interaction between opiates, 5-HT, and the DA system to explain autism, and more specifically stereotypies and SIB. Stereotypical behaviors observed among animals, head weaving, for example, are noticeably less frequent following lesions to the nigrostriatal and mesolimbic DA pathways and after the administration of DA antagonists (Lewis and Bodfish,

1998). Terminal fields that receive substantial amounts of DA innervation also contain large amounts of opioid peptides and receptors, suggesting an interaction effect between the opioid and DA systems (Angulo and McEwen, 1994). Another hypothesis, the DA supersensitivity hypothesis, proposes that repetitive behaviors and SIB result from low levels of DA in postsynaptic cells of the basal ganglia, resulting in the supersensitivity of the postsynaptic receptors. The presence of small amounts of DA subsequently produces activation. Support for this hypothesis includes animal models of self-injurious and stereotypic behaviors in which these behaviors are induced following the administration of DA agonists such as L-dopa (Lewis and Baumeister, 1982). Depriving animals of sensory stimulation and restricting their interactions with the environment in controlled laboratory experiments has been found to prevent DA innervation and subsequently produces spontaneous stereotypies (Martin *et al.*, 1991). Suomi and Harlow (1971) observed similar behaviors in nonhuman primates who had experienced early social deprivation. It appears that early deprivation or restricted environmental interaction results in a loss of DA innervation of important brain regions that results in DA receptor supersensitivity (Lewis and Bodfish, 1998).

Research linking structural parts of the brain to stereotypies have revealed some abnormalities in the cerebellum and the neuronal systems that are directly influenced by the cerebellum, including those that regulate attention, sensory modulation, autonomic activity, and behavior initiation (Courchesne *et al.*, 1988). Further research using functional magnetic resonance imaging may prove useful for specifying the underlying mechanisms involved.

Nonbiological theories for explaining stereotypies include behavioral theories in which the repetitive behaviors are reinforced by contingencies. These contingencies may be: (1) positive internal reinforcement such as sensory stimulation, (2) positive external reinforcement such as social attention, or (3) negative reinforcement—the removal of aversive stimuli. According to the sensory stimulation hypothesis, the stereotypy is maintained by access to reinforcing sensory and perceptual stimulation that may be a by-product of the stereotypic behavior itself (Lovaas *et al.*, 1987). Repetitive rocking or twirling, for example, may be maintained by vestibular stimulation or eye poking may be reinforced by visual sensations. Iwata *et al.* (1994) refer to the maintenance of these behaviors as *automatic reinforcement* because they are not socially mediated.

The positive reinforcement hypothesis focuses on the social consequences maintaining the stereotypic response. Attention or tangible rewards, such as toys or food, contingent upon these behaviors are hypothesized to reinforce the stereotypic response. The consequences need not necessarily include the receipt of desired rewards but can also involve the removal of aversive stimuli (Iwata *et al.*, 1994), in which case the respondent is reinforced negatively for manifesting stereotypic behaviors. Support for socially mediated consequences include observations of higher rates of rocking and hand-flapping among children with autism and other PDDs when confronted with a difficult task. These behaviors remit or are reduced upon the removal of the task. Rather than arguing for either the sensory stimulation or the behavioral hypothesis as causal explanations, both theories are supported empirically. Dawson *et al.* (1998) observed individuals with autism tended to engage in stereotypies for the purpose of sensory stimulation, as well as to remove aversive work or social stimuli.

A third etiological hypothesis for stereotypic behavior is the communication-based theory. Similar to behavioral hypotheses, stereotypic responses serve as a means to communicate one's needs, including eliciting a desired response from another, acquiring a desired object, or removal of an aversive environmental stimulus. This explanation may be more specific to children with communicative disorders that preclude them from using language productively. Such children have learned to use stereotypic behaviors as a means of communication to compensate for their language impairment. Support for this hypothesis includes the successful application of treatments designed to teach children sign language to communicate in lieu of engaging in stereotypic behaviors (Kennedy *et al.*, 2000). Others have suggested, on the other hand, that stereotypies are associated with cognitive impairment (Volkmar and Lord, 1998). However, Matson *et al.* (1996) found a significantly lower percentage of adults with severe or profound retardation (7%) exhibited stereotypies compared to adults with the dual diagnoses of severe or profound mental retardation and autism (75%). In addition, children with more severe manifestations of autism have been found to show a tendency to exhibit more severe stereotypical behaviors (Campbell *et al.*, 1990).

C. Assessment

Many of the assessment measures for autism include items pertaining specifically to stereotypic behavior. The ADI-R, for example, is a structured interview that assesses functioning in three domains mirroring the diagnostic criteria for autism, including the "restricted, repetitive, and stereotyped behaviors and interests" domain (Lord *et al.*, 1994). Items in this domain include repetitive motor movements and SIB. The CARS is another measure of autism that incorporates historical interview information from the parent and direct observation by a professional who rates the child's behavior in 15 domains (Schopler *et al.*, 1988). An item pertaining to "body use" measures the severity of stereotyped behaviors such as repetitive movements, rocking, spinning, and self-injury. An item pertaining to "object use" addresses inappropriate use of objects that may include stereotyped use. A third measure, the Autism Behavior Checklist (ABC) (Krug *et al.*, 1980, 1993), is a 57-item parent- or teacher-rating scale that is included as part of the Autism Screening Instrument for Educational Planning-2. One of the five subscales, "Body and Object Use," contains several items measuring stereotypic motor movements, such as whirling, rocking, spinning, and flapping, as well as stereotyped use of objects (e.g., spinning or banging objects).

Clinicians may need to be cautious when using these measures with younger children because of potential problems with overidentifying children with mental retardation as exhibiting autism (Rutter and Schopler, 1987; Wing and Gould, 1979). Both groups exhibit language delays, social impairments, and some evidence of restricted, repetitive, and stereotyped behaviors that can complicate the process of making differential diagnoses (Vig and Jedrysek, 1999; Wing and Gould, 1979). The problem with overidentification is further complicated by the fact that mental retardation can co-occur in 70–80% of children with autism. Some of the autism assessment measures, including the CARS and ABC, have also been criticized for failing to accurately diagnose children under the age of 3 (New York State Department of Health, 1999). Using the ADI-R and CARS, for example, has led to the overidentification of very young (under 2 years) and mentally retarded children (mental age < 18 months) as being autistic (DiLavore et al., 1995).

The Diagnostic Assessment for the Severely Handicapped II (DASH-II) is not a measure of autism specifically, but it does include two scales for movement disorders—stereotypies and SIB. Matson et al. (1997) evaluated the measure with 289 individuals with severe and profound mental retardation. The stereotypies and SIB scales correctly identified 32% of individuals with stereotypies, 94% with SIB, 75% of individuals exhibiting both, and 100% of controls, as based on DSM-IV criteria. The overall classification rate was 83%. In a second study of over 1000 individuals with severe or profound mental retardation, certain items were more characteristic of individuals with SMD and others for SIB. Repetitive body movements, limited sets of preferred activities, and repetitive words or sounds were most likely characteristic of SMD, whereas self-biting and picks at wounds differentiated the SIB group from the SMD and control groups. Although diagnostic, two to three items are not generally considered sufficient for diagnosing a syndrome per se.

D. Treatment

Treatments for stereotypies are generally behavior specific and based on behavioral functional analyses used to identify and remove or prevent reinforcing contingencies. Repeated episodes of the stereotypic response without contingent responses are necessary before the stereotypy is extinguished. Individuals who bite their arms, for example, might be outfitted with arm braces or casts to prevent them from garnering any self-stimulation from biting themselves. Aversive conditioning, including the application of contingent noxious smells and tastes (e.g., ammonia, lemon juice), have been used successfully to reduce stereotypies and SIB in children who are unresponsive to less restrictive contingent management approaches (Stoppelbein and Greening, 2005a,b). Treatment gains tend to be situation specific and, therefore, require that contingencies be applied across settings and caregivers. Less restrictive approaches should be exhausted before resorting to aversive therapies because of ethical concerns about using noxious stimuli in treatments.

References

American Psychiatric Association (APA) (2000). "Diagnostic and Statistical Manual of Mental Disorders," 4th ed., Text Revision (DSM-IV-TR). APA Press, Washington, DC.

Angulo, J. A., and McEwen, B. S. (1994). Molecular aspects of neuropeptide regulation and functioning in the corpus striatum and nucleus accumbens. *Brain Res. Rev.* **19,** 1–28.

Bush, G., Fink, M., Petrides, G., Dowling, F., and Francis, A. (1996). Catatonia. Part I: Rating scale and standardized examination. *Acta Psychiatr. Scand.* **93,** 129–136.

Campbell, M., Locascio, J., Choroco, M. C., Spencer, E. K., Malone, R. P., Kafantaris, V., and Overall, J. E. (1990). Stereotypies and tardive dyskinesia: Abnormal movements in autistic children. *Psychopharmacol. Bull.* **26,** 260–266.

Campbell, M., Anderson, L. T., Small, A. M., Adams, P., Gonzales, N. M., and Ernst, M. (1993). Naltrexone in autistic children: Behavioral symptoms and attentional learning. *J. Am. Acad. Child Adolesc. Psychiatry* **32,** 1283–1289.

Caroff, S. N., Mann, S. C., Campbell, E. C., and Sullivan, K. A. (2004). Epidemiology. *In* "Catatonia: From Psychopathology to Neurobiology" (S. N. Caroff, S. C. Mann, A. Francis, and G. L. Fricchione, Eds.), pp. 15–31. American Psychiatric Publishing, Washington, DC.

Courchesne, E., Yeung-Courchesne, R., Press, G. A., Hesselink, J. R., and Jernigan, T. L. (1988). Hypoplasia of cerebellar vermal lobules VI and VII in autism. *New Engl. J. Med.* **318,** 1349–1354.

Curzon, G. (1990). Stereotyped and other motor responses to 5-hydroxytryptamine receptor activation. *In* "Neurobiology of Stereotyped Behaviour" (S. J. Cooper and C. T. Dourish, Eds.), pp. 142–168. Oxford University Press, Oxford.

Dantzer, R. (1986). Behavioral, physiological and foundational aspects of stereotyped behavior: A review and re-interpretation. *J. Anim. Sci.* **62,** 1776–1786.

Dawson, J. E., Matson, J. L., and Cherry, K. E. (1998). An analysis of maladaptive behaviors in persons with autism, PDD-NOS, and mental retardation. *Res. Dev. Disabil.* **19,** 439–448.

Dhossche, D. M. (1998). Catatonia in autistic disorders. *J. Autism Dev. Disord.* **28,** 329–331.

Dhossche, D., Applegate, H., Abraham, A., Maertens, P., Bland, L., Bencsath, A., and Martinez, J. (2002). Elevated plasma gamma-aminobutyric acid (GABA) levels in autistic youngsters: Stimulus for a GABA hypothesis of autism. *Med. Sci. Monit.* **8,** PR1–PR6.

Dhossche, D. M. (2004). Autism as early expression of catatonia. *Med. Sci. Monit.* **10,** 31–39.

DiLavore, P. C., Lord, C., and Rutter, M. (1995). The pre-linguistic autism diagnostic observation schedule. *J. Autism Dev. Disord.* **25,** 355–379.

Fein, S., and McGrath, M. G. (1990). Problems in diagnosing bipolar disorder in catatonic patients. *J. Clin. Psychiatry* **51,** 203–205.

Fricchione, G. L. (2004). Brain evolution and the meaning of catatonia. In "Catatonia: From Psychopathology to Neurobiology" (S. N. Caroff, S. C. Mann, A. Francis, and G. L. Fricchione, Eds.), pp. 210–221. American Psychiatric Publishing, Washington, DC.

Geller, E., Ritvo, E. R., Freeman, B. J., and Yuwiler, A. (1982). Preliminary observations on the effect of fenfluramine on blood serotonin and symptoms in three autistic boys. *New Engl. J. Med.* **307**, 165–167.

Ghaziuddin, M., Quinlan, P., and Ghaziuddin, N. (2005). Catatonia in autism: A distinct subtype? *J. Intellect. Disabil. Res.* **49**, 102–105.

Guess, D., and Carr, E. (1991). Emergence and maintenance of stereotypy and self-injury. *Am. J. Ment. Retard.* **96**, 299–329.

Hare, D. J., and Malone, C. (2004). Catatonia and autistic spectrum disorders. *Autism* **8**, 183–195.

Iwamoto, E. T., and Way, E. L. (1977). Circling behavior and stereotypy induced by intranigral opiate microinjection. *J. Pharmacol. Exp. Ther.* **20**, 347–359.

Iwata, B. A., Pace, G. M., Dorsey, M. F., Zarcone, J. R., Vollmer, T. R., Smith, R. G., Rodgers, T. A., Lerman, D. C., Shore, B. A., Mazaleski, J. L., Goh, H. L., Cowdery, G. E., et al. (1994). The functions of self-injurious behavior: An experimental-epidemiological analysis. *J. Appl. Behav. Anal.* **27**, 215–240.

Kennedy, C. H., Meyer, K. A., Knowles, T., and Shukla, S. (2000). Analyzing the multiple functions of stereotypical behavior for students with autism: Implications for assessment and treatment. *J. Appl. Behav. Anal.* **33**, 559–571.

Krug, D. A., Arick, J. R., and Almond, P. J. (1980). Behavior checklist for identifying severely handicapped individuals with high levels of autistic behavior. *J. Child Psychol. Psychiatry* **21**, 221–229.

Krug, D. A., Arick, J. R., and Almond, P. J. (1993). "Autism Screening Instrument for Educational Planning," 2nd ed. Pro-Ed., Austin, TX.

LaGrow, S. J., and Repp, A. C. (1984). Stereotypic responding: A review of intervention research. *Am. J. Ment. Def.* **88**, 595–609.

Lewis, M. H., and Baumeister, A. A. (1982). Stereotyped mannerisms in mentally retarded persons: Animal models and theoretical analyses. In "International Review of Research in Mental Retardation" (N. R. Ellis, Ed.), Vol. 11, pp. 123–161. Academic Press, New York.

Lewis, M. H., and Bodfish, J. W. (1998). Repetitive behavior disorders in autism. *Mental Retardation and Developmental Disorders* **4**, 80–89.

Lord, C., Rutter, M., and Le Couteur, A. (1994). Autism diagnostic interview revised: A revised version of a diagnostic interview for caregivers of individuals with possible pervasive developmental disorders. *J. Autism Dev. Disord.* **24**, 659–685.

Lovaas, O. I., Newsom, C., and Hickman, C. (1987). Self-stimulatory behavior and perceptual reinforcement. *J. Appl. Behav. Anal.* **20**, 45–68.

Martin, L. J., Spicer, D. M., Lewis, M. H., Gluck, S. P., and Cork, L. C. (1991). Social deprivation of infant Rhesus monkeys alters the chemoarchitecture of the brain. I. Subcortical regions. *J. Neurosci.* **11**, 3344–3358.

Matson, J. L., Baglio, C. S., Smiroldo, B. B., Hamilton, M., Packlowskyj, T., Williams, D., and Kirkpatrick-Sanchez, S. (1996). Characteristics of autism as assessed by the diagnostic assessment for the severely handicapped-II (DASH-II). *Res. Dev. Disabil.* **17**, 135–143.

Matson, J. L., Hamilton, M., Duncan, D., Bamburg, J., Smiroldo, B., Anderson, S., and Baglio, C. (1997). Characteristics of stereotypic movement disorder and self-injurious behavior assessed with the diagnostic assessment for the severely handicapped (DASH-II). *Res. Dev. Disabil.* **18**, 457–469.

Matthews, L. H., Matthews, J. R., and Leibowitz, J. M. (2001). Tics, stereotypic movements, and habits. In "Handbook of Clinical Child Psychology" (C. E. Walker and M. C. Roberts, Eds.), 3rd ed., pp. 338–358. Wiley, New York.

McDougle, C., Naylor, S., Cohen, D., Aghajanian, G., Heninger, G., and Price, L. (1996). Effects of tryptophan depletion in drug-free adults with autistic disorder. *Arch. Gen. Psychiatry* **53,** 993–1000.

McKenna, P. J., Lund, C. E., and Mortimer, A. M. (1991). Motor, volitional and behavioral disorders of schizophrenia. Part 2: The "conflict of paradigms" hypothesis. *Br. J. Psychiatry* **158,** 328–336.

Morrison, K., and Rosales-Ruiz, J. (1997). The effect of object preference on task performance and stereotypy in a child with autism. *Res. Dev. Disabil.* **18,** 127–137.

Mortimer, A. M. (2004). Standardized instruments. *In* "Catatonia: From Psychopathology to Neurobiology" (S. N. Caroff, S. C. Mann, A. Francis, and G. L. Fricchione, Eds.), pp. 53–64. American Psychiatric Publishing, Washington, DC.

New York State Department of Health (1999). "Clinical Practice Guideline: Report of the Recommendations. Autism/Pervasive Developmental Disorders, Assessment and Intervention for Young Children (Age 0–3 years)." New York State Department of Health, Albany, New York.

Northoff, G. (2002). Catatonia and neuroleptic malignant syndrome: Psychopathology and pathophysiology. *J. Neural Transm.* **109,** 1453–1467.

Powell, S., Wechsler, E., Newman, H., and Lewis, M. H. (1997). Behavioral and biochemical effects of chronic fluoxetine treatment in an animal model of repetitive behavior disorder. *Soc. Neurosci.* (Abstracts) **23,** 802.

Realmuto, G. M., and August, G. J. (1991). Catatonia in autistic disorder: A sign of comorbidity or variable expression. *J. Autism Dev. Disord.* **21,** 517–528.

Rogers, D. (1991). Catatonia: A contemporary approach. *J. Neuropsychiatry Clin. Neurosci.* **3,** 334–340.

Rogers, D., Karki, C., Bartlett, C., and Pocock, P. (1991). The motor disorders of mental handicap: An overlap with the motor disorders of severe psychiatric illness. *Br. J. Psychiatry* **158,** 97–102.

Rojahn, J., Tasse, M. J., and Morin, D. (1998). Self-injurious behavior and stereotypies. *In* "Handbook of Child Psychopathology" (T. H. Ollendick and M. Hersen, Eds.), pp. 307–336. Plenum, New York.

Rutter, M., and Schopler, E. (1987). Autism and pervasive developmental disorders: Concepts and diagnostic issues. *J. Autism Dev. Disord.* **17,** 159–186.

Schopler, E., Reichler, R. J., and Renner, B. R. (1988). The Childhood Autism Rating Scale. Western Psychological Services, Los Angeles.

Schroeder, S. R., Rojahn, J., Mulick, J. A., and Schroeder, C. S. (1990). Self-injurious behavior. *In* "Handbook of Behavior Modification with the Mentally Retarded" (J. L. Matson, Ed.), pp. 141–180. Plenum, New York.

Starkstein, S., Petracca, G., Teson, A., Chemerinski, E., Merello, M., Migliorelli, R., and Leiguarda, R. (1996). Catatonia in depression: Prevalence, clinical correlates, and validation of a scale. *J. Neurol. Neurosurg. Psychiatry* **60,** 326–332.

Stöber, G. (2004). Genetics. *In* "Catatonia: From Psychopathology to Neurobiology" (S. N. Caroff, S. C. Mann, A. Francis, and G. L. Fricchione, Eds.), pp. 173–187. American Psychiatric Publishing Inc., Washington, DC.

Stoppelbein, L., and Greening, L. (2005a). Aromatic therapy. *In* "Encyclopedia of Behavior Modification and Cognitive Behavior Therapy: Clinical Child Applications" (M. Hersen, A. M. Gross, and R. S. Drabman, Eds.), pp. 674–675. Sage, Thousand Oaks, CA.

Stoppelbein, L., and Greening, L. (2005b). Lemon juice therapy. *In* "Encyclopedia of Behavior Modification and Cognitive Behavior Therapy: Clinical Child Applications" (M. Hersen, A. M. Gross, and R. S. Drabman, Eds.), Vol. II, pp. 889–890. Sage, Thousand Oaks, CA.

Suomi, S. J., and Harlow, H. F. (1971). Abnormal social behavior in young monkeys. *In* "The Exceptional Infant" (J. Hellmuth, Ed.), Vol. 2, pp. 483–529. Bruner-Mazel, New York.

Takaoka, K., and Takata, T. (2003). Catatonia in childhood and adolescence. *Psychiatry Clin. Neurosci.* **57,** 129–137.

Taylor, M. A. (1990). Catatonia: A review of a behavioral neurologic syndrome. *Neuropsychiatry Neuropsychol. Behav. Neurol.* **3,** 48–72.

Taylor, M. A. (2004). Clinical examination. In "Catatonia: From Psychopathology to Neurobiology" (S. N. Caroff, S. C. Mann, A. Francis, and G. L. Fricchione, Eds.), pp. 45–52. American Psychiatric Publishing, Washington, DC.

Ungvari, G. S., and Carroll, B. T. (2004). Nosology. In "Catatonia: From Psychopathology to Neurobiology" (S. N. Caroff, S. C. Mann, A. Francis, and G. L. Fricchione, Eds.), pp. 33–44. American Psychiatric Publishing, Washington, DC.

Ungvari, G. S., Leung, S. K., Ng, F. S., Cheung, H. K., and Leung, T. (2005). Schizophrenia with prominent catatonic features ('catatonic schizophrenia'). Part I: Demographic and clinical correlates in the chronic phase. *Prog. Neuropsychopharmacol. Biol. Psychiatry* **29,** 27–38.

van der Heijden, F., Tuinier, S., Arts, N. J. M., Hoogendoorn, M. L. C., Kahn, R. S., and Verhoeven, W. M. A. (2005). Catatonia: Disappeared or under-diagnosed? *Psychopathology* **38,** 3–8.

Vig, S., and Jedrysek, E. (1999). Autistic features in young children with significant cognitive impairment: Autism or mental retardation? *J. Autism Dev. Disord.* **29,** 235–248.

Volkmar, F. R., and Anderson, G. M. (1989). Neruochemical perspectives on infantile autism. In "Autism: Nature, Diagnosis, and Treatment" (G. Dawson, Ed.), pp. 208–224. Guilford Press, New York.

Volkmar, F. R., Cohen, D. J., and Paul, R. (1986). An evaluation of DSM-III criteria for infantile autism. *J. Am. Acad. Child Adolesc. Psychiatry* **25,** 190–197.

Volkmar, F. R., and Lord, C. (1998). Diagnosis and definition of autism and other pervasive developmental disorders. In "Autism and Pervasive Developmental Disorders" (F. R. Volkmar, Ed.), pp. 1–31. Cambridge University Press, Cambridge, UK.

Wing, L., and Gould, J. (1979). Severe impairments of social interaction and associated abnormalities in children: Epidemiology and classification. *J. Autism Dev. Disord.* **9,** 11–29.

Wing, L., and Shah, A. (2000). Catatonia in autistic spectrum disorders. *Br. J. Psychiatry* **176,** 357–362.

Zaw, F. K. M., Bates, G. D. L., Murali, V., and Bentham, P. (1999). Catatonia, autism, and ECT. *Dev. Med. Child Neurol.* **41,** 843–845.

PRADER–WILLI SYNDROME: ATYPICAL PSYCHOSES AND MOTOR DYSFUNCTIONS

Willem M. A. Verhoeven[*,†] and Siegfried Tuinier[*]

[*]Vincent van Gogh Institute for Psychiatry
Venray, The Netherlands
[†]Department of Psychiatry, Erasmus University Medical Center
Rotterdam, The Netherlands

I. Introduction
II. Development and Behavior
III. Psychotic Disorders
IV. Psychomotor Symptoms
V. Conclusions
   References

Prader–Willi syndrome (PWS) is the result of a lack of expression of genes on the paternally derived chromosome 15q11-q13 and can be considered as a hypothalamic disorder. Its behavioral phenotype is characterized by ritualistic, stereotyped, and compulsive behaviors as well as motor abnormalities. After adolescence, recurrent affective psychoses are relatively frequent, especially in patients with uniparental disomy. These psychotic states have a subacute onset with complete recovery and comprise an increase of psychomotor symptoms that show resemblance with catatonia. Some evidence has emerged that gamma-aminobutyric acid (GABA) dysfunctionality is involved in both PWS and catatonia. Treatment of these atypical psychoses should preferably include GABA mimetic compounds like lorazepam, valproic acid, and possibly topiramate.

I. Introduction

Prader–Willi syndrome (PWS) was first described in 1956 by Prader, Labhart, and Willi, and its prevalence is estimated to be 1 in 10 to 15.000. This multisystem disorder is accompanied by a variable degree of mental retardation, occurs in all races and both sexes, and arises from the lack of expression of genes on the paternally derived chromosome 15q11-q13. Candidate genes for PWS in this region are imprinted and silenced on the maternally inherited chromosome. PWS develops if the paternal alleles are defective, missing, or silenced. In 75%

of the cases there is a paternal deletion of 15q11-q13, maternal uniparental disomy (UPD) occurs in 22%, imprinting errors in 3% because of either a sporadic or inherited microdeletion in the imprinting center, and there is a paternal chromosomal translocation in 1% cases. Imprinting occurs partly through parent-off-origin allele-specific methylation of CpG residues, which is established either during or after gametogenesis and maintained throughout embryogenesis. Paternally expressed genes are particularly important for the development of the hypothalamus. Therefore, hypothalamic dysfunction is one of the key features of PWS (Nicholls and Knepper, 2001).

With respect to hypothalamus-related endocrine and metabolic dysfunctions, PWS is associated with growth hormone (GH) deficiency reflected by mild prenatal growth retardation and after birth by a short stature as well as a lack of pubertal growth spurt. In addition, the levels of sex hormones are decreased as a result of an abnormal function of LH-releasing hormone neurons (Goldstone, 2004). The reduction of the number of the oxytocin-containing cells may be related to hyperphagia because oxytocin has anorexigenic properties (Shapira et al., 2005; Swaab, 1997). Finally, aberrant control of body temperature and daytime hypersomnolence are often present (Nixon and Brouillette, 2002; Swaab, 2004).

Other characteristics reflect an involvement of the central nervous system (CNS) motor structures: central hypotonia in infancy, delayed psychomotor development, difficulties in motor aspects of language, and obsessive compulsive behavior. These abnormalities are thought to be related to cortical dysgenesis (Hashimoto et al., 1998; Leonard et al., 1993; Yoshii et al., 2002) and possibly to a defect in the gamma-aminobutyric acid (GABA)-type A receptor (Lucignani et al., 2004; Wagstaff et al., 1991). Civardi et al. (2004) demonstrated an overall corticospinal hypoexcitability most likely as a result of hyposensitivity of GABA receptors.

II. Development and Behavior

The developmental trajectory is characterized by pre- and postnatal hypotonia, neonatal feeding problems, delayed motor milestones, and poor coordination. Adults are mildly hypotonic with decreased upper body strength. Speech is often poorly articulated. In early childhood, a characteristic behavioral profile emerges with temper tantrums, impulsiveness, stubbornness, repetitive rituals, obsessive–compulsive and ritualistic behavior, as well as difficulties with changing routine resulting in irritability and aggression (Cassidy, 2001; Clarke et al., 2002; Dykens and Shah, 2003; Wigren and Hansen, 2003). Some patients show a symptomatic form of narcolepsy, including cataplexy (Mignot et al., 2002; Tobias *et al.*, 2002). Another specific behavioral abnormality is skin-picking, a stereotyped movement aimed at various parts of the skin of which the percentage of patients showing this disorder increases with age (Dimitropoulos *et al.*, 2001; Steinhausen *et al.*, 2004; Symons *et al.*, 1999). Some reports suggest that the behavioral abnormalities and cognitive disabilities are more pronounced in patients with a deletion as compared with those in whom UPD etiology is present (Roof *et al.*, 2000; Veltman *et al.*, 2004; Whittington *et al.*, 2004). Within the individuals with a deletion, the size of the deletion seems to be positively correlated with behavioral and psychological problems (Butler *et al.*, 2004), although no differences were reported by Varela *et al.* (2005).

Thus, some characteristics of PWS are clearly related to a hypothalamic dysfunction, whereas others point to a more general brain dysfunction. There seems to be a specific involvement of CNS motor structures expressed by stereotypies, repetitive and ritualistic behaviors, obsessive–compulsive features, as well as hypotonia and hyperactivity.

III. Psychotic Disorders

PWS patients may develop relapsing psychotic states that, in general, are characterized by a subacute onset and full recovery, as well as mood instability, polymorphous symptomatology, and a fluctuating course. All available reports on psychotic states in PWS patients are summarized in Table I.

In 1966, Kollrack and Wolf described a 20-year-old male patient who developed two psychotic episodes of a few weeks duration with prominent anxiety, paranoid ideation, affective instability, dysphoric mood, and psychomotor agitation that completely disappeared after a 10-day treatment with chlorprothixene and chlordiazepoxide. A diagnosis of schizophrenia, paranoid type was considered although a definite classification could not be made. The authors concluded that a diagnosis of "endogenous agitated depressive paranoidly coloured disorder" was the most appropriate diagnostic description.

Until the beginning of 1990s, only two questionnaire surveys including 40 (Bray *et al.*, 1983) and 35 (Whitman and Accardo, 1987) PWS patients respectively were published, which were mainly focused on behavioral and emotional symptoms. In the study by Bray *et al.* (1983), two patients could be identified with psychotic symptoms, particularly auditory hallucinations and fluctuating behavioral inhibition or disinhibition that were short-lived and self-limited. In their review on quantitative psychiatric manifestations, obtained by administering the Survey Diagnostic Instrument to the parents, Whitman and Accardo (1987) found in about one-third of the cases mild and/or recurrent psychotic symptoms such as auditory and visual hallucinations as well as strange and paranoid

TABLE I
REPORTS ABOUT PSYCHOTIC SYMPTOMS IN PRADER–WILLI SYNDROME

Authors	N	Symptoms	Duration and course	Treatment	Diagnosis
Kollrack and Wolff, 1966*	1	Anxiety, mood alterations, paranoid ideas, hallucinations	Recurrent short episodes with full recovery	Chlorprothixene + chlordiazepoxide	Paranoid-hallucinatory syndrome
Bray et al., 1983*	2 out of 40	Social withdrawal, hallucinations, uncontrollable anger	Short-lived and self-limiting	None	Emotional problems
Whitman and Accardo, 1987*	35 survey	Hallucinations, paranoid ideation	—	—	Behavioral symptoms
Tu et al., 1992*	3	Dysphoric mood, anxiety	—	Carbamazepine ($n = 1$)	—
Jerome, 1993*	1	Cyclic mood change, sleep disturbances, skin-picking behavior	24 h, 2–4 times per month	Lithium	Bipolar affective disorder
Bartolucci and Younger, 1994*	4 out of 9	Delusional thinking, hallucinations, dysphoric mood	Cyclic course, fluctuating behavioral traits	Low doses of antipsychotics, antidepressants or lithium	Psychotic disorder, atypical depression
Clarke, 1993*; Clarke et al., 1995,* 1998b*	8	Anxiety, abnormal believes, auditory and visual hallucinations, paranoid ideation	Acute onset, short lasting with full recovery, recurrent	Low doses of antipsychotics, carbamazepine ($n = 2$)	Atypical psychotic disorder

Takhar and Malla, 1997*	1	Periods of confusion, impaired attention and concentration, hallucinations, ideas of reference	Recurrent episodes with subacute onset and full recovery	Low doses of clozapine	Psychosis and drug-induced parkinsonism
Whittaker et al., 1997*	1	Delusional convictions, anxiety, hallucinations, perplexity	Acute onset, short lasting with full recovery	Low doses of antipsychotic	Acute confusional state, paranoid psychotic illness
Dhossche and Bouman, 1997*	1	Catatonic symptoms, anxiety, visual hallucinations, delusional ideas	Acute onset, short lasting with full recovery	Low doses of antipsychotic and lorazepam	Catatonia
Beardsmore et al., 1998	5	Delusions and/or hallucinations, depressive symptoms, anxiety, lability of mood	Recurrent episodes	—	Cycloid psychosis
Pereda Bikandi et al., 2001*	1	Anxiety, hallucinations, confusion, delusional convictions, mood disorders	Acute onset, short lasting with full recovery	Low doses of olanzapine	Atypical psychotic disorder
Vogels et al., 2003, 2004	10 out of 14	Hyper/hypoactivity, anxiety, dysphoric mood, delusions, hallucinations	Acute onset, recurrent episodes	—	Atypical psychotic disorder
Boer et al., 2002	6 and 6 out of 15	Hypokinesia, paranoid ideation, anxiety, confusion, delusions, delusional convictions, hallucinations	Subacute onset, recurrent episodes	Various antipsychotics carbamazepine ($n = 3$)	Psychotic disorder

*For references see: Verhoeven et al., 2003a,b.

ideation. Tu *et al.* (1992) described three patients with prominent anxiety and dysphoric mood of whom one was treated successfully with carbamazepine in a daily dose of 1200 mg. In none of the three patients a psychiatric diagnosis was specified. Jerome (1993) reported a 31-year-old female patient with long lasting cyclic mood changes who responded to treatment with lithium. In this case, a diagnosis of bipolar affective disorder was considered to be the most appropriate. In four out of nine patients, Bartolucci and Younger (1994) observed a cyclic pattern of psychopathological symptoms, especially mood and motor disturbances in both directions, auditory and visual hallucinations, delusional ideas usually accompanied by fear, and occasionally elated and mystical feelings. No clear psychiatric diagnosis and classification was made. In two single case reports, Whittaker *et al.* (1997) and Takhar and Malla (1997) described female patients aged 17 and 51 years, respectively, who developed a short lasting psychotic state with full recovery on treatment with low doses of antipsychotics. Characteristic symptoms included confusion, perplexity, anxiety, and delusional convictions. In both cases, a diagnosis of paranoid psychosis was made. Beardsmore *et al.* (1998) described in their report on affective psychoses in 23 PWS patients, 1 patient with a delusional disorder and 4 with a mixture of affective and psychotic symptoms. Clarke *et al.* (1993, 1995, 1998a) reported eight patients who developed an acute short-lasting psychosis with full recovery after treatment with low doses of antipsychotics. Psychotic symptomatology consisted of pan-anxiety, motility disturbances, abnormal beliefs not syntonic with mood, paranoid ideation, and hallucinations. The differential diagnosis included psychosis within the schizophrenic spectrum, atypical psychosis, and cycloid psychosis. Other investigators published 14 patients of whom 10 were diagnosed with atypical psychoses and 4 with a bipolar affective disorder. Apart from psychotic symptoms, the clinical picture included hyper- and hypoactivity (Descheemaeker *et al.*, 2002; Vogels *et al.*, 2003, 2004). Finally, Boer *et al.* (2002) reported a population-based study with 25 adult PWS patients of whom 15 had a complete neuropsychiatric examination. In six patients, psychotic symptoms were observed. A previous study described six patients with a psychotic disorder. Four of the case descriptions make special reference to motor disturbances, especially hypokinesia and disinhibition (Clarke *et al.*, 1998a).

Over the past years, 27 PWS patients (mean age: 31 years; 12 males, 15 females) were referred to for neuropsychiatric evaluation to the first author because of long lasting or recurrent behavioral problems and psychotic symptoms (mean age of onset: 24 years). Out of this group, 23 patients were published previously (Verhoeven *et al.*, 1998, 2000, 2002, 2003a,b). In all patients, a standard psychiatric examination was performed, and additional data about history and course were collected from all available sources. Tentative formal psychiatric diagnoses were established according to the clinical descriptions and diagnostic guidelines of the Interclassification of Diseases, 10th revision (ICD-10).

All but two patients (insufficient data in the medical record) had a history of mood instability paralleled by fluctuating behavioral problems. With respect to the actual psychopathology, six patients met the criteria for a bipolar affective disorder, in that they showed an episodic pattern of euphoria, hyperactivity and sleep disturbances, or depressed mood and inactivity. In the other 21 patients, the psychiatric symptomatology included emotional turmoil, anxieties, irritability, confusion, (rapid) mood swings, hallucinatory experiences, and paranoid ideation with a variable intensity and subacute onset. Therefore, a diagnosis of cycloid psychosis was considered to be the most appropriate (Table II). Treatment with one or a combination of mood stabilizers, such as lithium (0.8 mMol/l) and/or valproic acid (60 mg/l), resulted in most patients in full recovery and prevention of relapses of both psychotic episodes and mood and behavioral instability. Concerning the genetic etiology, three patients had a deletion, and a clinical diagnosis was made in four patients. Three patients were genetically confirmed but not differentiated. In the remaining 17 patients, a UPD was demonstrated.

In a population of PWS patients, the expected percentage of deletions and UPD is about 75 and 25, respectively (Vogels and Fryns, 2002), whereas in the present group, excluding those with a clinical diagnosis or a not specified genetic etiology, percentages of 15 and 85 are found. These results indicate a three- to fourfold overrepresentation of UPD patients in a group of PWS patients with a psychotic illness. This association is described in two other reports that deal with smaller groups (Boer et al., 2002; Vogels et al., 2003).

Focusing on the cross-sectional assessment of psychopathology, however, may conceal the history of affective instability in these patients. Descriptions of instability of mood and behavior in patients with intellectual disabilities have been published since the beginning of the last century and meet, depending on the presence of mood elevation, the criteria of either cyclothymia or unstable

TABLE II
Symptoms of 21 PWS Patients with Relapsing Atypical Psychoses

Symptoms and course	Number	Percentage
Auditory hallucinations	10	48
Perceptual disturbances	4	19
Paranoid ideation	16	76
Confusion	19	90
Anxieties	20	95
Mood swings	20	95
Emotional turmoil	17	80
Increased obsessive rituals	18	85
Agitation/hyperactivity	13	62
Subacute onset	21	100

mood disorder (Verhoeven and Tuinier, 2001). Since affective symptoms are frequently expressed as mood swings, irritability, impulsivity, aggression, and self injury, a diagnosis of depression is often overseen (Verhoeven et al., 2004). Although the literature on psychopathology in PWS frequently refers to affective symptoms, there is no information about the lifetime prevalence of affective disorders in this syndrome.

In the reported 27 patients, the psychosis was characterized by symptoms like confusion, auditory hallucinations, and paranoid behavior on the one hand and an increase of obsessive rituals, anxieties, and mood swings on the other and was preceded by affective instability for many years. Even though the actual psychopathology justifies a diagnosis of cycloid psychosis, longitudinal evaluation points toward an atypical bipolar disorder. It can, therefore, be hypothesized that the psychopathological phenotype of PWS patients with a UPD etiology comprises also an increased risk for a bipolar affective disorder (Verhoeven et al., 2003b). In these patients signs of cyclothymia or unstable mood disorder should be treated with mood stabilizers in order to prevent psychotic deterioration.

Formal categorical diagnoses in PWS patients with psychiatric symptoms are rather futile since the psychopathological symptom profile varies longitudinally and comprises an increase of preexistent behavioral abnormalities including motor symptoms. It is, therefore, advocated to use the term Prader–Willi psychiatric syndrome (Verhoeven et al., 2000).

Since the clinical picture is very polymorphous and fluctuating over time, the clinician easily ignores the motor symptoms that may indicate the existence of catatonia. Nonetheless, these symptoms have been described in several reports.

IV. Psychomotor Symptoms

There is an ongoing debate on how to classify or characterize the psychiatric symptomatology in patients with PWS. A great variety of diagnostic vignettes is applied as a result of the fluctuating and heterogeneous clinical picture. Most prominent, however, are affective and anxiety symptoms associated with an increase of the syndrome-specific motor dysfunctions like skin-picking, stereotypies, perseverations, compulsions, excitement, and aggression. This polymorphous symptomatology is sometimes attributed to a bipolar affective disorder or, in the acute phase, a cycloid psychosis although these diagnoses do not account for the prominent changes in motor-related symptoms. The latter points to the presence of a psychomotor syndrome associated with an affective psychosis. In a few PWS cases, catatonic symptoms have been described, but they have not been systematically assessed and, therefore, may have remained unrecognized (Dhossche, 2004).

It is not surprising that the psychomotor aspects of the recurrent psychoses in PWS are relatively neglected. A similar disregard has taken place in psychiatry in general. Over the past century, the prevalence of the diagnosis catatonic schizophrenia apparently dropped from 10–30% to 2–10% (Morrison, 1974; Stompe et al., 2002; Taylor and Fink, 2003). This decline has been explained by differences in patient samples, changes in diagnostic criteria, improvement of the diagnosis of neurological and somatic disorders, as well as the implementation of rehabilitation programs. Whether the introduction of antipsychotics has contributed to this phenomenon is still a matter of debate (Mahendra, 1981). More fundamental, however, could be the so-called "conflict of paradigms" that has emerged as a result of the separation of neurology and psychiatry after which motor disorders were no longer a major part of the psychiatric spectrum (Fink, 2001; Northoff, 1992; Pichot, 1997; Rogers, 1985). Interpretation became more influential at the expense of clinical observation and description. In a Dutch psychiatric hospital, the prevalence rate of catatonic schizophrenia over the last decades of the past century dropped from 7.8% to 1.3%. Evaluation of psychotic patients with a specific screening instrument for catatonia, however, revealed a much higher percentage (Van der Heijden et al., 2005). These figures suggest that the decline of catatonic symptoms is the result of selective observation.

Since the genes coding for the $GABA_A$-receptor subunits are localized within the deleted chromosomal segment in PWS patients, it can be hypothesized that some of the CNS abnormalities, like psychomotor dysfunctions, may arise from an altered gene expression or from a dysfunction of the $GABA_A$-receptor (Lucignani et al., 2004). An involvement of $GABA_A$ has been postulated in catatonia as well (Northoff, 2002; Northoff et al., 1999). This hypothesis emerged from the frequently observed therapeutic potential of the GABA-agonist lorazepam in acute catatonia (Fink, 2001; Gaind et al., 1994; Ungvari et al., 2001). In our patient group, treatment with the GABAergic anticonvulsant valproic acid appeared to result in a clinically relevant reduction of affective and psychomotor symptoms and prevention of relapse (Verhoeven et al., 1998, 2003a). One case report describes the effect of valproic acid in catatonia (Krüger and Bräunig, 2001). It has been suggested that treatment with topiramate, an anticonvulsant with among others GABAergic effects, may be effective in reducing stereotyped and compulsive behaviors in PWS patients (Shapira et al., 2002, 2004; Smathers et al., 2003).

V. Conclusions

PWS is associated with the occurrence of atypical psychoses after adolescence, especially in patients with UPD genetic etiology, which are characterized by a subacute onset, relapsing course, and the absence of residual symptoms. The clinical profile comprises affective symptoms and an increase of preexistent motor symptoms. Motoric abnormalities are part of the behavioral phenotype and should therefore not be considered as emotion-driven, goal-oriented behaviors. Most preferably, pharmacological treatment is targeted at an enhancement of the GABA function, resulting in an effect on affective and motor symptoms. It is advocated to abandon categorical diagnoses and establish a long-term prophylactic treatment regimen with GABA mimetic compounds.

References

Beardsmore, A., Dorman, T., Cooper, S. A., and Webb, T. (1998). Affective psychosis and Prader-Willi syndrome. *J. Intell. Disabil. Res.* **42,** 463–471.
Boer, H., Holland, A., Whittington, J., Butler, J., Webb, T., and Clark, D. (2002). Psychotic illness in people with Prader-Willi syndrome due to chromosome 15 maternal uniparental disomy. *Lancet* **359,** 135–136.
Butler, M. G., Bittel, D. C., Kibiryeva, N., Talebizadeh, Z., and Thompson, T. (2004). Behavioral differences among subjects with Prader-Willi syndrome and type I or type II deletion and maternal disomy. *Pediatrics* **113,** 565–573.
Cassidy, S. B. (2001). Prader-Willi syndrome. *In* "Management of Genetic Syndromes" (S. B. Cassidy and J. E. Allanson, Eds.), pp. 301–322. Wiley-Liss.
Civardi, C., Vicentini, R., Grugni, G., and Cantello, R. (2004). Corticospinal physiology in patients with Prader-Willi syndrome. *Arch. Neurol.* **61,** 1585–1589.
Clarke, D. J., Boer, H., Webb, T., Scott, P., Frazer, S., Vogels, A., Borghgraef, M., and Curfs, L. M. G. (1998a). Prader-Willi syndrome and psychotic symptoms: I. Case descriptions and genetic studies. *J. Intell. Disabil. Res.* **42,** 440–450.
Clarke, D. J., Boer, H., Whittington, J., Holland, A., Butler, J., and Webb, T. (2002). Prader-Willi syndrome, compulsive and ritualistic behaviours: The first population-based survey. *Br. J. Psychiat.* **180,** 358–362.
Descheemaeker, M. J., Vogels, A., Govers, V., Borghgraef, M., Willekens, D., and Swillen, A. (2002). Prader-Willi syndrome: New insights in the behavioural and psychiatric spectrum. *J. Intell. Disabil. Res.* **46,** 41–50.
Dhossche, D. M. (2004). Autism as early expression of catatonia. *Med. Sci. Monit.* **10,** 31–39.
Dimitropoulos, A., Feurer, I. D., Butler, M. G., and Thompson, T. (2001). Emergence of compulsive behavior and tantrums in children with Prader-Willi syndrome. *Am. J. Ment. Retard.* **106,** 39–51.
Dykens, E., and Shah, B. (2003). Psychiatric disorders in Prader-Willi syndrome. Epidemiology and management. *CNS Drugs* **17,** 167–178.
Fink, M. (2001). Catatonia: Syndrome or schizophrenia subtype? Recognition and treatment. *J. Neural Transm.* **108,** 637–644.
Gaind, G. S., Rosebush, P. I., and Mazurek, M. F. (1994). Lorazepam treatment of acute and chronic catatonia in two mentally retarded brothers. *J. Clin. Psychiat.* **55,** 20–23.
Goldstone, A. P. (2004). Prader-Willi syndrome: Advances in genetics, pathophysiology and treatment. *Trends Endocrin. Met.* **15,** 12–20.
Hashimoto, T., Mori, K., Yoneda, Y., Yamaue, T., Miyazaki, M., Harada, M., Miyoshi, H., and Kuroda, Y. (1998). Proton magnetic resonance spectroscopy of the brain in patients with Prader-Willi syndrome. *Pediatr. Neurol.* **18,** 30–35.

Krüger, S., and Bräunig, P. (2001). Intravenous valproic acid in the treatment of severe catatonia. *J. Neuropsych. Clin. N.* **13,** 303–304.

Leonard, C. M., Williams, C. A., Nicholls, R. D., Agee, O. F., Voeller, K. K., Honeyman, J. C., and Staab, E. V. (1993). Angelman and Prader-Willi syndrome: A magnetic resonance imaging study of differences in cerebral structure. *Am. J. Med. Genet.* **46,** 26–33.

Lucignani, G., Panzacchi, A., Bosio, L., Moresco, R. M., Ravasi, L., Coppa, I., Chiumello, G., Frey, K., Koeppe, R., and Fazio, F. (2004). GABA$_A$ receptor abnormalities in Prader-Willi syndrome assessed with positron emisson tomography and (^{11}C)flumazenil. *NeuroImage* **22,** 22–28.

Mahendra, B. (1981). Where have all the catatonics gone? *Psychol. Med.* **11,** 669–671.

Mignot, E., Lammers, G. J., Ripley, B., Okun, M., Nevsimalova, S., Overeem, S., Vankova, J., Black, J., Harsh, J., Bassetti, C., Schrader, H., and Nishino, S. (2002). The role of cerebrospinal fluid hypocretin measurement in the diagnosis of narcolepsy and other hypersomnias. *Arch. Neurol.* **59,** 1553–1562.

Morrison, J. R. (1974). Changes in subtype diagnosis of schizophrenia: 1920–1966. *Am. J. Psychiat.* **131,** 674–677.

Nicholls, R. D., and Knepper, J. L. (2001). Genome organization, function, and imprinting in Prader-Willi and Angelman syndromes. *Annu. Rev. Genom. Hum. G.* **2,** 153–175.

Nixon, G. M., and Brouillette, R. T. (2002). Sleep and breathing in Prader-Willi syndrome. *Pediatr. Pulm.* **84,** 209–217.

Northoff, G. (1992). Ansätze zu einer Neuropsychiatrie als Überbrückung der Trennung von Neurologie und Psychiatrie. *Schweizer Archiv für Neurologie und Psychiatrie* **143,** 27–38.

Northoff, G., Steinke, R., Czcervenka, C., Krause, R., Ulrich, S., Danos, P., Kroph, D., Otto, H. J., and Bogerts, B. (1999). Decreased density of GABA-A receptors in the left sensorimotor cortex in akinetic catatonia: Investigation of *in vivo* benzodiazepine receptor binding. *J. Neurol. Neurosur. PS.* **57,** 445–450.

Northoff, G. (2002). What catatonia can tell us about "top-down modulation": A neuropsychiatric hypothesis. *Behav. Brain Sci.* **25,** 555–604.

Pichot, P. (1997). Faut-il ressusciter la neuropsychiatrie? *La Presse Médicale* **26,** 1243–1247.

Prader, A., Labhart, A., and Willi, H. (1956). Ein Syndrom von Adipositas, Kleinwuchs, Kryptorchismus und Ologophrenie nach myatonieartigem Zustand im Neugeborenenalter. *Schweizerische Medizinische Wochenschrift* **86,** 1260–1261.

Rogers, D. (1985). The motor disorders of severe psychiatric illness: A conflict of paradigm. *Br. J. Psychiat.* **147,** 221–232.

Roof, E., Stone, W., Mac Lean, W., Feurer, I. D., Thompson, T., and Butler, M. G. (2000). Intellectual characteristics of Prader-Willi syndrome: Comparison of genetic subtypes. *J. Intell. Disabil. Res.* **44,** 25–30.

Shapira, N. A., Lessig, M. C., Murphy, T. K., Driscoll, D. J., and Goodman, W. K. (2002). Topiramate attenuates self-injurious behaviour in Prader-Willi syndrome. *Int. J. Neuropsychop.* **5,** 141–145.

Shapira, N. A., Lessing, M. C., Lewis, M. H., Goodman, W. K., and Driscoll, D. J. (2004). Effects of topiramate in adults with Prader-Willi syndrome. *Am. J. Ment. Retard.* **4,** 301–309.

Shapira, N. A., Lessig, M. C., He, A. G., James, G. A., Driscoll, D. J., and Liu, Y. (2005). Satiety dysfunction in Prader-Willi syndrome demonstrated by fMRI. *J. Neurosurg. Psychiat.* **76,** 260–262.

Smathers, S. A., Wilson, J. G., and Nigro, M. A. (2003). Topiramate effectiveness in Prader–Willi syndrome. *Pediatr. Neurol.* **28,** 130–133.

Steinhausen, H. C., Eiholzer, U., Hauffa, B. P., and Malin, Z. (2004). Behavioural and emotional disburbances in people with Prader-Willi syndrome. *J. Intell. Disabil. Res.* **48,** 47–52.

Stompe, T., Ortwein-Swoboda, G., Ritter, K., Schanda, H., and Friedmann, A. (2002). Are we witnessing the disappearance of catatonic schizophrenia? *Compr. Psychiat.* **43,** 167–174.

Swaab, D. F. (1997). Prader-Willi syndrome and the hypothalamus. *Acta Paediatr.* **423**(Suppl.), 50–54.

Swaab, D. F. (2004). Neuropeptides in hypothalamic neuronal disorders. *Int. Rev. Cytol.* **240,** 305–375.

Symons, F. J., Butler, M. G., Sanders, M. D., Feurer, I. D., and Thompson, T. (1999). Self-injurious behavior and Prader-Willi syndrome: Behavioral forms and body locations. *Am. J. Ment. Retard.* **104,** 260-269.

Taylor, M. A., and Fink, M. (2003). Catatonia in psychiatric classification: A home of its own. *Am. J. Psychiat.* **160,** 1233-1241.

Tobias, E. S., Tolmie, J. L., and Stephenson, J. B. P. (2002). Cataplexy in the Prader-Willi syndrome. *Arch. Dis. Child.* **87,** 170-171.

Ungvari, G. S., Kau, L. S., Wai-Kwong, T., and Shing, N. F. (2001). The pharmacological treatment of catatonia: An overview. *Eur. Arch. Psy. Clin. N.* **25,** 31-34.

Van der Heijden, F. M. M. A., Tuinier, S., Arts, N. J. M., Hoogendoorn, M. L. C., Kahn, R. S., and Verhoeven, W. M. A. (2005). Catatonia: Disappeared or under-diagnosed? *Psychopathology* **38,** 3-8.

Varela, M. C., Kok, F., Setian, M., Kim, C. A., and Koiffman, C. P. (2005). Impact of molecular mechanisms, including deletion size, on Prader-Willi syndrome phenotype: Study of 75 patients. *Clin. Genet.* **67,** 47-52.

Veltman, M. W. M., Thompson, R. J., Roberts, S. E., Thomas, N. S., Whittington, J., and Bolton, P. F. (2004). Prader-Willi syndrome. A study comparing deletion and uniparental disomy cases with reference to autism spectrum disorders. *Eur. Child Adoles. Psy.* **13,** 42-50.

Verhoeven, W. M. A., Curfs, L. M. G., and Tuinier, S. (1998). Prader-Willi syndrome and cycloid psychoses. *J. Intell. Disabil. Res.* **42,** 455-462.

Verhoeven, W. M. A., Tuinier, S., and Curfs, L. M. G. (2000). Prader-Willi psychiatric syndrome and velo-cardio-facial psychiatric syndrome. *Genet. Counsel.* **11,** 205-213.

Verhoeven, W. M. A., and Tuinier, S. (2001). Cyclothymia or unstable mood disorder? A systematic treatment evaluation with valproic acid. *J. Appl. Res. Intellect.* **14,** 147-154.

Verhoeven, W. M. A., Tuinier, S., and Curfs, L. M. G. (2002). Prader-Willi syndrome: A concise review of the genetic, pathophysiological and neuropsychiatric characteristics. In "Progress in Differentiated Psychopathology" (E. Franzek, E. Rüther, H. Beckmann, and G. S. Ungvori, Eds.), pp. 82-89. WKL Schriftenreihe.

Verhoeven, W. M. A., Tuinier, S., and Curfs, L. M. G. (2003a). Prader-Willi syndrome: Cycloid psychosis in a genetic subtype? *Acta Neuropsychiatrica* **15,** 32-37.

Verhoeven, W. M. A., Tuinier, S., and Curfs, L. M. G. (2003b). Prader-Willy Syndrome: The psychopathological phenotype in uniparental disomy. *J. Med. Genet.* **40,** e112.

Verhoeven, W. M. A., Sijben, A. E. S., and Tuinier, S. (2004). Psychiatric consultation in intellectual disability; dimensions, domains and vulnerability. *Eur. J. Psychiat.* **18,** 31-43.

Vogels, A., and Fryns, J. P. (2002). The Prader-Willi syndrome and the Angelman syndrome. *Genet. Cousel.* **13,** 385-396.

Vogels, A., Matthijs, G., Legius, E., Devriendt, K., and Fryns, J. P. (2003). Chromosome 15 maternal uniparental disomy and psychosis in Prader-Willi syndrome. *J. Med. Genet.* **40,** 72-73.

Vogels, A., De Hert, M., Descheemaeker, M. J., Govers, V., Devriendt, D., Legius, E., Prinzie, P., and Fryns, J. P. (2004). Psychotic disorders in Prader-Willi syndrome. *Am. J. Med. Genet.* **127,** 238-243.

Wagstaff, J., Knoll, J. H., Fleming, J., Kirkness, E. F., Martin-Gallardo, A., Greenberg, F., Graham, J. M., Jr., Menninger, J., Ward, D., and Venter, J. C. (1991). Localization of the gene encoding the $GABA_A$ receptor beta 3 subunit to the Angelman/Prader-Willi region of human chromosome 15. *Am. J. Hum. Genet.* **49,** 330-337.

Whittington, J., Holland, A., Webb, T., Butler, J., Clarke, D., and Boer, H. (2004). Cognitive abilities and genotype in a population-based sample of people with Prader-Willi syndrome. *J. Intell. Disabil. Res.* **48,** 172-187.

Wigren, M., and Hansen, S. (2003). Rituals and compulsivity in Prader-Willi syndrome: Profile and stability. *J. Intellect. Disabil. Res.* **47,** 428-438.

Yoshii, A., Krishnamoorthy, K. S., and Grant, P. E. (2002). Abnormal cortical development shown by 3D MRI in Prader-Willi syndrome. *Neurology* **59,** 644-645.

TOWARDS A VALID NOSOGRAPHY AND PSYCHOPATHOLOGY OF CATATONIA IN CHILDREN AND ADOLESCENTS

David Cohen

Department of Child and Adolescent Psychiatry, Hôpital Pitié-Salpêtrière, AP-HP
Université Pierre et Marie Curie, 47 bd de l'Hôpital, 75013 Paris, France

I. Introduction
II. Phenomenology of Catatonia in Young People
III. Towards a Broader Nosography for Child, Adolescent, and Adult Catatonia
IV. Models of Catatonia
V. Psychopathological Model for Catatonia
   A. Adherence to Delusional Ideas Resulting in Catatonia
   B. Resistance to Delusional Thinking or Conviction Resulting in Catatonia
   C. Hyperanxious or Hyperemotional States Resulting in Catatonia
VI. Discussion
VII. Conclusions
   References

Paraphrasing Taylor and Fink (2003), catatonia needs "a home of its own" in child and adolescent psychiatry. Limited but expanding literature supports that catatonia in children and adolescent can be identified reliably among other childhood conditions, is sufficiently common, treatable with the same specific treatments as adult catatonia (e.g., sedative drugs and electroconvulsive therapy), and can be worsened by other treatments (e.g., antipsychotics). Other findings in child and adolescent catatonia suggest that sex ratio and associated disorders may differ, and the proposed classification of Taylor and Fink (2003) needs modification. Adopting a broader diagnostic schedule may accommodate both child, adolescent, and adult catatonia. A *psychomotor automatism* variant should be included as a diagnosis, as well as specifiers for associated disorders such as acute nonpsychotic anxious state and pervasive developmental disorder. Duration of illness should be specified as acute or chronic. Regardless of associated psychiatric disorders, this chapter describes a new psychopathological model. Three main modalities of movement dysfunction in catatonic subjects are listed: (1) adherence to delusional ideas leading to a psychomotor automatism (De Clérambault, 1927); (2) resistance to delusional thinking or conviction; and finally (3) hyperanxious states. Case-vignettes illustrate the model, and future research directions are identified.

I. Introduction

Although Calmiel, in 1832, was the first to describe a sentinel report of malignant catatonia (Ainsworth, 1987), it was Kahlbaum (1874), who isolated the catatonic syndrome, characterized by the coexistence of psychic symptoms and motor symptoms resembling a muscular cramp. Kahlbaum (1874) stressed the frequent association of catatonia with depression or mania, as well as its link to organic conditions such as alcoholism, epilepsy, and syphilis. The catatonic syndrome became well known when Kraepelin and Bleuler, respectively, associated catatonia to dementia praecox and schizophrenia. Although both authors recognized that catatonic signs could appear in a variety of disorders, there was a subsequent tendency throughout the twentieth century to consider catatonia as a purely psychiatric condition associated with schizophrenia.

However, many reports published in the 1940s stressed that this view was too limited. For example, French psychiatrists Baruk (1959) and Ey (1950) pointed out that catatonia could also occur in organic conditions, manic episodes, melancholia, and hysteria; but their work remained unknown in the English medical literature. Similarly, research from the classic school of German psychiatry, particularly Kleist and Leonhard (Kleist, 1943), had little influence on international practice (Taylor and Fink, 2003). In the 1970s, the European position was endorsed by Morrison (1973), Abrams and Taylor (1976), and Gelenberg (1976), who seemed oblivious of the earlier French or German contributions. Finally, the broader view has been included in the Diagnostic and Statistical Manual, 4th ed. (DSM-IV), in which catatonia is still associated with schizophrenia, appears in a separate class as "catatonic disorder due to a general medical condition," and is a specifier of affective disorders "with catatonic features" (APA, 1994). In the contemporary International of Classification Diseases, 10th revision (ICD-10), catatonia is only associated with schizophrenia and stupor with melancholia. A diagnosis of a medical condition is also possible under "organic catatonic disorder" (WHO, 1994).

Aims in this chapter are to formulate a specific nosography for catatonia in young people and present a supporting psychopathological model. First, the phenomenology of catatonia in children and adolescents will be reviewed.

II. Phenomenology of Catatonia in Young People

The estimated prevalence of catatonia in young people varies widely, that is, between 0.6% and 17% (Cohen et al., 1999; Thakur et al., 2003) but seems consistently lower than estimated rates of catatonia in adults, which range from

7.6% to 38% of admissions to psychiatric inpatient facilities (Taylor and Fink, 2003). Catatonia has been poorly investigated in child and adolescent psychiatry. In a previous review (Cohen et al., 1999), only 42 cases were reported in the literature between 1977 and 1997. Takaoka and Takata (2003) listed 73 case-descriptions during the past 20 years (1982–2002). Most of the cases are adolescents. Systematic series over the last 5 years are limited to three: (1) Thakur et al. (2003) reported a consecutive series of 11 catatonic children and adolescents aged between 10 and 16 years in a pediatric clinic in Ranchi, India; (2) Wing and Shah (2000) reported 30 catatonic patients with a previous history of autism, in which catatonic features most often appeared between 10 and 19 years of age; and (3) Cohen et al. (2005) reported a prospective, consecutive series of 30 catatonic patients in two university clinics specialized in child and adolescent psychiatry in Paris, France. In an older study of childhood schizophrenia (Green et al., 1992), catatonic symptoms were present in more than one-third (12 of 38) of cases. Future studies in larger samples are needed to determine the true prevalence of catatonia in psychiatrically disturbed children and adolescents.

In a case-series (Cohen et al., 2005), more than half of the cases were of non-European origin (compared to approximately 20% of all inpatients in the study sites born to non-French native parents). A possible role of ethnic and/or cultural factors in the clinical expression of catatonia has been suggested in the adult literature (Leff, 1981; Varma, 1992). Given the high prevalence of catatonia (17%) in child and adolescent psychiatric inpatients in India, cultural factors may also play a role with child and adolescent inpatients (Thakur et al., 2003). However, selection bias may explain some of the differences in cultural prevalence, and future researchers should consider this in their sampling. Cross-cultural comparative studies of childhood catatonia would be very informative in clarifying the role of cultural factors in the expression of catatonia in children and adolescents.

Phenomenology and associated diagnoses of catatonia are similar to those reported in the adult literature but relative frequency of associated disorder differed, with schizophrenia being the most frequent diagnosis (Cohen et al., 1999). Moreover, the female-to-male ratio contrasts with that in adult studies. In young people, the majority of catatonic patients are boys (Cohen et al., 1999; Takaoka and Takata, 2003), and catatonic schizophrenia appears to be a clinically relevant but understudied subgroup affecting males (Cohen et al., 2005). In adults, women represent 75% of all cases of catatonia (Northoff et al., 1999; Rosebush et al., 1990). The difference in sex ratio might be a consequence of bias since affective disorders are overrepresented in samples of catatonic adults. Dhossche and Bouman (1997) compared frequencies of individual catatonic symptoms in pediatric cases (culled from the literature) versus adult cases. The estimated frequencies in adult catatonia were based on 463 catatonic cases pooled from seven studies. Symptoms in childhood versus adult catatonia were similar, except for incontinence that was not systematically reported in any of the adult studies (see Table I). Incontinence is not a DSM-IV symptom of catatonia but frequently reported, at least in children and adolescents.

In Table II, symptom frequencies in a case-series of 30 catatonic adolescents are shown (Cohen *et al.*, 2005). Stupor and acute onset are symptoms present in most of the subjects with a mood disorder diagnosis, but they can be encountered in patients with schizophrenia as well; thus, they cannot be regarded as specific of a diagnostic

TABLE I
FREQUENCIES OF DSM-IV CATATONIC SYMPTOMS IN CHILDHOOD VERSUS ADULT CATATONIA

	Children % ($N = 30$)	Adults % (mean)	95% CI
Mutism	87	78	68–88
Posturing/grimacing	52	66	50–82
Stupor	80	66	45–87
Staring	49	57	35–79
Negativism	38	49	34–64
Rigidity	38	40	20–60
Stereotypy	24	37	22–52
Waxy flexibility	62	35	14–56
Echolalia/echopraxia	14	19	11–27
Excessive motor activity	14	15	10–20
Automatic obedience	10	10	4–16
Incontinence	45	–	–

Adapted from Dhossche and Bouman (1997).

TABLE II
BUSH–FRANCIS CATATONIA RATING SCALE MODIFIED FOR USE IN CHILD AND ADOLESCENT PRACTICE: OCCURRENCE OF SYMPTOMS IN A SERIES OF 30 YOUNG PATIENTS WITH CATATONIA
(BUSH *ET AL.*, 1996; COHEN *ET AL.*, 2005)

Motor symptoms	(%)	Other symptoms	(%)
Catalepsy	67	Social withdrawal	90
Stupor	67	Mutism	80
Posturing	80	Mannerism	30
Waxy flexibility	60	Echophenomena	12
Staring	63	Incontinence	56
Negativism	80	Verbigeration[b]	33
Stereotypies	43	Refusal to eat	67
Psychomotor excitement	50	Social withdrawal	
Automatic compulsive movements[a]	53		
Muscular rigidity	67		

[a] Including grimacing.
[b] Meaningless and stereotyped repetition of words of note scoring is similar to that of Bush–Francis catatonia rating scale.

class. Automatic compulsive movements and stereotypies are highly suggestive (Cohen *et al.*, 2005; Kruger *et al.*, 2003) but not pathognomonic of schizophrenia.

A modified scale based on the Bush–Francis Catatonia Rating Scale (Bush *et al.*, 1996) was used. Six clinical items have been added based on Ey's (1950) clinical studies on catatonia. Four of these (i.e., incontinence, refusal to eat, catalepsy, and automatic movements) were present in 56, 67, 67, and 53% of the subjects. Future research should focus on the validity and reliability of this modified scale in other child and adolescent samples.

III. Towards a Broader Nosography for Child, Adolescent, and Adult Catatonia

In a recent review, Taylor and Fink (2003) stated that catatonia should have "a home of its own" in psychiatric nosography, based on the following propositions: (1) it can be distinguished from other behavioral syndromes; (2) it is sufficiently common; (3) it improves with specific symptomatic treatments such as sedative drugs and electroconvulsive therapy (ECT); and (4) it can be worsened by other treatments (e.g., antipsychotics). Five amendments have been proposed to Taylor and Fink's schedule in order to accommodate for findings in child and adolescent psychiatry and encompass a traditional French psychopathological entity (i.e., *psychomotor automatism*). Such a broadening may apply equally to youth and adult catatonia (Table III).

First, the addition of a fourth category of catatonia named psychomotor automatism was supported to help isolate the clinical presentation described later as the most frequent in adolescents (Cohen *et al.*, 2005). A typical case vignette is of a 16-year-old boy who had insidious onset of catatonic features, including automatic movements secondary to hallucinations—psychomotor automatism (De Clérambault, 1927), improves moderately when treated but continues to suffer a chronic course. This subtype is supported by a clinical study in adults. Kruger *et al.* (2003) performed a factor analysis of catatonic symptom distribution across four diagnostic groups; schizophrenia, pure mania, mixed mania, and major depression, and extracted a factor called "involuntary movements/mannerisms." In this study, patients with catatonic schizophrenia exhibited more frequently symptoms represented in this factor. Second, catatonia secondary to general medical conditions, toxic state, or neurologic disorder should be grouped together as a specifier. When a catatonic syndrome occurs in the course of a general medical condition, it is associated with a cerebral impairment. This was the case for one patient who exhibited neuro-lupus and catatonia. The organic diagnosis focus of treatment may be very specific—plasma exchange and immunosuppressive medication in this case (Périsse *et al.*, 2003). Third, child and adolescent psychiatric literature suggests that catatonia can also be associated

TABLE III
Proposed Categories for Diagnostic Classification of Catatonia in Children and Adolescents[a]

Classification element	Category[b]
Catatonia	
DSM code xxx.1	Nonmalignant catatonia
DSM code xxx.2	Delirious catatonia (or excited catatonia)
DSM code xxx.3	Malignant catatonia
DSM code xxx.4	Psychomotor automatism (the main symptom is compulsive automatic movements)
Specifier for associated disorder	
DSM code xxx.x1	Secondary to a mood disorder
DSM code xxx.x2	Secondary to a medical condition (including toxic state and neurologic disorder)
DSM code xxx.x3	Secondary to a psychotic disorder
DSM code xxx.x4	Secondary to an acute nonpsychotic anxious state
Specifier for symptom course	Acute
	Chronic
Specifier for history of PDD	With a history of PDD

[a] Adapted with permission from Taylor and Fink (2003).
[b] In bold, proposed modification in the field of child and adolescent practice.

with an acute nonpsychotic anxious state (Thakur et al., 2003; Ungvari et al., 1994). Finally, two new specifiers were proposed to add: one was related to the course of the disease (acute versus chronic), and the second was related to a history of pervasive developmental disorder (PDD), due to its relative frequency in youth cases of catatonia (Wing and Shah, 2000). Although in the largest prospective series (Cohen et al., 2005), all patients ($N = 5$) with a history of PDD exhibited catatonic schizophrenia, the literature indicates that mood disorders can be associated with catatonia in patients with a history of PDD as well (Ghaziuddin et al., 2005; Révuelta et al., 1994; Zaw and Bates, 1997). Despite speculations of disease continuity between autism and catatonia (Dhossche, 2004), much more research is needed to elucidate the relation between autism and catatonia. This specifier is focusing research on this neglected symptom dimension in autism and to find better treatments for these conditions.

IV. Models of Catatonia

The phenomenology of the syndrome shows that catatonic symptoms can be classified into motor (e.g., posturing, catalepsy, waxy flexibility), behavioral (e.g., negativism, mutism, automatic compulsive movements), affective (e.g., involuntary and uncontrollable emotional reactions, affective latence, flat affect, and withdrawal), and regressive symptoms (e.g., verbigeration, enuresis and encopresis, echophenomena). It is therefore extremely difficult to identify the subjective feelings experienced by a catatonic patient. For those who have an acute form of catatonia (e.g., some form of psychotic depression) (Cohen *et al.*, 1997; Northoff *et al.*, 1998), it is possible to ask patients retrospectively about such intrapsychic experience. For those who have a chronic form of catatonia (e.g., some forms of catatonic schizophrenia) (Cohen *et al.*, 1999), it is only possible with patients who do not show extreme mutism and negativism, when a careful interpersonal relation has been established.

Few models or hypotheses have been formulated, mostly biological, and all are somewhat unsatisfactory. Following the early experiments of bulbocapnine-induced catatonia in animals with neocortex in 1928 (De Jong and Baruk, 1930), interests have focused on the role of (1) $GABA_A$ receptor given the therapeutic efficacy of lorazepam, a $GABA_A$ receptor potentiator (Bush *et al.*, 1996; Rosebush *et al.*, 1990); (2) N-methyl-D-aspartate (NMDA) receptor given that some catatonic patients, nonresponsive to lorazepam, were treated successfully with amandatine or memantine, both NMDA receptor antagonists (Northoff *et al.*, 1997; Thomas, 2005). Other neurochemical systems involved are the dopamine and the serotonine systems, although direct involvement remains controversial (Northoff, 2002a).

The most interesting attempt to model the pathophysiology of catatonia has been proposed by Northoff who has directed works on catatonia for years in adult patients, including clinical (e.g., Northoff *et al.*, 1999), pharmacological (e.g., Northoff *et al.*, 1997), electrophysiological (e.g., Northoff *et al.*, 2000a), and neuroimaging studies (e.g., Northoff *et al.*, 2000b). On the basis of comparing available literature on catatonia and Parkinson's disease (Northoff, 2002b) and catatonia and neuroleptic malignant syndrome (Northoff, 2002a), he proposed a functional neuroanatomic model taking into account those similarities and differences regarding akinesia—a common feature in catatonia, Parkinson's disease and neuroleptic malignant syndrome—may reflect distinct modulation between cortico-cortical and cortico-subcortical relations. Northoff's model supports that clinical similarities between Parkinson's disease, neuroleptic malignant syndrome, and catatonia with respect to akinesia may be related to involvement of a dysregulation in a cortical-subcortical circuit (the so called "motor loop" between motor/premotor cortices and basal ganglia). Clinical differences in emotional and behavioral symptoms—that cannot be observed in either Parkinson's disease or neuroleptic malignant syndrome and define catatonia as a psychomotor syndrome—may be related with involvement of different cortical areas: orbitofrontal/parietal and premotor/motor cortices implying distinct kinds of modulation. This modulation is "vertical" in the case of the "motor loop" as opposed to "horizontal" in the case of psychic signs of catatonia. The affective, emotional, and behavioral symptoms in catatonia may be accounted for by dysfunction in orbitofrontal–prefrontal/parietal cortical connectivity reflecting "horizontal modulation" of cortico-subcortical relation. Despite its deep and elegant proposals, Northoff's model does not offer a more comprehensive psychological model of catatonia at the level of patient's intrapsychic experience. Based on the clinical experience of catatonia in adolescents, a psychopathological model for catatonia was proposed and developed, which tends to explain the psychomotor symptomatology of the syndrome and to parallel the main clinical categories.

V. Psychopathological Model for Catatonia

There is a conspicuous lack of more comprehensive psychological models of catatonia at the level of patient's intrapsychic experience. Findings in a few studies (Northoff *et al.*, 1998; Rosebush and Mazurek, 1999) as well as author's own experiences give some insight into the subjective experience of catatonic patients. First, akinetic patients with catatonia appear unable to experience pain or fatigue despite prolonged posturing. This point is supported by possible complication of catatonia in skin injury lesions, even in young patients (Cohen *et al.*, 1999). Second, they appeared unaware of the objective position of their body or the consequences of their movements. Third, most of them reported intense and uncontrollable emotions, including one patient who had a blockade of his will with contradictory and ambivalent thoughts. Finally, all patients of Northoff *et al.* (1998) series remembered very well the persons who treated them on admission confirming that catatonic patients have no major deficit in memory and/or general awareness. This point was also highlighted by Rosebush and Mazurek (1999). The same experience has been shared with young patients except when a history of autism with no language does not permit retrospective psychological investigation. Similarly, except when catatonia is associated with a neurologic disorder, catatonic patients do not have abnormal neurological examination (Cohen *et al.*, 2005; Rosebush and Mazurek, 1999). Thus, neurological foundations are usually preserved in catatonic patients. Catatonic symptoms should be regarded as functional and also understood at the level of subjects' experience resulting in catatonic motor dysfunction.

Based on the clinical experience of catatonia in adolescents, a psychopathological model for catatonia was proposed, which tends to explain the psychomotor symptoms of the syndrome, regardless of associated psychiatric disorders. First, a review of the main cognitive dimensions involved in voluntary human locomotion and movement is given as an aid in defining the specifics of catatonic psychopathology.

Figure 1 summarizes a cognitive model of voluntary human movements. Assuming an intact gross motor system, voluntary movement results from intentionality (or will), the behavioral planification, and the emotional context

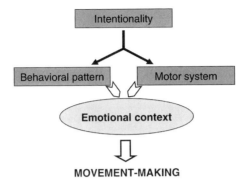

FIG. 1. A cognitive model of human voluntary movement (modified from Bloulac *et al.*, 2004).

(associated with the whole procedure), which in conjunction result in a voluntary movement. The intentionality belongs to the realm of human conscious activities, and its implication is not necessary for all type of movements (such as breathing). The behavioral planification per se is a complex procedure that involves, at minimum, attentional capacity, motivation skills, memory schemes and motor control, and computing (Bloulac *et al.*, 2004). A major but forgotten component (regarding catatonia) relates to the emotional context that is always present during a voluntary movement. The entire human repertoire can be encountered, such as sadness, anxiety, fear, angriness, joy, and so on. Finally, the implication of the "motor system" is a truism but not under the scope of this chapter since there is evidence of normal neurological examination in catatonic subjects. However, its contribution to catatonic states may be suspected in cases of malignant catatonia secondary to extrapyramidal effects of neuroleptic prescriptions (Northoff, 2002a; Taylor and Fink, 2003).

Our psychopathological model that refers to the cognitive model of human locomotion described earlier, distinguishes three main modalities of subjective experiences that result in movement dysfunction in catatonic subjects: (1) adherence to delusional ideas that leads to a psychomotor automatism (De Clérambault, 1927), (2) resistance to delusional thinking or conviction, and (3) hyperanxious or hyperemotional states.

A. ADHERENCE TO DELUSIONAL IDEAS RESULTING IN CATATONIA

De Clérambault (1927) described the principle of psychomotor automatism. The most typical features of psychomotor automatism are automatic movements secondary to hallucinations and adherence to delusional ideas. The following case-vignette offers an illustration (Cohen *et al.*, 1999).

A was 16-year old when admitted to the inpatient unit for compulsive water drinking. At age 5, A started to be treated with psychotherapy for a nonautistic PDD. Although his intellectual skills were in the low average range, he remained in a regular school until age of 14 years. Then, he entered a special education program in which he received vocational training. One year later, he was sexually abused by a male peer. A began to isolate himself and started drinking a lot of water. A was first hospitalized for 2 months with symptoms described as compulsive behaviors: rituals to eat and drink and ordering rituals. He explained that he needed to drink a certain number of glasses of water in order to alleviate his anxiety. Fluoxetine (20 mg/day) and cyamepromazine (50 mg/day) were started together with behavioral therapy for water control. As he was feeling better, he left the hospital for the summer holidays.

Three months later, A was readmitted to the unit in emergency because of life-threatening water intoxication. Few days before hospitalization, his parents said he was drinking up to 14 liters of water a day, shortly before admission. A was confused, complaining of headache and nausea, due to a cerebral edema confirmed by brain-imaging. Water restriction was sufficient to improve his confusion, but A exhibited catatonic signs, including stereotypies imitating the movement of drinking (even when water was not available), negativism, and catatonic excitement. He had numerous hallucinations including auditory orders to drink and to do specific movements resembling compulsive behaviors. He also had delusions with fears of HIV infection, or poisoning. A diagnosis of catatonic schizophrenia was made, and antipsychotics were administered. Chlorpromazine (250 mg/day) and then haloperidol (15 mg/day) failed to improve the patient's condition and instead induced numerous extrapyramidal effects. Thioridazine (350 mg/day) subsequently showed notable efficacy. A was able to attend behavioral therapy for water control and left the inpatient unit after 8 months. Follow-up was arranged at a day care hospital.

During a 2-year follow-up period, no catatonic symptom returned, while residual signs of schizophrenia, such as marked social withdrawal and lack of initiative, persisted. Pharmacotherapy was kept unchanged, but behavioral therapy was no longer necessary. A is now 22-year old and about to work in a supervised environment. Case 1 is typical of psychomotor automatism since he was totally adherent to his delusional voices and hallucinations during the acute phase resulting in automatic movements of drinking.

B. Resistance to Delusional Thinking or Conviction Resulting in Catatonia

Resistance to delusional thinking may also manifest as catatonia. In this scenario, subjects do not agree or follow voices and hallucinations giving orders, but desperately resist via bodily symptoms. These motor symptoms can be motor rituals (very similar to automatic movements) designed to alleviate anxiety, or catalepsy and immobility designed to prevent any disaster. Case 2 is an exemplar (Cohen et al., 1999).

C was a 13-year-old boy admitted to the unit for a paralytic catatonia. C had been treated since he was 4-year old for Asperger disorder. He had followed the regular curriculum of a supportive private school and attended psychotherapy twice a week until 11-year old when it was stopped as he was doing pretty well. A few months prior to admission, C developed severe compulsions and was diagnosed with obsessive–compulsive disorder. Clomipramine was started. His symptoms, however, were the beginning of his catatonic state. On admission, the catatonic syndrome was extreme due to a delay before hospitalization. C exhibited catalepsy with waxy flexibility and posturing lasting sometimes several hours. He also had muscular rigidity, negativism, stereotyped movements of the mouth (grimacing), hands, and shoulders, enuresis, and extreme acrocyanosis. C was not mute but exhibited blunted affect, disorganized thoughts, neologisms, and inappropriate smiles. He disclosed prominent delusions and hallucinations including voices of orders, tactile and olfactive hallucinations, and visions of skeletons. His compulsions appeared to be related to his delusions, and he was able to explain that he needed to stay immobile because he was terrified by voices ordering him to commit suicide.

Catatonic schizophrenia was diagnosed and antipsychotics were started after discontinuation of clomipramine. Catatonia did not respond to chlorpromazine (300 mg/day) or haloperidol (15 mg/day) and lorazepam (5 mg/day). A relay with thioridazine (200 mg/day) provoked a severe dystonic reaction that subsided after 1 week of medication discontinuation. After 4 months, ECT was proposed to the parents because of complications including skin injury lesions, but they refused the treatment. Amisulpride (800 mg/day) was associated with pack therapy[1] twice a week. This new therapeutic regimen led to moderate symptomatic improvement. After 6 months, C was able to leave the hospital and enter a day care unit.

Now 18-year-old C still exhibits residual symptoms both of the catatonic and schizophrenic spectrum. His treatment includes amisulpride (800 mg/day), biperidene (8 mg/day), psychotherapy twice a week, milieu therapy in a day hospital, and family therapy. Case 2 is typical since he desperately resisted to his delusional voices and hallucinations asking for suicide during the acute phase resulting in catalepsy, negativism, and immobility.

---

[1] Pack therapy: Envelopment in damp sheets for 1 hour sessions with patient expressing cenesthesic sensations and spontaneous fantasies.

C. Hyperanxious or Hyperemotional States Resulting in Catatonia

This might be the most frequent subjective experience reported by stuporous and akinetic catatonic patients. Many authors have described hyperanxious experience in adults (Northoff et al., 1998; Rosebush et al., 1990) and adolescents (Thakur et al., 2003; Ungvari et al., 1994). This experience is usually associated with psychotic states, but psychosis is not required (Ungvari et al., 1994). Case 3 is an exemplar of acute stuporous state associated with hyperemotionality and described how a brief neuroleptic medication induced a malignant form of catatonia (Cohen et al., 1997).

M was a 15-year-old girl. One month prior to admission, M started to be sad, developed early-morning wakening, and lost her appetite without any apparent cause. Severe psychomotor retardation worried her family. Her rapidly worsening mood symptoms prompted hospitalization. Delusional ideas concerning her own existence and a state of depressive stupor appeared a few days before her admission. When M was admitted to the inpatient psychiatric unit, she had lost 7 kg, showed severe psychomotor retardation, and exhibited catatonic posturing. Cyamepromazine was started (75 mg/day). The next day, she was totally stuporous and refused food and water. Physical examination revealed a muscular rigidity, a trismus, a brief period of hyperthemia (38 °C), and an alteration of consciousness. Laboratory tests were all within normal limits, except creatin phospho-kinase (CPK) that had a concentration increased 12 times. Bacteriological samples, EEG, ECG, cerebrospinal fluid analysis, and head CT were normal. Within 48 hours after rehydratation, neuroleptic retrieve, and Tropatepine[2] (20 mg/day), awareness was recovered, rigidity disappeared, and muscular enzymes decreased. She started speaking and moving spontaneously, though catatonic posturing and retardation remained. Delusion consisted of the absolute conviction that she was already dead, waiting to be buried, had no more teeth, had no more hair, and had a malformed uterus.[3] The family was proposed ECT. Clinical improvement was notable after the first ECT in which she was then able to eat normally, speak fluently, and had no more delusive ideas. After the sixth treatment, she appeared to be lightly disinhibited and concerned with sexual matters. Although she had no other manic symptoms, it was decided to stop ECT. Rapid relapse led us to six more ECT sessions. Except headache following ECT, a few days of dysinhibition, and 3 weeks of mild confusion and disturbed memory, she showed no other secondary effects. Treatment at discharge associated amineptine (90 mg/day) and amisulpiride (600 mg/day).

---

[2] An anticholinergic drug available in Europe only.
[3] In the French nosography, this delusional state has the name of Cotard's syndrome.

After the first ECT and later on, M's clinical status revealed that during the acute phase she was totally conscious but overcome by extreme anxiety as feeling she was already dead. It was wondered if her delusional ideas might have a meaning in her personal history. So far, she was asked to express the free associations that came in mind when they were evoked. Concerning the disappearance of her teeth, she felt surprised to think of her brother-in-law. She added that she would be ashamed to be given dental care by him and she started to cry every night since her sister's wedding and departure. On the idea of a reproductive malformation, she thought of guilty feelings associated with repeated masturbation that she had historically practiced from childhood until the beginning of her puberty. At 6 months follow-up, M had no more psychiatric symptoms. Memory impairments disappeared within 3 months, and she was able to go back to school 2 months after discharge. A manic episode occurred 1 year after the catatonic, warranting a mood stabilizer (carbamazepine 600 mg/day). She followed a psychodynamic therapy for 3 years and stayed 2 years in a residential setting for college student with psychiatric diagnosis. At 8 years follow-up, M continues to be euthymic, is planning to get married, and finishing her studies in social sciences.

Case 3 is typical of extreme emotional state resulting in catatonia. Anxiety was very high, and delusional ideas (e.g., being already dead) testify to the level of emotional involvement. Although comparing human behaviors to animal behaviors is an epistemological jump, whether these hyperanxious states resulting in catatonia may be related to the terror immobility reflex (e.g., freezing) described in the behavioral repertoire of many animals (Panksepp, 1998) is still not clear. According to Gallup and Maser (1977), comparing tonic immobility in animals and catatonic state in human may have an evolutionary significance as a survival mechanism. Catatonia may "represent fragments of primitive defences against predators that now misfire under conditions of exaggerated stress." They suggested that such conditions occur "in situations in which the person is frozen with fear." The fear can be external. But it is more frequently internal in human (as shown in case 3).

VI. Discussion

The proposed psychopathological model for catatonia is a modest attempt to give meaning and rationality to a psychomotor experience. It is based on patients' testimony of their subjective experiences during catatonic states when they accepted to share this experience during encounters. Figure 2 summarizes the proposal and attempts to link the main clinical presentation of the catatonic syndrome as exposed in the first part of this chapter.

Although this model may seem simplistic, it has clinical and heuristic validity. First, it fits well with the idea that the correlation of clinical form of catatonia and

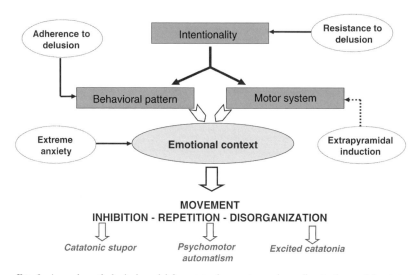

FIG. 2. A psychopathological model for catatonic symptoms. According to the model, catatonic signs are a consequence of: (1) a dysfunction of intentionality (e.g., when a patient resists to delusional ideas), (2) a dysfunction of the planification of behavioral patterns (e.g., when a patient strictly follows voices or hallucinations ordering movements—psychomotor automatism), and (3) a dysfunction of emotional regulation marked by extreme emotional involvement (e.g., in stuporous anxious states). Although direct correlation are not possible, these dysfunctions in movement making may explain the variety of catatonic motor signs (catatonic stupor, psychomotor automatism, excited catatonia).

associated psychiatric disorder is weak, although some direction may be highlighted: stupor and psychotic depression (Rosebush et al., 1990; Taylor and Fink, 2003), automatic movements secondary to psychomotor automatism and schizophrenia (Cohen et al., 2005; Kruger et al., 2003), and excited catatonia and mania (Kruger et al., 2003; Taylor and Fink, 2003). Second, the model allows for the diversity of clinical expressions and presentations of the catatonic syndrome. Third, the model gives a major role to subjective experience including emotional context compared to motor dysfunction. Finally, the model fits well with the fact that in many animals extreme fears can induce an immobilization reflex leading to a state resembling that of stupor.

However, the model has some major limitations. First, the model does not integrate data on functional anatomy of motor activities including regulation and planification. On the other hand, the neuronal basis of catatonia is not known and awaits further neuroscientific inquiry. To this respect, Northoff's proposals are interesting (described earlier). Second, a major dimension of human thinking, that is, temporality, has not been considered. This dimension is not included within the cognitive model of voluntary human movement proposed, although movement execution also includes a temporal paradigm (e.g., initiation, execution, and then termination). Therefore, one can hypothesize that some catatonic patients also experience an alteration of temporality. There are some evidences showing that this dimension might play a key role in the subjective experience of catatonic patients. On one hand, in severe retardation, one can expect a suspension of temporality leading to stupor. In case 3, M had the conviction she was already dead, in some way "out of time." Similarly, Northoff reported that many catatonic patients are not anymore aware of the consequence of their movements. Of course, this could be a consequence of an absence of awareness of their body position. It could also be interpreted as a lack of temporal awareness and anticipation. On the other hand, major increase of thinking speed may also result in a complete disorganization or a blockade of the sense of temporality. A 14-year-old girl with psychotic mania exhibited stupor and mutism. The diagnosis of mania was not possible at admission and was made when careful evaluation of functioning during the prior week was made with her mother. After recovery, thanks to a lithium trial, she explained that during the phase of stupor and mutism "thinking and ideas were that fast that I had no time to do anything even eating, talking, moving."

Finally, the model does not encompass patients with no language, such as those with autistism, since one does not have access to subjective experiences. In these subjects, it is to stress that the developmental course should be considered for diagnosis of comorbid catatonia, as suggested by Wing and Shah (2000) so that catatonia will not be overdiagnosed.

VII. Conclusions

Catatonia is an infrequent but severe condition in young people. While symptomatology and associated disorders are similar to those reported in the adult literature, findings differ in regard to the female-to-male ratio and the relative frequencies of associated disorders. There is a need for research in the field of catatonic schizophrenia in young people, as it appears to be a clinically relevant subgroup frequently resistant to therapeutic approaches. The inclusion of catatonia as a specific syndrome in the psychiatric nosography of both juvenile and adult practices may help its recognition and stimulate research in the field. Research should include psychopathological issues regardless the clinical or phenomenological presentation of the syndrome. In the psychopathological speculations that have not included the temporal dimension of movement and thinking, there are three distinguished modalities of movement dysfunction in catatonic subjects: (1) adherence to delusional ideas that leads to a psychomotor automatism (De Clérambault, 1927), (2) resistance to delusional thinking or conviction, and finally (3) hyperanxious states.

References

Abrams, R., and Taylor, M. A. (1976). Catatonia: A prospective clinical study. *Arch. Gen. Psychiatry* **33,** 579–581.
Ainsworth, P. (1987). A case of lethal catatonia in a 14-year-old girl. *Br. J. Psychiatry* **150,** 110–112.
American Psychiatric Association (APA) (1994). "Diagnostic and Statistical Manual IV." APA Press, Washington DC.
Baruk, H. (1959). "Traité de psychiatrie." Masson, Paris.
Bloulac, B., Burbaud, P., Cazalets, J. P., and Gross, C. (2004). Fonctions motrices. *In* "Encyclopédie Médico-Chirugicale." 17–002-D-10.
Bush, G., Fink, M., Petrides, G., and Francis, A. (1996). Catatonia I. Rating scale and standardized examination. *Acta Psychiatr. Scand.* **93,** 129–136.
Cohen, D., Cottias, C., and Basquin, M. (1997). Cotard's syndrome in a 15 year old girl. *Acta Psychiatr. Scand.* **95,** 164–165.
Cohen, D., Flament, M., Dubos, P. F., and Basquin, M. (1999). The catatonic syndrome in young people. *J. Am. Acad. Child Adolesc. Psychiatry* **38,** 104–106.
Cohen, D., Nicolas, J. D., Flament, M., Périsse, D., Dubos, P. F., Bonnot, O., Speranza, M., Graindorge, C., Tordjman, S., and Mazet, P. (2005). Clinical relevance of chronic catatonic schizophrenia in children and adolescents: Evidence from a prospective naturalistic study. *Schizophr. Res.* **76,** 301–308.
De Clérambault, G. (1927). Psychose à base d'automatisme et syndrome d'automatisme. *Annales Médico-Psychologiques* **1,** 193–239.
De Jong, H., and Baruk, H. (1930). "La catatonie expérimentale par la bulbocapnine." Masson, Paris.
Dhossche, D. (2004). Autism as early expression of catatonia. *Med. Sci. Monit.* **10,** RA31–RA39.
Dhossche, D., and Bouman, N. (1997). Catatonia in an adolescent with Prader-Willi Syndrome. *Ann. Clin. Psychiatry* **4,** 247–253.
Ey, H. (1950). "Etudes psychiatriques." Désclée de Brouwer & Cie, Paris.
Gallup, G. G., and Maser, J. D. (1977). Tonic immobility: Evolutionary underpinnings of human catalepsy and catatonia. *In* "Psychopathology: Experimental Models" (J. D. Maser and L. P. Seligman, Eds.), pp. 334–357. Freeman, San Francisco.
Gelenberg, A. J. (1976). The catatonic syndrome. *Lancet* **i,** 1339–1341.
Ghaziuddin, M., Quilan, P., and Ghaziuddin, N. (2005). Catatonia in autism: A distinct subtype? *J. Intellect. Disabil. Res.* **49,** 102–105.
Green, W., Padron-Gayol, M., Hardesty, A., and Bassiri, M. (1992). Schizophrenia with childhood onset: A phenomenological study of 38 cases. *J. Am. Acad. Child Adolesc. Psychiatry* **31,** 968–976.
Kahlbaum, K. (1874). "Die Katatonie oder das Spannungsirresein." Verlag August Hirshwald, Berlin.
Kleist, K. (1943). Die Katatonien. *Nervenarzt* **16,** 1–10.
Kruger, S., Bagby, R. M., Hoffler, J., and Braunig, P. (2003). Factor analysis of the catatonia rating scale and catatonic symptom distribution across four diagnostic groups. *Compr. Psychiatry* **44,** 472–482.
Leff, J. (1981). "Psychiatry Around the World: A Ranscultural View." Marcel Dekker Inc., New York.
Morrison, J. R. (1973). Catatonia: Retarded and excited types. *Arch. Gen. Psychiatry* **28,** 39–41.
Northoff, G. (2002a). Catatonia and neuroleptic malignant syndrome: Psychopathology and pathophysiology. *J. Neural. Transm.* **109,** 1453–1467.
Northoff, G. (2002b). What catatonia can tell us about "top-down modulation": A neuropsychiatric hypothesis. *Behav. Brain Sci.* **25,** 555–604.
Northoff, G., Eckert, J., and Fritze, J. (1997). Glutamatergic dysfunction in catatonia? Successful treatment of three acute akinetic catatonic patients with the NMDA antagoniste amandatine. *J. Neurol. Neurosurg. Psychiatry* **62,** 404–406.

Northoff, G., Koch, A., Wenke, J., Eckert, J., Böker, H., Pflug, B., and Bogerts, B. (1999). Catatonia as a psychomotor syndrome: A rating scale and extrapyramidal motor symptoms. *Mov. Disord.* **14,** 404–416.

Northoff, G., Krill, W., Wenke, J., Gille, J., Eckert, J., Russ, M., Pester, U., Diekmann, S., Pflug, B., and Bogerts, B. (1998). Major differences in subjective experience of akinesia in catatonic and parkinsonic patients. *Cogn. Neuropsychiatry* **3,** 161–178.

Northoff, G., Pfennig, A., Krug, M., Danos, P., Leschinger, A., Schwarz, A., and Bogerts, B. (2000a). Delayed onset of late movement-related cortical potentials and abnormal response to lorazepam in catatonia. *Schizophr. Res.* **44,** 193–211.

Northoff, G., Richter, A., Gessner, M., Baumgart, F., Leschinger, A., Danos, P., Tempelmann, C., Kötter, R., Stephan, K., Hagner, T., Bogerts, B., Scheich, H., *et al.* (2000b). Right parietal dysfunction in akinetic catatonia: A combined study of neuropsychology and regional cerebral blood flow. *Psychol. Med.* **30,** 583–596.

Panksepp, J. (1998). The sources of fear and anxiety in the brain. *In* "Affective Neuroscience: The Foundations of Human and Animal Emotions," pp. 206–225. Oxford University Press, New York.

Périsse, D., Amoura, Z., Cohen, D., Saintigny, P., Mekhloufi, F., Mazet, P., and Piette, J. C. (2003). Effectiveness of plasma exchange in an adolescent with systemic lupus erythematosus and catatonia. *J. Am. Acad. Child Adolesc. Psychiatry* **42,** 497–499.

Révuelta, E., Bordet, R., Piquet, T., Ghawche, F., Destee, A., and Goudemand, M. (1994). Catatonie aiguë et syndrome malin des neuroleptiques: Un cas au cours d'une psychose infantile. *Encephale* **20,** 351–354.

Rosebush, P. I., and Mazurek, M. F. (1999). Catatonia: Re-awakening to a forgotten disorder. *Movt. Disord.* **14,** 395–397.

Rosebush, P. I., Hildebrand, A. M., Furlong, B. G., and Mazurek, M. F. (1990). Catatonic syndrome in a general psychiatric population: Frequency, clinical presentation, and response to lorazepam. *J. Clin. Psychiatry* **51,** 357–362.

Takaoka, K., and Takata, T. (2003). Catatonia in childhood and adolescence. *Psychiatry Clin. Neurosci.* **57,** 129–137.

Taylor, M. A., and Fink, M. (2003). Catatonia in psychiatric classification: A home of its own. *Am. J. Psychiatry* **160,** 1233–1241.

Thakur, A., Jagadsen, K., Dutta, S., and Sinha, V. K. (2003). Incidence of catatonia in children and adolescents in a pediatric psychiatric clinic. *Aust. N. Z. J. Psychiatry* **37,** 200–203.

Thomas, C. (2005). Memantine and catatonic schizophrenia. *Am. J. Psychiatry* **162,** 326.

Ungvari, G. S., Leung, C. M., and Lee, T. S. (1994). Benzodiazepines and the psychopathology of catatonia. *Pharmacopsychiatry* **27,** 242–245.

Varma, V. K. (1992). Transcultural psychiatry. *In* "Postgraduate Psychiatry" (J. N. Vyas and N. Ahuja, Eds.), pp. 611–635. Churchill Livingstone, New Delhi.

Wing, L., and Shah, A. (2000). Catatonia in autistic spectrum disorders. *Br. J. Psychiatry* **176,** 357–362.

World Health Organization (1994). Organisation, Mondiale de la Santé. "Classification internationale des maladies," 10ème ed. Masson, Paris.

Zaw, Z. F., and Bates, G. D. L. (1997). Replication of zolpidem test for catatonia in an adolescent. *Lancet* **349,** 1914.

SECTION III
BIOLOGY

IS THERE A COMMON NEURONAL BASIS FOR AUTISM AND CATATONIA?

Dirk Marcel Dhossche,* Brendan T. Carroll,[†,‡] and Tressa D. Carroll[§]

*Department of Psychiatry and Human Behavior
University of Mississippi Medical Center, Jackson, Mississippi 39216, USA
[†]Department of Psychiatry, University of Cincinnati, Cincinnati, Ohio 05267, USA
[‡]Psychiatry Service, VA Medical Center, Chillicothe, Ohio 43162, USA
[§]The Neuroscience Alliance, West Jefferson, Ohio 43162, USA

I. Introduction
II. Motor Theory of Language and Autism
III. Are Autism and Catatonia Neurobiological Syndromes?
IV. Catatonia as a Neurobiological Syndrome
V. Autism as a Neurobiological Syndrome
VI. Is There a Common Neuronal Basis for Autism and Catatonia?
VII. Future Directions: Parallel Family-Based Studies in Autism and Catatonia
VIII. Implications
    References

Neuronal bases for autism and catatonia are unknown although integrative theories may soon become feasible as research in autism and catatonia advances. Catatonia and autism may both qualify as neurobiological syndromes in their own right. There is emerging evidence that catatonia may be a common syndrome in autism. Although the relation between autism and catatonia is unclear, coexpression of autism and catatonia may be due to abnormalities in common neuronal circuitries. This possibility constitutes another level of complexity to neurobiological inquiry, but also provides an opportunity to advance our understanding of both disorders. There is a great potential benefit in studying the relation between catatonia and autism in order to focus future research on subtype-specific causes and treatments. Future research avenues are outlined.

I. Introduction

Neuronal substrates of catatonia and autism are yet to be clearly defined. Integration of the limited but expanding knowledge base of neuropathological, biochemical, and neurophysiologic findings has been tried for autism and catatonia separately. However, there is emerging evidence of an association between autism and catatonia (Dhossche, 1998; Wing and Shah, 2000). Following Kraemer (1996), this suggests the following possibilities: (1) autism may be a risk factor for catatonia; (2) autism and catatonia are expressions of the same disorder; (3) autism and catatonia represent different stages of the same disorder; (4) autism and catatonia are separate but related disorders, due to linked genes or environmental risk factors; and (5) comorbid states of autism and catatonia are chance affairs. Regardless of the nature of the association, coexpression of autism and catatonia may be due to abnormalities in common neuronal circuitries. This possibility constitutes another level of complexity to neurobiological inquiry but also provides an opportunity to advance our understanding of both disorders. Advances in catatonia research may benefit autism research and vice versa.

In this chapter, neurobiological findings in autism and catatonia are reviewed, evidence that autism and catatonia are neurobiological syndromes is discussed, and commonalities between brain mechanisms leading to autism and catatonia are sought. In the next section, special attention is given to the putative neural basis of concurrent motor abnormalities and language deficits that are prominent in most people with autism and catatonia.

II. Motor Theory of Language and Autism

> The Motor Theory is a theory of the origin and functioning of language. The theory is that the structures of language (phonological, lexical and syntactic) were derived from and modelled on the pre-existing complex neural systems which had evolved for the control of body movement...Speech is essentially a motor activity (a stream of articulatory gestures). Language made use of the elementary pre-set units of motor action to produce equivalent phonological units (phonemic categories)...The syntactic processes and structures of language were modelled on the motor "syntax."
>
> Robin Allott (2001)

Arguably, the most conspicuous and mysterious feature in many people with autism is absence of speech despite a normal anatomical speech apparatus. The cause is unknown probably in part because of our incomplete understanding of how infants learn language (Kuhl, 2004).

How can deficiencies of psychomotor function and language acquisition be connected? Empirical studies in autistic children suggest an association between these two areas. For example, high-functioning children with autism and children with specific language disorders had more neuromotor problems than control children (Noterdaeme et al., 2002). In another study of autistic children (Eisenmajer et al., 1998), early language delay was associated with more autistic symptoms, developmental milestone delays, and lower receptive language abilities. Stone and Yoder (2001) found that language outcome at age 4 in children with autism spectrum disorders (ASDs) was predicted, after controlling for language skills at age 2, by motor imitation ability (and number of hours of speech therapy). Joint attention, play level, and environmental variables were not predictive. In a study of 2-year-old children with autism (Stone *et al.*, 1997), imitation of body movements was associated with expressive language skills, and imitation of actions with objects was associated with play skills.

The association between deficient motor function or imitation ability and language skills suggests a common or related brain substrate. A more in-depth discussion of neural networks that may be involved can be found elsewhere (Belmonte *et al.*, 2004; Williams *et al.*, 2001). The Motor Theory of Language formulated by Allott (2001) offers an intuitive understanding on the intricate and evolutionary relation between language acquisition and motor systems. Allott (2001) describes: "Language is the capacity of one individual to alter, through structured sound emission, the mental organisation of another individual...The theory is that language was constructed on the basis of a previously existing complex system, the neural motor system. The programs and procedures which evolved for the construction and execution of simple and sequential motor movements formed the basis of the programs and procedures going to form language."

The Motor Theory emphasizes the cross-modal aspects of language, language in use as a continual neural restructuring, the vital importance of the primitive motor-elements, as well as the interlocking of motor control and language. Other functions may also be deficient if it is true that the autistic brain is characterized by abnormal neural information processing. An explanation is also provided for the social and emotional dysfunction in autism: "Language, vision, and motor control all go together; language is the supreme medium of empathy and language almost certainly plays the major role in making possible consciousness, self-awareness; language allows one to construct a model of one's world, to create rational expectancies, particularly about the behavior of others. For someone with no reliable pattern of expectation about the 'world,' every moment of life becomes like wandering through a Chamber of Horrors, unknown and unexpected horrors."

It is the challenge of future studies to unravel pertinent neuronal mechanisms and to stop or reverse them in their earliest stages.

III. Are Autism and Catatonia Neurobiological Syndromes?

Fujii and Ahmed (2004) have proposed five criteria for a neurobiogical syndrome:

1. A neurobiological syndrome is a constellation of symptoms that are reliably associated with disturbance that is functional, structural, neurochemical, or neuropathological in a circumscribed structural location or neural circuit.
2. Similar neurobiological disturbances (that is, in the location or in the neural circuit) that are secondary to different etiologies would result in similar cognitive or behavioral symptoms.
3. Smaller amounts of similar neurobiological disturbances are associated with milder symptoms.
4. Additional symptoms such as cognitive, mood, psychiatric, or other associated neurological symptoms are related to other networks simultaneously affected by underlying neurochemical or neuropathological processes.
5. Aside from treating the underlying disease process, treatment for the associated symptoms of a neurobiological disorder of different etiologies is similar.

There are several candidate neurobiological syndromes such as psychosis (e.g., hallucinations and delusions), catatonia, autism, certain types of depression (e.g., melancholia), and so on. Although any attempt for validation of these putative syndromes suffers from our limited knowledge of the relation between brain function and behavior, especially at the cellular and molecular level, this approach has heuristic value for refining current classification systems such as Diagnostic and Statistical Manual (DSM) and International Classification of Diseases (ICD), focusing research, and tailoring treatments on an individual basis. Most importantly, clinical presentation would be more firmly tied with neuropathology and etiology than in models such as DSM or ICD.

Fujii and Ahmed (2004) have applied their criteria to psychosis (e.g., hallucinations and delusions). They found that there is converging evidence indicating that psychotic disorders of different etiologies are associated with abnormalities to frontal and temporal areas; milder forms of psychosis may be present in people with schizotypal personality disorder with similar but more subtle brain abnormalities than in people with schizophrenia; other common comorbid symptoms of schizophrenia-like psychosis such as depression share similar pathologies; and finally antipsychotics, including the atypical antipsychotics, reduce psychotic symptoms, regardless of diagnosis or etiology, possibly through similar mechanisms of action.

IV. Catatonia as a Neurobiological Syndrome

Catatonia is a brain disease with a cyclic, alternating course, in which the mental symptoms are, consecutively, melancholia, mania, stupor, confusion, and eventually dementia. One or more of these symptoms may be absent from the complete series of psychic "symptom-complexes." In addition to the mental symptoms, locomotor neural processes with the general character of convulsions occur as typical symptoms.

K. L. Kahlbaum (1874)

Could catatonia qualify as a distinct neurobiological syndrome according to the criteria of Fujii and Ahmed (2004)? If so, this would be contrary to Kraepelin's view of catatonia as an exclusive subtype of dementia praecox or schizophrenia, but in line with Kahlbaum's original description of catatonia as a separate brain disorder with a cyclic, alternating, and ultimately progressive course (Kahlbaum, 1874).

In fact, Taylor and Fink (2003) have made this argument. After reviewing the evidence for the diagnostic validity of catatonia, they concluded: "Catatonia can be distinguished from other behavioral syndromes by a recognizable cluster of clinical features. Catatonia is sufficiently common to warrant classification as an independent syndrome. It can be reliably identified, has a typical course when appropriately treated, responds to specific treatments, and is worsened by other treatments. It is associated with many pathophysiologic processes and most often with mood disorder. These findings, which are consistent with established methods of defining distinct diagnostic groupings, support consideration of catatonia as an individual category in psychiatric diagnostic systems."

In support of the first criterion of Fujii and Ahmed (2004), putative neural circuitries for catatonia have been described by Northoff (2002) and Carroll *et al.* (2005) involving multiple focal sites such as the anterior cingulate gyrus, the thalamus (mediodorsal), the basal ganglia, the medial frontal cortex, the inferior orbital frontal cortex, and the parietal cortex. Other studies also implicate the pons and upper brainstem, and abnormalities of the cerebellum. Neurochemistry studies supported by functional brain imaging have provided insight into types of cerebral dysfunction responsible for producing the catatonic syndrome. Possible neurochemical etiologies for medical catatonias include glutaminergic antagonism, gamma-aminobutyric acid (GABA) antagonism, serotonergic actions, and dopamine antagonism.

Catatonia has been described as a feature of many general medical and neurological conditions (Carroll and Goforth, 2004), including metabolic disturbances, endocrinopathies, viral infections (including HIV), typhoid fever, heat stroke, and autoimmune disease. All these conditions are both commonly associated with delirium and catatonia. Drug intoxications and withdrawals may also induce catatonia. Neurologic conditions associated with catatonia include postencephalitic states, parkinsonism, bilateral globus pallidus disease, thalamic and parietal lobe lesions, frontal lobe disease, and general paresis. Diffuse disease processes associated with medical catatonia support the notion that pathway dysfunction rather than focal (site-specific) dysfunction cause catatonia.

The third criterion for a neurobiological syndrome states that milder neurobiological disturbances cause milder symptoms. Transient drug intoxications with rapid resolution of catatonia illustrate this criterion.

Fourth, catatonia is characterized by concurrent motor, emotional, cognitive, and behavioral symptoms as discussed by Northoff (2002). Finally, several studies have found that catatonia responds to benzodiazepines and electroconvulsive therapy (ECT), regardless of etiology (Fink and Taylor, 2003). Future research should continue to examine the validity of catatonia by expanding data on the pathophysiology, biochemistry, and treatment catatonia.

V. Autism as a Neurobiological Syndrome

> However, it is the third question "What causes autism" that is the most problematic and reminds me of Mark Twain's comment, "I was gratified to be able to answer promptly, and I did. I said I don't know."
>
> John F. Mantovani (2003)

Although brain substrates of autistic symptoms are unknown, previous attempts to describe a coherent anatomical or pathophysiological theory of autism have emphasized the involvement of different brain areas sometimes depending on whether the authors considered social, cognitive, or affective abnormalities as the primary deficit in autism.

More integrative analyzes have included motility and perceptual disturbances. For example, Maurer and Damasio (1982) using anatomical and etiological inferences based on a wide range of autistic symptoms concluded that dysfunction of phylogenetically older parts of the frontal and temporal lobes best accounted for the clinical manifestations of autism. Gualtieri (1991) on the other hand noted similarities between autism and the Kluver-Bucy syndrome, a syndrome described in rhesus monkeys after bilateral temporal lobectomies. This suggested to him that deep temporal and frontal lobes may be involved.

Evidence from behavioral, imaging, and postmortem studies indicates that the frontal lobes, as well as other brain regions such as the cerebellum and limbic system, develop abnormally in children with autism. Patients with ASDs had significantly decreased metabolism in both the anterior and posterior cingulate gyri, compared to controls (Haznedar et al., 2000). Increases in prefrontal gray matter have been reported in young autistic boys (Carper and Courchesne, 2000; Carper and Courchesne, 2005). Others have found hypoperfusion in the prefrontal areas of autistic individuals as compared to normals (Wilcox et al., 2002). There is evidence that the volume of the left hippocampus was larger in both the parents of children with autistic disorder, as well as in autistic adults, compared to controls (Rojas et al., 2004). This suggests that some hippocampal abnormalities in autism may be genetically determined. Familial patterns of prefrontal abnormalities in autism have not yet been studied.

Developmental processes have been recognized as key to understanding autism (Belmonte *et al.*, 2004): "Looking beyond surface features is a challenge for autism research: primary dysfunctions can be masked by the evolution of compensatory processing strategies which normalize behaviour, and also by the induction of activity-dependent secondary dysfunctions which disrupt behavior in new ways." The authors lament the lack of data on the primary dysfunctions: "It is quite remarkable and difficult to fathom that we currently have more functional imaging data about how the autistic brain processes a face or a theory of mind than we do about the way it processes, say, location, colour, orientation, or spatial frequency; at what level of processing do the perceptual and cognitive abnormalities begin?"

Although developmental factors complicate neurobiological studies in autism, autism is still a candidate neurobiological syndrome. Autism has been associated with various etiologies. A wide range of severity of autism has been described. It is likely that core disturbances in autism affect associated behavioral/cognitive networks. Treatments for hyperactivity, aggression, anxiety, catatonia, and obsessive features in people with autism reflect the treatments for the primary conditions, although they seem less efficacious.

VI. Is There a Common Neuronal Basis for Autism and Catatonia?

Taking clinical similarities as a criterion to infer relationships among common pathophysiological substrates has merit. The idea is congruent with the concept of a neurobiological syndrome due to impairment of specific neural substrates in different etiological contexts (Fujii and Ahmed, 2004). We have previously discussed the role of age-of-onset on the clinical expression of a disorder and explored if this may explain the differences in symptom profile between autism and catatonia. Even if autism and catatonia are in fact different disorders, the co-occurrence of autism and catatonia in some individuals suggests some relation between the neuronal bases for autism and catatonia.

Another similarity between autism and catatonia may be disturbance of will, volition, or motivation. In autism, there are defects of social interaction and communication with a restricted repertoire of activity and interest. Autistic people misperceive the external world as threatening, hostile, or overwhelming. These cognitive and affective interpretations have been important to help understand autism. However, in catatonia, a different concept, that is, the disturbance of will, has proven to be equally useful. Catatonic negativism and withdrawal from the environment are seen as a form of abulia (the loss of will), while positivism and excitement are seen as a form of excess of will. Temple Grandin has reported these alterations with excess and absence of will (Sacks, 1995). Perseveration, rituals, compulsions, shallow affect, lack of empathy, negative reactions to changes in routine or surroundings, and defect in planning have been typically described in patients with acute and chronic frontal lobe syndromes. Shared dysfunction in frontal cortex areas could explain the presence of abnormal volition in autism and catatonia.

There may be other similarities between autism and catatonia, but definitive conclusions need to be deferred to future research with designs allowing head-to-head comparisons of patients with autism and catatonia. Such parallel studies are discussed in the next section.

Of some interest is the role of the cerebellum in autism and catatonia. The cerebellum remains an understudied yet important area in both disorders. During phylogenesis, the growth of the cerebellum parallels that of the cerebrum. It makes up 10% of the brain's total weight and contains half of its neurons. The neocerebellum is linked to motor, premotor, and supplementary motor regions, as well as primary somatosensory cortices and associative and paralimbic cerebral areas, including the prefrontal cortex and the hypothalamus. The fiber connections imply that there are almost unlimited possibilities for cooperation between the cerebellum and the rest of the brain. The cerebellum has been associated with motor control, but studies have extended its contribution to nonmotor functions. The role of the cerebellum in normality and pathological conditions, such as autism and catatonia, is unknown but is likely associated with higher mental tasks in addition to motor functions (Kern, 2003; Levishon *et al.*, 2000).

Unfortunately, testing for cerebellar function involves identifying neurological soft signs on examination. Specifically, these include but are not limited to testing of: (1) hypotonia, (2) pendular reflexes, (3) dysmetria, and (4) asynergia. These focused tests are not performed as part of the standard neurological examination and are not usually performed on catatonic and autistic patients who are often uncooperative. The cerebellum tends to compensate for injuries early in development. Consequently, testing for cerebellar abnormalities does not always detect cerebellar damage.

Evidence for cerebellar involvement in catatonia is limited to imaging studies, given the limited number of neuroanatomical studies in catatonia. A reduction in cerebellar volume in chronic schizophrenia (Okugawa *et al.*, 2003), catatonic schizophrenia (Joseph *et al.*, 1985; Wilcox, 1991), and autism (Courchesne *et al.*, 1988) has been shown. Several other lines of inquiry also suggest cerebellar dysfunction in autism.

One of the most consistent findings in autism is the selective vulnerability of the cerebellar Purkinje cell (Kemper and Bauman, 1993; Ritvo *et al.*, 1986). Nine brains of patients with well-documented autism (six children and three young adult males) have been studied systematically (Bauman *et al*, 1997). All brains showed a marked reduction of Purkinje cells and a variable decrease in granular cells throughout the cerebellar hemispheres. There was no significant gliosis and no retrograde of inferior olivary neurons, suggesting that the abnormalities were acquired early in development at or before the 30th week of gestation. The neuroanatomical abnormalities observed postmortem in the cerebellum could theoretically contribute to some of the disordered information processing characteristics of the autistic syndrome and explain some of the atypical behavior in autism. In fact, abnormal activation of the cerebellum during motor and attention task was found in a functional imaging study of autistic patients (Allen and Courchesne, 2003). Findings in an electrophysiological study of autism suggest that frontal and parietal spatial attention dysfunction may be due to cerebellar deficits (Townsend *et al.*, 2001).

Another study found that increased frontal lobe cortex volume correlates with the degree of cerebellar abnormality in a subset of patients with autism (Carper and Courchesne, 2000, 2005). In a study of children with congenital ataxia (Ahsgren *et al.*, 2005), neuropsychiatric evaluations showed that 50% of such children had varying levels of autistic symptoms. This is a higher percentage than is found in most syndromes that have a subgroup of children with autistic traits, again supporting a role of the cerebellum in autism.

Meanwhile, Lee *et al.* (2003) postulated a cerebello-limbic circuit for autism. They identified that cerebellar lesions may lead to disturbances of: (1) auditory memory and language sequencing (left cerebellar lesions), (2) spatial and visual sequential memory (right cerebellar lesions), (3) mutism and speech disturbances in children and adults after posterior fossa tumor resection (Sherman *et al.*, 2005) as well as nonsurgical cerebellar involvement (Mewasingh *et al.*, 2003), and sometimes in association with visual disturbances (Daniels *et al.*, 2005), and (4) irritability and other behavioral disturbances (vermal lesions) in adults (Schmamhmann and Sherman, 1998) and children (Richter *et al.*, 2005). Cerebellar signs include difficultly in imitation, difficulty in termination of movements and abnormalities of tone. The concept of a cerebello-limbic loop may be important in autism, in conjunction with cerebello-frontal influences (Carper and Courchesne, 2000). However, firm conclusions must await future, focused investigations.

VII. Future Directions: Parallel Family-Based Studies in Autism and Catatonia

Studies may need to employ head-to-head comparisons of patients with autism and catatonia in order to substantiate common biological, neuropsychological, or genetic risk factors. No prior studies have directly compared autism and catatonia. In this section, the use of parallel studies of autism and catatonia, using a family design, is explored. Under the null hypothesis, no overlap would be expected between these markers in first-degree relatives (parents and siblings) of probands with autism and catatonia. Such a design offers a great advantage over studying these disorders separately by reducing methodological variance. Neurocognitive features can be taken as an example of a putatively common familial marker in autism and catatonia.

Genetic risk may be expressed in subtle neuropsychological abnormalities in unaffected relatives of people with neuropsychiatric disorders. Executive function is an umbrella term, typically associated with frontal-lobe functioning, and used to encompass the processes that underlie goal-directed behavior, for example, planning, working memory, inhibition of prepotent responses, and cognitive flexibility. Previous studies have shown that individuals with autism show impaired performance on tests of executive function (Hughes et al., 1994; Ozonoff et al., 1993). A significant proportion of parents of autistic children also show impaired executive function (Hughes et al., 1997) supporting impaired executive function as a familial marker of the broader autism phenotype.

Weak central coherence or the information-processing bias favoring part/detail processing over processing of wholes/meaning has also been examined as a cognitive phenotype in autistic probands and their relatives (Happe and Briskman, 2001). People with autism and their first-degree relatives seem to perform in a superior way on certain cognitive tasks that benefit from failure to perceive gestalt, for example, the Embedded Figures Test (Witkin et al., 1971). This cognitive anomaly has been coined weak central coherence (Frith, 1989). Correlates of superior disembedding skills in autism probands and their relatives are unknown.

Several studies (Addington and Addington, 1997; Rund, 1993; Tam et al. 1998) have shown that the performance on executive and attentional tests between bipolar and schizophrenic patients is similar, so that executive and attentional deficits might not be specific to schizophrenia. Attentional and executive impairments as assessed by the Stroop Color-Word Test have also been found both in patients with schizophrenia and in their unaffected first-degree relatives, suggesting that they might be considered as familial vulnerability markers (Zalla et al., 2004). Other components of executive function as assessed by the Verbal Fluency Test, the Wisconsin Card Sorting Test and the Trail Making Test do not seem to be different between people with schizophrenia or bipolar disorder (and their family members) and controls.

No family studies of executive function in catatonia are found in the literature. However, executive function has been examined in a small group ($N = 8$) of patients with catatonic schizophrenia (Bark et al., 2005). Several prefrontal cortical tests were administered and compared between patients with catatonic schizophrenia, patients with paranoid schizophrenia ($N = 19$), and healthy controls ($N = 26$). The authors hypothesized that ventral prefrontal cortical function tests (Iowa Gambling Task and Object Alternation Task) would show impairment in patients with catatonic schizophrenia compared to paranoid schizophrenics and controls. An earlier imaging study (Northoff et al., 2004) showed an association between catatonic symptoms and dysfunction in the ventromedial prefrontal cortex. Abnormalities in the Iowa Gambling Task and Object Alternation Task were found in patients with catatonic schizophrenia, but not those with paranoid schizophrenia or in controls.

The study supports the presence of a specific deficit in decision-making in catatonic schizophrenia with an inability to shift from initial high-risk preferences to a more advantageous low-risk strategy. Such deficits were not observed in patients with paranoid schizophrenia and may translate clinically in behavioral anomalies, such as perseverations, stereotypies, echolalia/echopraxia, negativism, and automatic obedience, in which patients show an inability to shift to a different motor behavior. The authors believe that this supports differential pathology of the ventral prefrontal cortex in catatonic schizophrenia compared to paranoid schizophrenia. Neuropsychological (Abbruzzese et al., 1997) and imaging (Callicott et al., 2000; Manoach et al., 2000) studies suggest dysfunction of the dorsolateral prefrontal cortex in paranoid schizophrenia. Future studies should also assess similar tests in other types of schizophrenia, in catatonia not associated with schizophrenia, in acute catatonia, and in family members of these various groups. Specific impairments in executive functions may reflect a valid endophenotype with specific genetic risk factors.

VIII. Implications

Theoretical and practical implications of shared neurobiological substrates of autism and catatonia would be considerable. Searching for common features provides an opportunity to learn more about developmental influence on clinical expression, psychopathology, and pathophysiology of early-onset disorders.

Practical advances concern a widening of therapeutic armamentarium. This is important because autism is increasingly diagnosed, but in many children, current treatments are only minimally effective. Favorable responses to a variety of psychotropic agents and behavioral interventions have been reported in autism, but none leading to major symptom reductions. Finding a shared neuronal basis for autism and catatonia would bring closer the remote possibility that established, for example, ECT (Dhossche and Stanfill, 2004) and future anti-catatonic treatments could be beneficial at some point along the course of the development of certain types of autism.

Acknowledgments

Research support from the Thrasher Research Fund, Salt Lake City, Utah, is gratefully acknowledged. The authors thank Tina Fore, Medical Librarian, Chillicothe VA Medical Center, Chillicothe, Ohio, for locating research articles, and Allison East, Ph.D., Cara Reeves, M.A., and David Taylor for proofreading the text. The work is dedicated to the late Michael Boerger, Ph.D., for inspiration.

References

Abbruzzese, M., Ferri, S., and Scarone, S. (1997). The selective breakdown of frontal functions in patients with obsessive-compulsive disorder and in patients with schizophrenia: A double dissociation experimental finding. *Neuropsychol.* **35,** 907–912.

Addington, J., and Addington, D. (1997). Attentional vulnerability indicators in schizophrenia and bipolar disorder. *Schizophr. Res.* **23,** 197–204.

Ahsgren, I., Baldwin, I., Goetzinger-Falk, C., Erikson, A., Flodmark, O., and Gilberg, C. (2005). Ataxia, autism, and the cerebellum: A clinical study of 32 individuals with congenital ataxia. *Dev. Med. Child Neurol.* **47,** 193–198.

Allen, G., and Courchesne, E. (2003). Differential effects of developmental cerebellar abnormality on cognitive and motor functions in the cerebellum: An fMRI study of autism. *Am. J. Psychiatry* **160,** 262–273.

Allott, R. (2001). Autism and the motor theory of language. *In* "The Great Mosaic Eye: Language and Evolution," pp. 93–113. Book Guild, Lewes, UK.

Bark, R., Dieckmann, S., Bogerts, B., and Northoff, G. (2005). Deficit in decision making in catatonic schizophrenia: An exploratory study. *Psychiatr. Res.* **134,** 131–141.

Bauman, M., Filipek, P., and Kemper, T. (1997). Early infantile autism. *In* "The Cerebellum and Cognition" (J. Schmahmann, Ed.), pp. 367–386. Academic Press, New York.

Belmonte, M., Cook, E., Anderson, G., Rubenstein, J., Greenough, W., Beckel-Mitchener, A., Courchesne, E., Boulanger, L., Powell, S., Levitt, P., Perry, E., Jiang, Y., et al. (2004). Autism as a disorder of neural information processing: Directions for research and targets for therapy. *Mol. Psychiatry* **9,** 646–663.

Callicott, J., Bertolino, A., Mattay, V., Duyn, J., Coppola, R., Goldberg, T., and Weinberger, D. (2000). Physiological dysfunction of the dorsolateral prefrontal cortex in schizophrenia revisited. *Cereb. Cortex* **10,** 1079–1092.

Carper, R., and Courchesne, E. (2000). Inverse correlation between frontal lobe and cerebellum sizes in children with autism. *Brain* **123,** 836–844.

Carper, R., and Courchesne, E. (2005). Localized enlargement of the frontal cortex in early autism. *Biol. Psychiatry* **57,** 126–133.

Carroll, B., and Goforth, H. (2004). Catatonia due to general medical conditions. *In* "Catatonia: From Psychopathology to Neurobiology" (S. Caroff, S. Mann, A. Francis, and G. L. Fricchione, Eds.). American Psychiatric Press, Washington, DC.

Carroll, B., Thomas, C., Jayanti, K., Hawkins, J. W., and Burbage, C. (2005). Treating persistent catatonia when benzodiazepines fail. *Curr. Psychiatry* **4,** 56–64.

Courchesne, E., Yeung-Courchesne, R., Press, G., Hesselink, J., and Jernigan, T. (1988). Hypoplasia of cerebellar vermis lobules VI and VII in autism. *N. Engl. J. Med.* **318,** 1349–1354.

Daniels, S., Moores, L., and Di Fazio, M. (2005). Visual disturbance associated with postoperative cerebellar mutism. *Pediatr. Neurol.* **32,** 127–130.

Dhossche, D. (1998). Catatonia in Autistic Disorders (brief report). *J. Autism Dev. Disord.* **28,** 329–331.
Dhossche, D., and Stanfill, S. (2004). Could ECT be effective in autism? *Med. Hypotheses* **63,** 371–376.
Eisenmajer, R., Prior, M., Leekam, S., Wing, L., Ong, B., Gould, J., and Welham, M. (1998). Delayed language onset as a predictor of clinical symptoms in pervasive developmental disorders. *J. Autism Dev. Disord.* **28,** 527–533.
Fink, M., and Taylor, M. (2003). Catatonia. *In* "A Clinician's Guide to Diagnosis and Treatment." University Press, Cambridge.
Frith, U. (1989). "Autism: Explaining the Enigma." Blackwell, Oxford.
Fujii, D., and Ahmed, I. (2004). Is psychosis a neurobiological syndrome? *Can. J. Psychiatry* **49,** 713–718.
Gualtieri, C. (1991). The functional neuroanatomy of psychiatric treatments. *Psychiatr. Clin. North Am.* **14,** 113–124.
Happe, F., and Briskman, J. (2001). Exploring the cognitive phenotype of autism: Weak "central coherence" in parents and siblings of children with autism: I. Experimental tests. *J. Child Psychol. Psychiatry* **42,** 299–307.
Haznedar, M., Buchsbaum, M., Wei, T.-C., Hof, P., Cartwright, C., Bienstock, C., and Hollander, E. (2000). Limbic circuitry in patients with autism spectrum disorders studied with positron emission tomography and magnetic resonance imaging. *Am. J. Psychiatry* **157,** 1994–2001.
Hughes, C., Russell, J., and Robbins, T. (1994). Evidence for central executive dysfunction in autism. *Neuropsychologia* **32,** 477–492.
Hughes, C., Leboyer, M., and Bouvard, M. (1997). Executive function in parents of children with autism. *Psychol. Med.* **27,** 209–220.
Joseph, A., Anderson, W., and O' Leary, D. (1985). Brainstem and vermis atrophy in catatonia. *Am.. J. Psychiatry* **142,** 352–354.
Kahlbaum, K. (1874). "Die Katatonie Oder Das Spannungsirresein." Verlag August Hirshwald, Berlin.
Kemper, T., and Bauman, M. (1993). The contribution of neuropathologic studies to the understanding of autism. *Neurol. Clin.* **11,** 175–187.
Kern, J. (2003). Purkinje cell vulnerability and autism: A possible etiological connection. *Brain and Development* **25,** 377–382.
Kraemer, H. (1996). Biostatistical recommendations. *J. Autism Dev. Disord.* **26,** 150–154.
Kuhl, P. (2004). Early language acquisition: Cracking the speech code. *Nat. Rev. Neurosci.* **5,** 831–843.
Lee, D., Lopez-Alberda, R., and Bhattacharjee, M. (2003). Childhood autism: A circuit syndrome? *Neurologist* **9,** 99–109.
Levishon, L., Cronin-Golomb, A., and Schmahmann, J. (2000). Neuropsychological consequences of cerebellar tumour resection in children. Cerebellar cognitive affective syndrome in a paediatric population. *Brain* **123,** 1041–1050.
Manoach, D., Gollub, R., Benson, E., Searl, M., Goff, D., Halpern, E., Saper, C., and Rauch, S. (2000). Schizophrenic subjects show aberrant fMRI activation of dorsolateral prefrontal cortex and basal ganglia during working performance. *Biol. Psychiatry* **48,** 99–109.
Mantovani, J. (2003). Not knowing (Editorial). *Dev. Med. Child Neurol.* **45,** 75.
Maurer, R., and Damasio, A. (1982). Childhood autism from the point of view of behavioral neurology. *J. Autism Dev. Disord.* **12,** 195–205.
Mewasingh, L., Kadhim, H., Christophe, C., Christiaens, F., and Dan, B. (2003). Nonsurgical cerebellar mutism (anarthia) in two children. *Pediatr. Neurol.* **28,** 59–63.
Northoff, G. (2002). What catatonia can tell us about "top-down" modulation: A neuropsychiatric hypothesis. *Behav. Brain Sci.* **25,** 555–604.
Northoff, G., Kotter, R., Baumgart, F., Danos, P., Boeker, H., Kaulisch, T., Schlagenhauf, F., Walter, H., Heinzel, A., Witzel, T., and Bogerts, B. (2004). Orbitofrontal cortical dysfunction in akinetic catatonia: A functional magnetic resonance imaging study during negative emotional stimulation. *Schizophr. Bull.* **30,** 405–427.

Noterdaeme, M., Mildenberger, K., Minow, F., and Amorosa, H. (2002). Evaluation of neuromotor deficits in children with autism and children with a specific speech and language disorder. *Eur. Child Adolesc. Psychiatry* **11,** 219–225.

Okugawa, G., Sedvall, G., and Agartz, I. (2003). Smaller cerebellar vermis but not hemisphere volumes in patients with chronic schizophrenia. *Am. J. Psychiatry* **160,** 1614–1617.

Ozonoff, S., Strayer, D., McMahon, W., and Filloux, F. (1993). Executive function in autism: An information processing approach. *J. Child Psychol. Psychiatry* **35,** 1015–1032.

Richter, S., Schoch, B., Kaiser, O., Groetschel, H., Dimitrova, A., Hein-Kropp, C., Maschke, M., Gizewski, E., and Timman, D. (2005). Behavioral and affective changes in children and adolescents with chronic cerebellar lesions. *Neurosci. Lett.* **381,** 102–107.

Ritvo, E., Freeman, B., Scheibel, A., Duong, T., Robinson, H., Guthrie, D., and Ritvo, A. (1986). Lower Purkinje cell counts in the cerebella of four autistic subjects: Initial findings of the UCLA-NSAC Autopsy Research Report. *Am. J. Psychiatry* **143,** 862–866.

Rojas, D., Smith, J., Benkers, T., Camou, S., Reite, M., and Rogers, S. (2004). Hippocampus and amygdala volumes in parents of children with autistic disorder. *Am. J. Psychiatry* **161,** 2038–2044.

Rund, B. (1993). Backward-masking performance in chronic and nonchronic schizophrenics, affectively disturbed patients, and normal control subjects. *J. Abnorm. Psychol.* **102,** 74–81.

Sacks, O. (1995). "An Anthropologist on Mars: 7 Paradoxical Tales," pp. 244–296, Vintage Books, New York.

Schmamhmann, J., and Sherman, J. (1998). The cerebellar cognitive affective syndrome. *Brain* **121,** 561–579.

Sherman, J., Sheehan, J., Elias, W., and Jane, J. (2005). Cerebellar mutism in adults after posterior fossa surgery: A report of 2 cases. *Surg. Neurol* **63,** 476–479.

Stone, W., Ousley, O., and Littleford, C. (1997). Motor imitation in young children with autism: What's the object? *J. Abnorm. Child Psychol.* **25,** 475–485.

Stone, W., and Yoder, P. (2001). Predicting spoken language level in children with autism spectrum disorders. *Autism* **5,** 341–361.

Tam, W., Sewell, K., and Deng, H. (1998). Information processing in schizophrenia and bipolar disorder: A discriminant analysis. *J. Nerv. Ment. Dis.* **186,** 597–603.

Taylor, M., and Fink, M. (2003). Catatonia in psychiatric classification: A home of its own. *Am. J. Psychiatry* **160,** 1–9.

Townsend, J., Westerfield, M., Leaver, E., Makeig, S., Jung, T., Pierce, K., and Courchesne, E. (2001). Event-related brain response abnormalities in autism: Evidence for impaired cerebello-frontal spatial attention networks. *Brain Res.Cogn. Brain Res.* **11,** 127–145.

Wilcox, J. (1991). Cerebellar atrophy and catatonia. *Biol. Psychiatry* **29,** 733–734.

Wilcox, J., Tsuang, M., Ledger, E., Algeo, J., and Schnurr, T. (2002). Brain perfusion in autism varies with age. *Neuropsychobiology* **46,** 13–16.

Williams, J., Whiten, A., Suddendorf, T., and Perrett, D. (2001). Imitation, mirror neurons and autism. *Neurosci. Biobehav. Rev.* **25,** 287–295.

Wing, L., and Shah, A. (2000). Catatonia in autistic spectrum disorders. *Br. J. Psychiatry* **176,** 357–362.

Witkin, H., Oltman, P., Raskin, E., and Karp, S. (1971). "A Manual for the Embedded Figure Test." Consulting Psychologists Press, California.

Zalla, T., Joyce, C., Szoke, A., Schurhoff, F., Pillon, B., Komano, O., Perez-Diaz, F., Bellivier, F., Alter, C., Dubois, B., Rouillon, F., Houde, O., *et al.* (2004). Executive dysfunctions as potential markers of familial vulnerability to bipolar disorder and schizophrenia. *Psychiatr. Res.* **121,** 207–217.

SHARED SUSCEPTIBILITY REGION ON CHROMOSOME 15 BETWEEN AUTISM AND CATATONIA

Yvon C. Chagnon

Laval University Robert-Giffard Research Center, Beauport, Québec, Canada

I. Introduction
II. Methods
III. Results
IV. Conclusions
References

We have compiled significant linkage results from 20 genome scans for the autism syndrome disorder (ASD) and 2 for catatonia in schizophrenia (SZ). Localization of the markers has been updated across the studies using the same cytological (Genetic Location Database), physical (National Center for Biological Information), and genetic (Marshfield) maps. Eight autosomal chromosomes (1, 2, 3, 7, 9, 13, 15, and 17) showed significant linkages with ASD, and one with catatonia (15). Chromosome 15 was further characterized for SZ genome scans ($N = 4$) since catatonia was observed in SZ patients, for candidate genes for ASD and catatonia, and for the numerous chromosomal rearrangement and abnormalities associated to ASD. From these results, we observed that four potential susceptibility regions for ASD could be observed on chromosome 15 at 15q11-q13, 15q14-q21, 15q22-q23, and 15q26, respectively. All the four regions were shared between ASD and SZ, with 15q15-q21 being also shared with catatonia. Strong candidate genes, such as gamma-aminobutyric acid receptor B3, A5, and G3, have shown associations with ASD at 15q11-q13 susceptibility region where the majority of the chromosomal rearrangements are also found. On the other hand, negative association results were observed at 15q14-q21 susceptibility region for catatonia with the genes encoding the zinc transporter SLC30A4, the cholinergic receptor nicotinic alpha polypeptide 7, and the delta-like 4 *Drosophila*. Further, fine mapping and candidate gene analyses are needed to highlight potential common genes between ASD and catatonia for this chromosome.

I. Introduction

Autism spectrum disorder (ASD) is a severe neurodevelopmental disorder characterized by delayed or absent speech, impairments in social interaction and communication, and repetitive behaviors and restricted interests. Few common psychopathological dimensions of ASD are observed in other mental disorders, such as schizophrenia (SZ) and bipolar disorder (BP), one of which being catatonia (Abrams and Taylor, 1976; Dhossche, 1998; Realmuto and August, 1991; Wing and Shah, 2000). Moreover, family studies have reported increased rates of SZ-like and affective disorders (Larsson *et al.*, 2005). Parental SZ-like psychosis and affective disorder were significant risk factors for ASD in offspring, in a nationwide Danish case-control study of 698 children diagnosed with ASD between 1972 and 1999 (Larsson *et al.*, 2005). Relative risks were 3.44, 95% CI 1.48–7.95 and 2.91, 95% CI 1.65–5.14, for parental SZ-like disorder and affective disorder, respectively. Other significant variables were breech presentation (RR = 1.63), low Apgar score at 5 minutes (RR = 1.89), and gestational age at birth less than 35 weeks (RR = 2.45).

Previous studies have typically not separated out catatonic subtypes of SZ, affective disorder, or other psychotic disorders. A subtype of unsystematic SZ is characterized by periodic catatonia where acute psychotic episodes are followed by remission. These disorders have shown strong heritabilities with some 15–20 genes expected to be involved in ASD (Spence, 2004), while a major gene effect is predicted for catatonia (Stöber *et al.*, 1995). From the literature, it is obvious that the same genes and biological pathways could be shared among different mental disorders. For example, the serotonin transporter gene has been associated to different mental disorders including ASD (Yirmiya *et al.*, 2001). Another example is the catechol-*O*-methyltransferase gene for which positive associations have been reported with SZ (Chen *et al.*, 2004; Egan *et al.*, 2001; Glatt *et al.*, 2003a; Sazci *et al.*, 2004; Wonodi *et al.*, 2003) and schizotypy (Avramopoulos *et al.*, 2002), with BP (Shifman *et al.*, 2004), with anxiety (Enoch *et al.*, 2003; McGrath *et al.*, 2004), with anorexia nervosa (Gabrovsek *et al.*, 2004), and with the 22q11.2 deletion syndrome (Bearden *et al.*, 2004).

We have compiled the results from published genome scans for ASD and for catatonia in SZ, and observed a shared chromosomal susceptibility region on chromosome 15. For this chromosome, we then also compiled significant genome scan results with SZ for comparison with catatonia SZ, as chromosomal rearrangements and abnormalities associated with ASD, and candidate gene analyses for ASD and catatonia.

II. Methods

Relevant genome scan papers have been identified by a search in the PubMed database using the key words "linkage OR genome scan" and "autism OR catatonia OR schizophrenia." Relevant references found therein were also used. The criterion for inclusion of the results of a genome scan was a Lod score of 3.0 and greater, corresponding to a p value of 0.0001 and smaller. Only chromosomes significant susceptibility regions for ASD have been retained, while all linkage results for catatonia were included since the low number of studies. The chromosomal location of all the linked markers has been updated using the same and most recent version of the physical map in megabases (Mb) from the National Center for Biological Information (NCBI built 35.1). One Mb corresponds to 10^6 bases or nucleotides of DNA. Because the location on the physical map was not available for all markers, we also used a genetic map in centimorgan (cM) units from Marshfield (Broman et al., 1998) to determine the relative position of the marker. One cM corresponds roughly to 1 Mb. Additionally, the cytological locations have been updated using the predictive locations (GMAP) from the Genetic Location Database (Collins et al., 1996). For chromosome 15, we have also reported chromosomal rearrangements and abnormalities, positive or negative candidate gene analyses associated to ASD or catatonia, and significant genome scans for SZ for comparison with catatonia SZ.

III. Results

The Table I presents the results from the 20 genome scans related to ASD ($N = 18$) and to catatonia SZ ($N = 2$). We observed that eight autosomal chromosomes (1, 2, 3, 7, 9, 13, 15, and 17) showed significant linkages with ASD. Linkages with catatonia SZ were also observed on chromosomes 15 and 22. Four genome scans with significant results for SZ were available for chromosome 15 and none for chromosome 22. On chromosome 1, two susceptibility loci for ASD are observed at 1p21.2 and 1q24.1, respectively. The calcium-regulated potassium channel (*KCNN3*) gene is located in the second region at 1q21.3 and less then 3 Mb from the Asperger syndrome locus (see NCBI). *KCNN3* had shown a possible association with SZ but not with a subgroup of SZ showing catatonia (Stöber et al., 2000a). In contrast, another study reported negative association results between *KCNN3* and SZ (Glatt et al., 2003b). Additional studies are needed to confirm the association of *KCNN3* with SZ and eventually with catatonia SZ.

The Fig. 1 presents some of the chromosomal rearrangements and associated syndrome related to ASD for chromosome 15. Most of the rearrangements/abnormalities are observed at 15q11-q13 where the Prader–Willi/Angelman syndrome loci are located. Two more telomeric regions have also been reported with a deletion at 15q22-q23 (Smith et al., 2000), and a duplication of the 15q25-qter region at 15p producing a true trisomy (Bonati et al., 2005). The figure presents a compendium of all chromosomal rearrangements, linkage results, and candidate gene analysis for this chromosome. From these, it could be seen that

TABLE I
SIGNIFICANT GENOME SCAN RESULTS OBSERVED FOR ASD AND SHARED BY OTHER MENTAL DISORDERS (MD). PHYSICAL (NCBI) AND GENETIC (MARSHFIELD) ARE REPORTED

Chromosome	Location	Markers	NCBI	Marshfield	Statistic	MD	References
1	1p21.2	*D1S1631*	105,372,673	136.9	$Z_{max} = 3.4$	ASD	Risch et al., 1999
	1q21.3	*KCNN3*	151,655,827	na	ns	Catatonia SZ	Stöber et al., 2000a
	1q24.1	*D1S484*	157,580,640	169.7	$Z_{max} = 3.6$	ASD (AS)	Ylisaukko-Oja et al., 2004
2	2q31.1	*D2S335*	172,392,096	175.9	NPL = 3.3	ASD	Buxbaum et al., 2001
	2q31.3	*D2S2188*	175,430,218	180.8	MLS = 4.8	ASD	IMGSAC, 2001a
	2q32.1	*D2S364*	182,860,040	186.2	NPL = 3.3	ASD	Buxbaum et al., 2001
3	3q27.1	*D3S3037*	na	190.4	Lod = 4.3	ASD	Auranen et al., 2002
7	7q32.1	*D7S530*	128,216,773	134.6	MLS = 3.6	ASD	IMGSAC, 1998
	7q	*D7S684*	137,521,669	147.2	MLS = 3.6	ASD	IMGSAC, 1998
	7q34	*D7S1824*	139,465,749	149.9	$Z = 3.0$	ASD	Alarcon et al., 2002
	7q36.2	*D7S2462*	152,999,336	169.8	Lod = 3.7	ASD	Auranen et al., 2002
9	9p22.3	*D9S157*	17,618,382	32.2	MLS = 3.1	ASD	IMGSAC, 2001a
	9q34.3	*D9S1826*	135,674,269	159.6	MLS = 3.6	ASD	IMGSAC, 2001a
		D9S158	136,325,009	161.7	MLS = 3.2	ASD	IMGSAC, 2001a
13	13q22.3	*D13S800*	72,772,693	55.3	MML = 3.0	ASD	IMGSAC, 2001b
15	15q11.2	*D15S128*	022,681,893	6.1	$Z = 4.0$	SZ	Freedman et al., 2001
					Lod = 0.7	ASD	Philippe et al., 1999
	15q12	*UBE3A*	023,231,137	6.1	$p = 0.004$	ASD	Nurmi et al., 2001
					ns	ASD	Veenstra-VanderWeele et al., 1999
	15q	*ATP10A*[a]	023,659,963	na	$p = 0.03, 0.03$	ASD	Nurmi et al., 2003
	15q11.1	*GABRB3*	024,570,020	na	Lod = 4.7	ASD	Shao et al., 2003
					$p = 0.001$	ASD	Cook et al., 1998
					$p = 0.01$ to 0.04	ASD	McCauley et al., 2004
					ns	ASD	Maestrini et al., 1999

	Marker	Position	cM	Statistic	Disorder	Reference
	GABRA5	24,724,386	na	$p = 0.03$	ASD	McCauley et al., 2004
	GABRG3	24,799,413	na	$p = 0.02, 0.03$	ASD	Menold et al., 2001
15q12	D15S217	25,747,918	na	$Z = 1.8$	ASD	Bass et al., 2000
15q11.2	CHRNA7[b]	30,110,018	na	Lod = 5.3	SZ/P50	Freedman et al., 1997
15q13.3	D15S118	34,024,135	32.6	$Z = 4.0$	SZ	Freedman et al., 2001
15q15.3				Lod = 1.1	ASD	Philippe et al., 1999
	D15S1042	34,050,903	32.6	$Z = 2.6, 3.9$	Catatonia SZ	Stöber et al., 2000b, 2002
	D15S1012	36,794,835	36.0	$Z = 2.8$	Catatonia SZ	Stöber et al., 2000b
15q21.1	SLC30A4	43,602,294	na		Catatonia SZ	Kury et al., 2003
	D15S659	44,161,300	43.5	$Z = 3.9$	Catatonia SZ	Stöber et al., 2002
15q21.2	D15S117	56,266,889	51.2	MLS = 2.6	ASD	Lamb et al., 2005
15q22.3	D15S125	64,880,007	64.2	MLS = 2.6	ASD	Lamb et al., 2005
	PKM2 (PK3)	70,278,439	na	$p = 0.02*$	SZ	Stone et al., 2004
15q26.2	D15S652	90,318,371	90.0	$Z = 2.3$	ASD	Risch et al., 1999
15q26.3	D15S1014	95,803,594	107.7	Lod = 4.6	SZ/BP	Maziade et al., 2005
17q12	D17S1299	36,607,512	62.0	MLS = 3.6	ASD	Cantor et al., 2005
17q21.1	D17S2180	45,940,225	66.9	MLS = 4.1	ASD	Cantor et al., 2005
22q13	D22S1169	47,722,962	60.6	$Z = 1.9$	Catatonia SZ	Stöber et al., 2000b

*Genome-wide adjusted.
[a]Previously named ATP10C.
[b]Location of D15S1360 within CHRNA7.

na: not available; ns: not significant. LDB: Cytological locations: http://cedar.genetics.soton.ac.uk/public_html/; NCBI: physical location in nucleotides: http://www.ncbi.nlm.nih.gov/; and Marshfield: genetic distance in centimorgans: http://research.marshfieldclinic.org/genetics/ MD: mental disorders; ASD: autism spectrum disorder; AS: Asperger syndrome; and SZ: schizophrenia. MLS: multipoint Lod score; NPL: nonparametric Lod score; and MML: maximum multipoint heterogeneity LOD score. KCN3: calcium-activated potassium channel; PFKFB2: 6-phosphofructo-2-kinase/fructose-2,6-bisphosphatase 2; UBE3A: E6-AP ubiquitin-protein ligase; ATP10A: ATPase, Class V, type 10A; GABRB3, GABRA5, and GABRG3: gamma-aminobutyric acid receptor B3, A5, G3; CHRNA7: alpha 7-nicotinic receptor; and SLC30A4: zinc transporter.

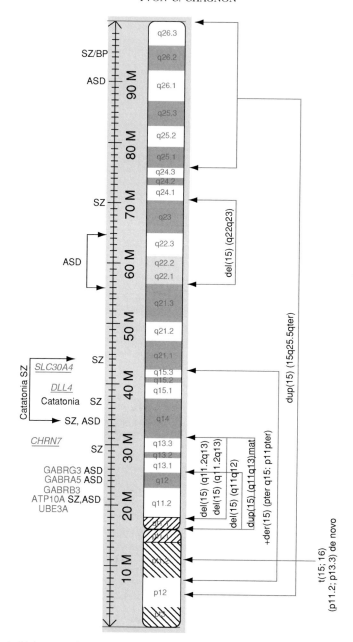

Fig. 1. Linkage results and candidate genes (above) and chromosomal rearrangements/abnormalities (below) observed on chromosome 15 in relation to ASD and catatonia. Genes with negative association results are also shown (underlined).

four different chromosomal regions potentially carried susceptibility loci for ASD. It could also be seen that all the four regions are shared with SZ and one region also with catatonia SZ (Fig. 1). The first region is at 15q11-q13 between approximately 20 Mb and 30 Mb. Originally, 15q11-q13 was associated with ASD because of the numerous chromosomal abnormalities observed for this region in autistic patients (Fig. 1). Significant linkage and candidate gene association results for ASD have also been reported at 15q11-q13 (Table I). Linkages with the marker *D15S128* located at 23 Mb have been observed for both ASD and SZ. The E6-AP ubiquitin-protein ligase (*UBE3A*) gene located at 15q12 and 23 Mb was associated with the Angelman syndrome (Veenstra-VanderWeele et al., 1999) and with ASD (Nurmi et al., 2001). The marker *D15S122*, located in the 5'UTR region of *UBE3A*, was linked to BP (Maziade et al., 2005). The *UBE3A* appeared then to be related to four syndromes and mental disorders (Angelman, autism, SZ, BP).

Associations and linkages were also observed with different gamma-aminobutyric acid (GABA) receptors. A significant linkage (Shao et al., 2003) and a positive association (Buxbaum et al., 2002; Cook et al., 1998) was observed between ASD and the GABA receptor B3 (*GABRB3*) using the 155CA-2 microsatellite marker. An association between ASD and *GABRB3* was also observed using two SNPs located in the intron 7 of *GABRB3* (McCauley et al., 2004). Finally, the *GABRB3* 155CA-2 marker had shown no association with BP (Papadimitriou et al., 2001b). For its part, the GABA receptor 5 (*GABRA5*) was associated to ASD for six individual SNPs and their haplotypes (McCauley et al., 2004). *GABRA5* was also associated with SZ (Devon et al., 2001; Papadimitriou et al., 2001a), and BP (Otani et al., 2005; Papadimitriou et al., 1998, 2001b). Finally, two SNPs in the GABA receptor G3 showed association with ASD (Menold et al., 2001).

In addition to *UBEA3* and the GABA receptors, *ATP10A* previously named *ATP10C*, showed association with ASD using two SNPs and a haplotype block (Nurmi et al., 2003). Also, it is not clear at this time if all these genes are involved individually in ASD, or as a group or as subgroups of them interacting with each other in an epistatic effect. These associations could also come from linkage disequilibrium between the markers. However, high-local recombination or, conversely, low-linkage disequilibrium have been reported for this chromosomal region (McCauley et al., 2004). For instance, the sex-averaged genetic map estimates that the rate for this interval is 4 cM/Mb, compared to a genome-wide average of 1.3 cM/Mb (McCauley et al., 2004).

The second chromosomal region at 15q14-q21 is located between 32 and 44 Mb, very close to 15q11-q13, and could be part eventually of the same susceptibility region. However, some features could distinguish these two regions. First, most of the chromosomal abnormalities are observed at 15q11-q13 in contrast to 15q14-q21 (see Fig. 1). Second, linkages with catatonia are located exclusively at

15q14-q21. For instance, markers *D15S1012* at 37 Mb, *D15S1042* at 34 Mb, and *D15S659* at 44 Mb are all linked to catatonia in SZ patients, while the marker *D15S118* at 34 Mb is linked to both ASD and SZ. The zinc transporter gene *SLC30A4*, located at 15q21.1 and 44 Mb, had shown altered expression patterns in postmortem analysis of the brains of SZ patients (Mirnics et al., 2000). However, no genetic variants within the coding and the putative promoter region of *SLC30A4* was found in affected individuals from large pedigrees of catatonia SZ patients showing a perfect co-segregation of a chromosomal segment between marker *D15S1042* and *D15S659* (Kury et al., 2003). Similarly, no association with catatonia has been observed with the cholinergic receptor nicotinic alpha polypeptide 7 (*CHRN7*) and the delta-like 4 Drosophila (*DLL4*) genes, both located at 15q14 and 31.0 and 39.0 Mb, respectively (McKeane et al., 2005; Meyer et al., 2002).

The third chromosomal region is located at 15q22-q23, between 56 and 70 Mb (Fig. 1). A deletion of this region has been associated to ASD and significant linkage results have been observed for both ASD and SZ. It has to be noted that the linkage with ASD has been observed in nonmale affected sib-pairs only which suggests possible parent of origin specific effects (Lamb et al., 2005). As many as 140 genes are located in this region, according to NCBI. Among them, the *N*-methyl-D-aspartate (NMDA) receptor-regulated 2 gene (*NARG2*) is located at 15q22.2 and 58.6 Mb. NMDA receptors play an important role in the transition from proliferation of neuronal precursors to differentiation of neurons (Sugiura et al., 2001). Postmortem brain abnormalities of the glutamate neurotransmitter system have been observed in ASD (Purcell et al., 2001). To our knowledge, *NARG2* has not been studied yet in relation to ASD, catatonia, or SZ.

The fourth and last region encompassed the q terminal end of the chromosome 15 at 15q26 and 90–108 Mb (Fig. 1). ASD and SZ share this relatively short region. The marker *D15S652* linked to ASD is also related to the major depressive disorder 2 susceptibility locus (see NCBI). Fifty-five genes are located in this region, according to NCBI. The desmuslin gene (*DMN*) is located at 15q26.3 and 97.5 Mb. *DMN* is mainly expressed in muscular tissues where it is involved in different myopathies and is also expressed in the brain (Mizuno et al., 2001a,b). *DMN* encodes an intermediate filament protein that interacts with the desmin, also an intermediate filament protein, and with the alpha form of the dystrobevin (Mizuno et al., 2001b). Interestingly, the beta form of the dystrobevin has been shown to be bound by the dystrobevin binding protein 1 (DTNBP1), for which the gene is located at 6p22.3. *DTNBP1* gene has been associated to SZ in multiple studies (Funke et al., 2004; Kohn et al., 2004; Schwab et al., 2003; Straub et al., 2002; Van Den Bogaert et al., 2003; van den Oord et al., 2003), and possibly to BP in a subset of cases with psychosis (Raybould et al., 2005).

Finally, susceptibility regions for ASD were observed at 2q31.1-q32.1, 3q27, 7q32.1-q36.2, 9p22.3 and 9q34.3, 13q22.3, and 17q12-q21.1 (see Table I).

However, no significant linkage with catatonia was observed on these chromosomes, neither between ASD or catatonia nor the two sexual X and Y chromosomes.

IV. Conclusions

Four potentially different susceptibility chromosomal regions for ASD, SZ, and catatonia have been identified on chromosome 15 by putting together on a same physical map scale, the significant results from different genome scans. All the four regions are shared by ASD and SZ, while the second region was also shared by catatonia SZ. This supports the hypothesis that common intermediate phenotypes are shared between these disorders. Five candidate genes have shown positive associations with ASD in the first region, while three have shown negative results in the second region with catatonia. As such, this chromosome becomes a prime target for direct investigations of common intermediate phenotypes in relevant cohorts. Catatonia could be a shared syndrome in ASD and SZ, and may be useful as an intermediate phenotype. Future studies in ASD and SZ, and eventually other mental disorders, such as BP and attention deficit/hyperactivity deficiency, should assess and distinguish the catatonic syndrome within study samples. Findings in this review suggest consecutive susceptibility regions on chromosome 15. The identification and validation of additional candidate genes on chromosome 15 by fine mapping and association studies should be a priority particularly in the second region with catatonia.

References

Abrams, R., and Taylor, M. A. (1976). Catatonia. A prospective clinical study. *Arch. Gen. Psychiat.* **33,** 579–581.
Alarcon, M., Cantor, R. M., Liu, J., Gilliam, T. C., and Geschwind, D. H. (2002). Evidence for a language quantitative trait locus on chromosome 7q in multiplex autism families. *Am. J. Hum. Genet.* **70,** 60–71.
Auranen, M., Vanhala, R., Varilo, T., Ayers, K., Kempas, E., Ylisaukko-Oja, T., Sinsheimer, J. S., Peltonen, L., and Jarvela, I. (2002). A genomewide screen for autism-spectrum disorders: Evidence for a major susceptibility locus on chromosome 3q25–27. *Am. J. Hum. Genet.* **71,** 777–790.
Avramopoulos, D., Stefanis, N. C., Hantoumi, I., Smyrnis, N., Evdokimidis, I., and Stefanis, C. N. (2002). Higher scores of self reported schizotypy in healthy young males carrying the COMT high activity allele. *Mol. Psychiatr.* **7,** 706–711.
Bass, M. P., Menold, M. M., Wolpert, C. M., Donnelly, S. L., Ravan, S. A., Hauser, E. R., Maddox, L. O., Vance, J. M., Abramson, R. K., Wright, H. H., Gilbert, J. R., Cuccaro, M. L., *et al.* (2000). Genetic studies in autistic disorder and chromosome 15. *Neurogenetics* **2,** 219–226.

Bearden, C. E., Jawad, A. F., Lynch, D. R., Sokol, S., Kanes, S. J., McDonald-McGinn, D. M., Saitta, S. C., Harris, S. E., Moss, E., Wang, P. P., Zackai, E., Emanuel, B. S., *et al.* (2004). Effects of a functional COMT polymorphism on prefrontal cognitive function in patients with 22q11.2 deletion syndrome. *Am. J. Psychiatr.* **161,** 1700–1702.

Bonati, M. T., Finelli, P., Giardino, D., Gottardi, G., Roberts, W., and Larizza, L. (2005). Trisomy 15q25.2-qter in an autistic child: Genotype-phenotype correlations. *Am. J. Med. Genet. A* **133,** 184–188.

Broman, K. W., Murray, J. C., Sheffield, V. C., White, R. L., and Weber, J. L. (1998). Comprehensive human genetic maps: Individual and sex-specific variation in recombination. *Am. J. Hum. Genet.* **63,** 861–869.

Buxbaum, J. D., Silverman, J. M., Smith, C. J., Kilifarski, M., Reichert, J., Hollander, E., Lawlor, B. A., Fitzgerald, M., Greenberg, D. A., and Davis, K. L. (2001). Evidence for a susceptibility gene for autism on chromosome 2 and for genetic heterogeneity. *Am. J. Hum. Genet.* **68,** 1514–1520.

Buxbaum, J. D., Silverman, J. M., Smith, C. J., Greenberg, D. A., Kilifarski, M., Reichert, J., Cook, E. H., Jr., Fang, Y., Song, C. Y., and Vitale, R. (2002). Association between a GABRB3 polymorphism and autism. *Mol. Psychiatr.* **7,** 311–316.

Cantor, R. M., Kono, N., Duvall, J. A., Alvarez-Retuerto, A., Stone, J. L., Alarcon, M., Nelson, S. F., and Geschwind, D. H. (2005). Replication of autism linkage: Fine-mapping peak at 17q21. *Am. J. Hum. Genet.* **76,** 1050–1056.

Chen, X., Wang, X., O'Neill, A. F., Walsh, D., and Kendler, K. S. (2004). Variants in the catechol-*O*-methyltransferase (COMT) gene are associated with schizophrenia in Irish high-density families. *Mol. Psychiatr.* **9,** 962–967.

Collins, A., Frezal, J., Teague, J., and Morton, N. E. (1996). A metric map of humans: 23,500 loci in 850 bands. *Proc. Natl. Acad. Sci. USA* **93,** 14,771–14,775.

Cook, E. H., Jr., Courchesne, R. Y., Cox, N. J., Lord, C., Gonen, D., Guter, S. J., Lincoln, A., Nix, K., Haas, R., Leventhal, B. L., and Courchesne, E. (1998). Linkage-disequilibrium mapping of autistic disorder, with 15q11–13 markers. *Am. J. Hum. Genet.* **62,** 1077–1083.

Devon, R. S., Anderson, S., Teague, P. W., Muir, W. J., Murray, V., Pelosi, A. J., Blackwood, D. H., and Porteous, D. J. (2001). The genomic organisation of the metabotropic glutamate receptor subtype 5 gene, and its association with schizophrenia. *Mol. Psychiatr.* **6,** 311–314.

Dhossche, D. (1998). Brief report: Catatonia in autistic disorders. *J. Autism Dev. Disord.* **28,** 329–331.

Egan, M. F., Goldberg, T. E., Kolachana, B. S., Callicott, J. H., Mazzanti, C. M., Straub, R. E., Goldman, D., and Weinberger, D. R. (2001). Effect of COMT Val108/158 Met genotype on frontal lobe function and risk for schizophrenia. *Proc. Natl. Acad. Sci. USA* **98,** 6917–6922.

Enoch, M. A., Xu, K., Ferro, E., Harris, C. R., and Goldman, D. (2003). Genetic origins of anxiety in women: A role for a functional catechol-*O*-methyltransferase polymorphism. *Psychiatr. Genet.* **13,** 33–41.

Freedman, R., Coon, H., Myles-Worsley, M., Orr-Urtreger, A., Olincy, A., Davis, A., Polymeropoulos, M., Holik, J., Hopkins, J., Hoff, M., Rosenthal, J., Waldo, M. C., *et al.* (1997). Linkage of a neurophysiological deficit in schizophrenia to a chromosome 15 locus. *Proc. Natl. Acad. Sci. USA* **94,** 587–592.

Freedman, R., Leonard, S., Olincy, A., Kaufmann, C. A., Malaspina, D., Cloninger, C. R., Svrakic, D., Faraone, S. V., and Tsuang, M. T. (2001). Evidence for the multigenic inheritance of schizophrenia. *Am. J. Med. Genet.* **105,** 794–800.

Funke, B., Finn, C. T., Plocik, A. M., Lake, S., DeRosse, P., Kane, J. M., Kucherlapati, R., and Malhotra, A. K. (2004). Association of the DTNBP1 locus with schizophrenia in a U.S. population. *Am. J. Hum. Genet.* **75,** 891–898.

Gabrovsek, M., Brecelj-Anderluh, M., Bellodi, L., Cellini, E., Di Bella, D., Estivill, X., Fernandez-Aranda, F., Freeman, B., Geller, F., Gratacos, M., Haigh, R., Hebebrand, J., *et al.* (2004).

Combined family trio and case-control analysis of the COMT Val158Met polymorphism in European patients with anorexia nervosa. *Am. J. Med. Genet. B. Neuropsychiatr. Genet.* **124,** 68–72.

Glatt, S. J., Faraone, S. V., and Tsuang, M. T. (2003a). Association between a functional catechol-*O*-methyltransferase gene polymorphism and schizophrenia: Meta-analysis of case-control and family-based studies. *Am. J. Psychiatry* **160,** 469–476.

Glatt, S. J., Faraone, S. V., and Tsuang, M. T. (2003b). CAG-repeat length in exon 1 of KCNN3 does not influence risk for schizophrenia or bipolar disorder: A meta-analysis of association studies. *Am. J. Med. Genet. B. Neuropsychiatr. Genet.* **121,** 14–20.

IMGSAC (1998). A full genome screen for autism with evidence for linkage to a region on chromosome 7q. *Hum. Mol. Genet.* **7,** 571–578.

IMGSAC (2001a). A genomewide screen for autism: Strong evidence for linkage to chromosomes 2q, 7q, and 16p. *Am. J. Hum. Genet.* **69,** 570–581.

IMGSAC (2001b). An autosomal genomic screen for autism. *Am. J. Med. Genet.* **105,** 609–615.

Kohn, Y., Danilovich, E., Filon, D., Oppenheim, A., Karni, O., Kanyas, K., Turetsky, N., Korner, M., and Lerer, B. (2004). Linkage disequilibrium in the DTNBP1 (dysbindin) gene region and on chromosome 1p36 among psychotic patients from a genetic isolate in Israel: Findings from identity by descent haplotype sharing analysis. *Am. J. Med. Genet. B. Neuropsychiatr. Genet.* **128,** 65–70.

Kury, S., Rubie, C., Moisan, J. P., and Stober, G. (2003). Mutation analysis of the zinc transporter gene SLC30A4 reveals no association with periodic catatonia on chromosome 15q15. *J. Neural. Transm.* **110,** 1329–1332.

Lamb, J. A., Barnby, G., Bonora, E., Sykes, N., Bacchelli, E., Blasi, F., Maestrini, E., Broxholme, J., Tzenova, J., Weeks, D., Bailey, A. J., and Monaco, A. P. (2005). Analysis of IMGSAC autism susceptibility loci: Evidence for sex limited and parent of origin specific effects. *J. Med. Genet.* **42,** 132–137.

Larsson, H. J., Eaton, W. W., Madsen, K. M., Vestergaard, M., Olesen, A. V., Agerbo, E., Schendel, D., Thorsen, P., and Mortensen, P. B. (2005). Risk factors for autism: Perinatal factors, parental psychiatric history, and socioeconomic status. *Am. J. Epidemiol.* **161,** 916–925.

Maestrini, E., Lai, C., Marlow, A., Matthews, N., Wallace, S., Bailey, A., Cook, E. H., Weeks, D. E., and Monaco, A. P. (1999). Serotonin transporter (5-HTT) and gamma-aminobutyric acid receptor subunit beta3 (GABRB3) gene polymorphisms are not associated with autism in the IMGSAC families. The International Molecular Genetic Study of Autism Consortium. *Am. J. Med. Genet.* **88,** 492–496.

Maziade, M., Roy, M., Chagnon, Y., Cliche, D., Fournier, J., Montgrain, N., Dion, C., Lavallee, J., Garneau, Y., Gingras, N., Nicole, L., Pires, L., *et al.* (2005). Shared and specific susceptibility loci for schizophrenia and bipolar disorder: A dense genome scan in Eastern Quebec families. *Mol. Psychiatry* **10,** 486–499.

McCauley, J. L., Olson, L. M., Delahanty, R., Amin, T., Nurmi, E. L., Organ, E. L., Jacobs, M. M., Folstein, S. E., Haines, J. L., and Sutcliffe, J. S. (2004). A linkage disequilibrium map of the 1 Mb 15q12 GABA(A) receptor subunit cluster and association to autism. *Am. J. Med. Genet. B. Neuropsychiatr. Genet.* **131,** 51–59.

McGrath, M., Kawachi, I., Ascherio, A., Colditz, G. A., Hunter, D. J., and De Vivo, I. (2004). Association between catechol-*O*-methyltransferase and phobic anxiety. *Am. J. Psychiatr.* **161,** 1703–1705.

McKeane, D. P., Meyer, J., Dobrin, S. E., Melmed, K. M., Ekawardhani, S., Tracy, N. A., Lesch, K. P., and Stephan, D. A. (2005). No causative DLL4 mutations in periodic catatonia patients from 15q15 linked families. *Schizophr. Res.* **75,** 1–3.

Menold, M. M., Shao, Y., Wolpert, C. M., Donnelly, S. L., Raiford, K. L., Martin, E. R., Ravan, S. A., Abramson, R. K., Wright, H. H., Delong, G. R., Cuccaro, M. L., Pericak-Vance, M. A., *et al.* (2001). Association analysis of chromosome 15 GABAA receptor subunit genes in autistic disorder. *J. Neurogenet.* **15,** 245–259.

Meyer, J., Ortega, G., Schraut, K., Nurnberg, G., Ruschendorf, F., Saar, K., Mossner, R., Wienker, T. F., Reis, A., Stober, G., and Lesch, K. P. (2002). Exclusion of the neuronal nicotinic acetylcholine receptor alpha7 subunit gene as a candidate for catatonic schizophrenia in a large family supporting the chromosome 15q13-22 locus. *Mol. Psychiatr.* **7,** 220-223.

Mirnics, K., Middleton, F. A., Marquez, A., Lewis, D. A., and Levitt, P. (2000). Molecular characterization of schizophrenia viewed by microarray analysis of gene expression in prefrontal cortex. *Neuron* **28,** 53-67.

Mizuno, Y., Puca, A. A., O'Brien, K. F., Beggs, A. H., and Kunkel, L. M. (2001a). Genomic organization and single-nucleotide polymorphism map of desmuslin, a novel intermediate filament protein on chromosome 15q26.3. *BMC Genet.* **2,** 8.

Mizuno, Y., Thompson, T. G., Guyon, J. R., Lidov, H. G., Brosius, M., Imamura, M., Ozawa, E., Watkins, S. C., and Kunkel, L. M. (2001b). Desmuslin, an intermediate filament protein that interacts with alpha-dystrobrevin and desmin. *Proc. Natl. Acad. Sci. USA* **98,** 6156-6161.

Nurmi, E. L., Bradford, Y., Chen, Y., Hall, J., Arnone, B., Gardiner, M. B., Hutcheson, H. B., Gilbert, J. R., Pericak-Vance, M. A., Copeland-Yates, S. A., Michaelis, R. C., Wassink, T. H., *et al.* (2001). Linkage disequilibrium at the Angelman syndrome gene UBE3A in autism families. *Genomics* **77,** 105-113.

Nurmi, E. L., Amin, T., Olson, L. M., Jacobs, M. M., McCauley, J. L., Lam, A. Y., Organ, E. L., Folstein, S. E., Haines, J. L., and Sutcliffe, J. S. (2003). Dense linkage disequilibrium mapping in the 15q11-q13 maternal expression domain yields evidence for association in autism. *Mol. Psychiatr.* **8,** 624-634570..

Otani, K., Ujike, H., Tanaka, Y., Morita, Y., Katsu, T., Nomura, A., Uchida, N., Hamamura, T., Fujiwara, Y., and Kuroda, S. (2005). The GABA type A receptor alpha5 subunit gene is associated with bipolar I disorder. *Neurosci. Lett.* **381,** 108-113.

Papadimitriou, G., Dikeos, D., Daskalopoulou, E., Karadima, G., Avramopoulos, D., Contis, C., and Stefanis, C. (2001a). Association between GABA-A receptor alpha 5 subunit gene locus and schizophrenia of a later age of onset. *Neuropsychobiology* **43,** 141-144.

Papadimitriou, G. N., Dikeos, D. G., Karadima, G., Avramopoulos, D., Daskalopoulou, E. G., Vassilopoulos, D., and Stefanis, C. N. (1998). Association between the GABA(A) receptor alpha5 subunit gene locus (GABRA5) and bipolar affective disorder. *Am. J. Med. Genet.* **81,** 73-80.

Papadimitriou, G. N., Dikeos, D. G., Karadima, G., Avramopoulos, D., Daskalopoulou, E. G., and Stefanis, C. N. (2001b). GABA-A receptor beta3 and alpha5 subunit gene cluster on chromosome 15q11-q13 and bipolar disorder: A genetic association study. *Am. J. Med. Genet.* **105,** 317-320.

Philippe, A., Martinez, M., Guilloud-Bataille, M., Gillberg, C., Rastam, M., Sponheim, E., Coleman, M., Zappella, M., Aschauer, H., Van Maldergem, L., Penet, C., Feingold, J., *et al.* (1999). Genome-wide scan for autism susceptibility genes. Paris Autism Research International Sibpair Study. *Hum. Mol. Genet.* **8,** 805-812.

Purcell, A. E., Jeon, O. H., Zimmerman, A. W., Blue, M. E., and Pevsner, J. (2001). Postmortem brain abnormalities of the glutamate neurotransmitter system in autism. *Neurology* **57,** 1618-1628.

Raybould, R., Green, E. K., MacGregor, S., Gordon-Smith, K., Heron, J., Hyde, S., Caesar, S., Nikolov, I., Williams, N., Jones, L., O'Donovan, M. C., Owen, M. J., *et al.* (2005). Bipolar disorder and polymorphisms in the dysbindin gene (DTNBP1). *Biol. Psychiatr.* **57,** 696-701.

Realmuto, G. M., and August, G. J. (1991). Catatonia in autistic disorder: A sign of comorbidity or variable expression? *J. Autism Dev. Disord.* **21,** 517-528.

Risch, N., Spiker, D., Lotspeich, L., Nouri, N., Hinds, D., Hallmayer, J., Kalaydjieva, L., McCague, P., Dimiceli, S., Pitts, T., Nguyen, L., Yang, J., *et al.* (1999). A genomic screen of autism: Evidence for a multilocus etiology. *Am. J. Hum. Genet.* **65,** 493-507.

Sazci, A., Ergul, E., Kucukali, I., Kilic, G., Kaya, G., and Kara, I. (2004). Catechol-*O*-methyltransferase gene Val108/158Met polymorphism, and susceptibility to schizophrenia: Association is more significant in women. *Brain. Res. Mol. Brain. Res.* **132,** 51-56.

Schwab, S. G., Knapp, M., Mondabon, S., Hallmayer, J., Borrmann-Hassenbach, M., Albus, M., Lerer, B., Rietschel, M., Trixler, M., Maier, W., and Wildenauer, D. B. (2003). Support for association of schizophrenia with genetic variation in the 6p22.3 gene, dysbindin, in sib-pair families with linkage and in an additional sample of triad families. *Am. J. Hum. Genet.* **72,** 185–190.

Shao, Y., Cuccaro, M. L., Hauser, E. R., Raiford, K. L., Menold, M. M., Wolpert, C. M., Ravan, S. A., Elston, L., Decena, K., Donnelly, S. L., Abramson, R. K., Wright, H. H., *et al.* (2003). Fine mapping of autistic disorder to chromosome 15q11-q13 by use of phenotypic subtypes. *Am. J. Hum. Genet.* **72,** 539–548.

Shifman, S., Bronstein, M., Sternfeld, M., Pisante, A., Weizman, A., Reznik, I., Spivak, B., Grisaru, N., Karp, L., Schiffer, R., Kotler, M., Strous, R. D., *et al.* (2004). COMT: A common susceptibility gene in bipolar disorder and schizophrenia. *Am. J. Med. Genet. B. Neuropsychiatr. Genet.* **128,** 61–64.

Smith, M., Filipek, P. A., Wu, C., Bocian, M., Hakim, S., Modahl, C., and Spence, M. A. (2000). Analysis of a 1-megabase deletion in 15q22-q23 in an autistic patient: Identification of candidate genes for autism and of homologous DNA segments in 15q22-q23 and 15q11-q13. *Am. J. Med. Genet.* **96,** 765–770.

Spence, S. J. (2004). The genetics of autism. *Semin. Pediatr. Neurol.* **11,** 196–204.

Stöber, G., Franzek, E., Lesch, K. P., and Beckmann, H. (1995). Periodic catatonia: A schizophrenic subtype with major gene effect and anticipation. *Eur. Arch. Psychiatry Clin. Neurosci.* **245,** 135–141.

Stöber, G., Meyer, J., Nanda, I., Wienker, T. F., Saar, K., Jatzke, S., Schmid, M., Lesch, K. P., and Beckmann, H. (2000a). hKCNN3 which maps to chromosome 1q21 is not the causative gene in periodic catatonia, a familial subtype of schizophrenia. *Eur. Arch. Psychiatry Clin. Neurosci.* **250,** 163–168.

Stöber, G., Saar, K., Ruschendorf, F., Meyer, J., Nurnberg, G., Jatzke, S., Franzek, E., Reis, A., Lesch, K. P., Wienker, T. F., and Beckmann, H. (2000b). Splitting schizophrenia: Periodic catatonia-susceptibility locus on chromosome 15q15. *Am. J. Hum. Genet.* **67,** 1201–1207.

Stöber, G., Seelow, D., Ruschendorf, F., Ekici, A., Beckmann, H., and Reis, A. (2002). Periodic catatonia: Confirmation of linkage to chromosome 15 and further evidence for genetic heterogeneity. *Hum. Genet.* **111,** 323–330.

Stone, W. S., Faraone, S. V., Su, J., Tarbox, S. I., Van Eerdewegh, P., and Tsuang, M. T. (2004). Evidence for linkage between regulatory enzymes in glycolysis and schizophrenia in a multiplex sample. *Am. J. Med. Genet. B. Neuropsychiatr. Genet.* **127,** 5–10.

Straub, R. E., Jiang, Y., MacLean, C. J., Ma, Y., Webb, B. T., Myakishev, M. V., Harris-Kerr, C., Wormley, B., Sadek, H., Kadambi, B., Cesare, A. J., Gibberman, A., *et al.* (2002). Genetic variation in the 6p22.3 gene DTNBP1, the human ortholog on the mouse dysbindin gene, is associated with schizophrenia. *Am. J. Hum. Genet.* **71,** 337–348.

Sugiura, N., Patel, R. G., and Corriveau, R. A. (2001). N-methyl-D-aspartate receptors regulate a group of transiently expressed genes in the developing brain. *J. Biol. Chem.* **276,** 14,257–14,263.

Van Den Bogaert, A., Schumacher, J., Schulze, T. G., Otte, A. C., Ohlraun, S., Kovalenko, S., Becker, T., Freudenberg, J., Jonsson, E. G., Mattila-Evenden, M., Sedvall, G. C., Czerski, P. M., *et al.* (2003). The DTNBP1 (dysbindin) gene contributes to schizophrenia, depending on family history of the disease. *Am. J. Hum. Genet.* **73,** 1438–1443.

van den Oord, E. J., Sullivan, P. F., Jiang, Y., Walsh, D., O'Neill, F. A., Kendler, K. S., and Riley, B. P. (2003). Identification of a high-risk haplotype for the dystrobrevin binding protein 1 (DTNBP1) gene in the Irish study of high-density schizophrenia families. *Mol. Psychiatr.* **8,** 499–510.

Veenstra-VanderWeele, J., Gonen, D., Leventhal, B. L., and Cook, E. H., Jr. (1999). Mutation screening of the UBE3A/E6-AP gene in autistic disorder. *Mol. Psychiatr.* **4,** 64–67.

Wing, L., and Shah, A. (2000). Catatonia in autistic spectrum disorders. *Br. J. Psychiatry* **176,** 357–362.

Wonodi, I., Stine, O. C., Mitchell, B. D., Buchanan, R. W., and Thaker, G. K. (2003). Association between Val108/158 Met polymorphism of the COMT gene and schizophrenia. *Am. J. Med. Genet. B. Neuropsychiatr. Genet.* **120,** 47–50.

Yirmiya, N., Pilowsky, T., Nemanov, L., Arbelle, S., Feinsilver, T., Fried, I., and Ebstein, R. P. (2001). Evidence for an association with the serotonin transporter promoter region polymorphism and autism. *Am. J. Med. Genet.* **105,** 381–386.

Ylisaukko-Oja, T., Nieminen-von Wendt, T., Kempas, E., Sarenius, S., Varilo, T., von Wendt, L., Peltonen, L., and Jarvela, I. (2004). Genome-wide scan for loci of Asperger syndrome. *Mol. Psychiatr.* **9,** 161–168.

SECTION IV
TREATMENT

CURRENT TRENDS IN BEHAVIORAL INTERVENTIONS FOR CHILDREN WITH AUTISM

Dorothy Scattone* and Kimberly R. Knight[†]

*Department of Pediatrics, Division of Child Development and Behavioral Medicine
The University of Mississippi Medical Center, Jackson, Mississippi 39216, USA
[†]Department of Pediatrics, Mississippi Child Development Institute
The University of Mississippi Medical Center, Jackson, Mississippi 39216, USA

I. Current Trends in Behavioral Interventions for Children with Autism in the United States
II. Language
   A. Sign Language and the Picture Exchange Communication System
   B. Assesment of Basic Language and Learning Skills
III. Social Skills
   A. Pivotal Response Training
   B. Priming
   C. Social Stories
IV. Self-Management
V. Conclusions
   References

This chapter focuses on behavioral interventions involving skill acquisition for children and adolescents with autism over the last 12 years. Literature and outcome data are reviewed with respect to three categories: language, social skills, and self-management. Generalization of results is somewhat problematic, as all of the interventions discussed consist of relatively small sample sizes and utilize single case design methodology. However, expansion and refinement of research methodology over the last decade, with more researchers replicating previous studies, may make broader application of research findings possible.

I. Current Trends in Behavioral Interventions for Children with Autism in the United States

Over the past decade, behavioral treatments have been widely accepted in the United States and are often preferred methods in both private clinics and school systems. Many of these treatments focus on skill acquisition, specifically in language (Sundberg and Partington, 1998) and social skills development (Pierce and Schreibman, 1995). Their objective is to reduce aggression, self-injury, and other maladaptive behaviors while teaching appropriate replacement behaviors (Scattone *et al.*, 2002). Some of these procedures require lengthy training and the presence of an expert, making them somewhat difficult to implement. Others promote independence and can be implemented relatively easily by a teacher or parent at greater cost efficiencies (Pierce and Schreibman, 1997; Schreibman and Whalen, 2000).

Over the last 10–12 years, interventions in these areas have become more sophisticated, incorporating the use of video technology, computers, and naturalistic procedures. Some of these methods have been shown to generalize across settings, or with other individuals not directly participating in the study, as well as increasing independence for the individual with autism. Most of this research involves a single case study or a small sample consisting of only three or four individuals, making it difficult to project conclusions to the population at large. However, as an increasing number of researchers replicate these studies, it will likely be possible to apply findings more broadly.

This chapter reviews behavioral interventions focused on skill acquisition. The published studies discussed are divided into three categories: language, social skills, and self-management. All studies were published between 1993 and 2004 and involve only children who were diagnosed with an autism spectrum disorder (ASD).

II. Language

A. Sign Language and the Picture Exchange Communication System

The development of language and communication skills is a complex area of study and the subject of a growing body of research. Some of this research has focused on the use of sign language and pictures to promote functional communication and increased vocalizations (Charlop-Christy *et al.*, 2002). The success of sign language has been well documented with nonvocal individuals, including children with mental retardation (Partington *et al.*, 1994) and, most recently, with children suffering from autism (Bartman and Freeman, 2003; Tincani, 2004). Some researchers suggest that signs may serve as a prompt and can facilitate speech, especially if the signs are iconic (Sundberg and Partington, 1998). For example, the sign for "ball" suggests placing hands into the shape of a ball and looks similar to the item it represents; the sign for car pantomimes hands on a steering wheel.

Signs represent a topography-based communication system that is similar to speech (Michael, 1985); each motor movement differs as does each spoken word. Some researchers suggest that sign language programs are often most successful when first signs taught are requests for a child's preferred items and activities (e.g., juice, toys, cookies, tickle, swing) and often fail when first signs taught are somewhat abstract (i.e., more, please, thank you) or are simply labels of items (Sundberg and Partington, 1998). This may be primarily due to motivational factors. If the motivation for juice is strong, it may be easier to teach the child to sign "juice" under conditions of motivation rather than to simply label "juice" when the child does not want a drink.

There are many advantages of using signs as an augmentative communication system, including possible improvement of speech and development of motor imitation. If motor imitation is strong, signs may be easy to acquire. Among the disadvantages are the need for specialized training, assistance by another individual to shape each sign, and extensive effort on the part of the listener, who must use signs to communicate as well (Sundberg and Partington, 1998).

An alternative method, the Picture Exchange Communication System (PECS), represents a selection-based communication system and was developed to teach nonvocal individuals to communicate by exchanging picture icons with a communication partner (Bondy and Frost, 2002). The pictures are typically kept in a book and may either be handed to a partner or secured with Velcro onto a board and pointed to. Pictures may be arranged to form sentences. For example, the icon for "I want" may be placed next to the icon for "toy."

The picture system facilitates ease of communication between the "listener" or teacher and the "speaker" or child with autism. Also, pictures only require simple matching initially, the response form is the same for each communication attempt (i.e., scanning and pointing), and pictures are portable and low cost. However, the picture system is problematic in two respects. It is necessary to carry the pictures at all times for communication to occur, and as the complexity of communication increases, the pictures often become more abstract (Sundberg and Partington, 1998; Tincani, 2004). For example, the icon for "beautiful" is often a picture of someone signing the word beautiful.

Deciding which method to select for a particular child may be difficult since both methods have been shown effective in producing independent communication (Charlop-Christy *et al.*, 2002; Sundberg and Sundberg, 1990). For children with autism, the goal is often speech acquisition. Thus, research has begun to compare these two methods in terms of ease of acquisition and facilitation of speech. For example, Tincani (2004) compared PECS with sign language training on the ease of acquisition of requesting preferred items and activities for two children between the ages of 5 and 6, one with autism and the other with pervasive developmental disorder—not otherwise specified. Both children could imitate some words when prompted but neither communicated functionally using speech.

After a stimulus preference assessment was conducted and a list of 10–12 preferred items (e.g., edibles, drinks, toys) was compiled for each child, sign language and PECS training were conducted. Sign language training was conducted in the manner described by Sundberg and Partington (1998). The experimenter held up an item and named the item while simultaneously signing. The trainer sat behind the child and shaped the child's hands to form the sign. The child was given access to the item even if it was not independently signed for. Eventually, the time between holding up the item, and the vocal word, and signed prompt were delayed several seconds in an effort to encourage spontaneous communication and to avoid prompt dependency. Only reinforcing items were used to teach signs; an item was determined reinforcing if the child reached for it when held up. If the child did not reach for the item, another item on the list was selected.

PECS training was conducted in a manner similar to the procedure for signs and as stated in the manual (Bondy and Frost, 2002). An adult sat in front of the child and held up a reinforcing item. A second adult, sitting behind the child, physically prompted the child to exchange the card for the item. Gradually, the distances between the communication partner and the child, and the communication book and the child, were increased. The second adult physically prompted the child to go to the book and hand the card to the adult. Finally, the child was taught to select a picture from an array.

The best treatment was determined by alternating the treatments, exposing each participant to PECS and signs. For one participant, PECS was easier to acquire and produced more independent requests for preferred items and activities, whereas signs produced more independent requests for the other participant. For the participant using PECS, a decrease in vocalizations was observed, which coincided with exchanging pictures. However, when a reinforcement procedure was added to the PECS, vocalizations increased for this participant. Overall, signs produced more vocalizations for both participants, suggesting that signs may have functioned as a prompt for spoken words (Tincani, 2004).

B. Assesment of Basic Language and Learning Skills

The assessment of basic language and learning skills (ABLLS) is a curriculum-based language intervention and skills tracking system, which is designed to identify language and other critical skills requiring intervention (Partington and Sundberg, 1998). The ABLLS can aid in the development of individual educational plans for children with autism by determining a starting point for intervention, providing specific skills to teach, and the order in which they should be taught. The ABLLS focuses on 25 content areas, including cooperation, motor imitation, receptive language, requesting, labeling, group instruction, grooming, academics, visual performance, and so on, and breaks up each content area into small step-by-step instruction. With the ABLLS, the acquisition of new skills can be visually displayed on the tracking system. One can determine the starting point based on the skills the child has already acquired and progress in a systematic manner, eliminating the need for guesswork in developing a plan for intervention (Partington and Sundberg, 1998).

III. Social Skills

A. Pivotal Response Training

Although children with autism often demonstrate profound impairments in social interaction, many interventions have been developed that target this core area, including pivotal response training (PRT), Priming, and Social Stories. PRT is a naturalistic teaching strategy that focuses on using the child's motivation to teach complex behaviors such as language (Koegel et al., 1987a), symbolic play (Stahmer, 1995), and social responding (Pierce and Schreibman, 1995). Koegel and Frea (1993) suggest that once these pivotal behaviors are affected, other collateral behavior changes may occur as well. This procedure is considered "naturalistic" in that it usually occurs in the natural environment (e.g., a playground), utilizes shared control, and contains multiple exemplars (Pierce and Schreibman, 1997). According to Koegel et al. (1987b), motivation is critical to learning these complex skills; without motivation, learning and generalization may not occur. Motivation may be enhanced by having the target child select activities, varying the tasks, reinforcing language use, and interspersing easy and difficult demands (Pierce and Schreibman, 1997).

Pierce and Schreibman (1997) used PRT with typical peers in an effort to teach two children with autism to engage in a variety of social behaviors, including initiating play and conversations. Both participants were largely unresponsive in social situations and engaged in a variety of inappropriate social behaviors (e.g., echolalic speech, repetitive noises). The teacher selected four peers, three to serve as peer tutors and one untrained peer to assess generalizability.

Peers were instructed to use the following strategies during free-play activities: (1) ensure that the target child is attending before delivering a prompt, (2) let the target child choose activities to ensure motivation, (3) vary toys frequently, (4) model appropriate social behavior by giving verbal statements (e.g., "This game is fun"), (5) verbally reinforce conversational attempts, (6) encourage conversation, (7) narrate play, (8) extend conversations by asking questions, (9) model turn taking, and (10) teach responsivity to multiple cues by commenting about object properties, for example, saying "big blue ball" when playing with a blue ball (Pierce and Schreibman, 1997). During PRT, toys were placed in the middle of the classroom floor, and the children were instructed to play together. A trained professional provided feedback to peers and teacher training.

The intervention was highly effective in increasing spoken words for the participants, and with peers not included in the study, generalized conversation skills occurred. Further, the quality of each target student's vocalizations increased. For example, both children produced longer sentences and more frequent verbal interactions. However, the intervention required extensive time to train peers (i.e., 2 months) and the ongoing presence of an expert (Pierce and Schreibman, 1997).

B. Priming

Priming is an antecedent manipulation that is low in demand, provides a high level of reinforcement, and conducted just before an activity that is difficult or challenging for the child (Zanolli *et al.*, 1996). This procedure was applied to a social-skills training methodology by teaching two preschool children with autism to make spontaneous initiations (Zanolli *et al.*, 1996). Initially, typical peers were trained to initiate comments and hand preferred items to target students. After peer training, 5-min activity sessions were conducted in the classroom with only one peer and the participant. At the start of the session, the teacher asked the children to greet each other and play. Peers provided brief access to a reinforcer each time a target student initiated (e.g., looking, smiling, touching) with them.

A multiple baseline design across activities was employed to assess the level of social initiations. Unprompted initiations to peers increased for both participants after the priming procedure was implemented. A variety of initiations not directly taught were observed as well. Initially, teacher prompting was often necessary for delivery of the reinforcer to a target student. However, near the conclusion of the study, peers were responding independently. Response generalization occurred during a group-activity phase, although generalization across activities was not observed.

Priming was recently examined using videotaped instruction to reduce disruptive transition behavior (e.g., biting, pinching, screaming, hitting) of three children between the ages of 3 and 6 (Schreibman and Whalen, 2000). Priming consisted of previewing a 1–4 min videotaped presentation of the target activity before it was to take place. Routes in a mall or store were problem areas depicted in the videos for two of the participants, while the routine before leaving home was depicted for the third. The routes for the public settings ended with a store or store department that was reinforcing to the participant (e.g., toy department) because the children often tantrummed immediately at the entrance and continued until they reached preferred areas, making shopping difficult if not impossible.

A multiple baseline design was employed to examine effects of video priming on tantrumous behavior. A reduction in disruptive behavior was immediately observed for one participant, and a gradual reduction occurred for the other two participants after previewing the videotape. In addition, generalization occurred for two of the participants in untreated settings. A benefit of using a video priming methodology, such as the one described here, is the short preparation time and absence of lengthy training sessions, making its implementation a potentially good choice for parents or teachers.

C. SOCIAL STORIES

Social Stories are short individualized stories designed to teach children with autism a certain skill, event, concept, or social behavior via a written format (Gray, 1998, 2000). They are most often developed by parents or teachers and written in first person formats, as if the children themselves are describing the situations. Social stories describe what is happening and why, who will participate, and when an event or activity will occur. The stories further provide the appropriate responses individuals are expected to make. Social stories consist of a predetermined ratio of sentences that describe, direct, and provide the perspective of others, with the goal being description over direction. Specific procedures for writing social stories are outlined in Gray's book (2000).

Social stories have been used for a variety of purposes, including sharing, greetings (Swaggert *et al.*, 1995), reducing tantrums (Kuttler *et al.*, 1998), social initiations (Norris and Dattilo, 1999; Scattone *et al.*, 2005), following directions, and reducing perseverative speech and inappropriate behaviors (Brownell, 2002; Scattone *et al.*, 2002).

The first published study on social stories was undertaken to improve greeting behavior for an 11-year-old girl and play skills for two 7-year-old boys with autism (Swaggert *et al.*, 1995). Each social story was five pages long and contained picture icons of target behaviors and individuals mentioned in the story combined with a social-skills training methodology or a response-cost system. Each story was individualized to fit that particular student's specific situation and was read to them just before the activity. In addition, one of the participants had the opportunity to earn a reinforcer (i.e., cola) contingent on appropriate behavior. An AB design was employed with each participant, and a reduction in undesirable behaviors occurred for all participants after exposure to their story.

The social stories intervention was also examined when used in isolation (i.e., without verbal or physical prompts, pictures, or token economies). Scattone *et al.* (2002) developed social stories to reduce disruptive behaviors of three students with autism between the ages of 8 and 12. The students were all capable of speech communication, and two were able to read. Target behaviors included chair tipping, shouting, and staring assessed via a multiple baseline design across participants. Each participant either read or was read his social story just prior to data collection. Decreases in disruptive behaviors occurred for two of the three participants when the intervention was implemented with chair tipping resulting in an immediate decrease.

Reading was a preferred activity for the participant who tipped his chair, and he was often observed reciting the rules of his story to other members of the classroom. This observation suggests that reading ability, and reading as a preferred activity in general, may be beneficial when using this intervention. In addition, social stories may serve to establish rule-governed behavior for some children by incorporating rules into their repertoire and having them apply those rules to a given social situation.

Social stories were combined with video feedback to improve the social interactions of five students with autism and their typical peers (Thiemann and Goldstein, 2001). The students were between the ages of 6 and 12 and had basic functional communication skills. A multiple baseline design was employed across social communication skills (securing attention, initiating comments, initiating requests, contingent responses). Four social stories with picture prompts were developed, one for each skill. The target child read one of the stories just before the social interaction, and comprehension was assessed immediately after via a series of predetermined questions. Peers provided assistance during this process until the participant reached 80% accuracy in comprehension.

After holding short social activity sessions, videotaped feedback was provided to the group. The video was played back, and corrective feedback and discussion on target skills were provided at various parts in the video. All participants demonstrated improved social interactions compared to baseline performance. However, behaviors were not maintained for three of the five participants after the social story was discontinued.

Social stories have also been developed to facilitate social initiations to typical peers, without the addition of picture prompts or extensive adult supervision. For example, Scattone *et al.* (in press) investigated the effectiveness of social stories on increasing appropriate social interactions in three boys with autism between the ages of 8 and 12. All the boys had functional language but were not observed to either initiate or respond much to their peers during free time activities. Three individual social stories, compiled in a book-like format, were developed to provide each participant with the appropriate social initiations and responses he was expected to make during that activity.

A multiple baseline design across participants was used to assess changes in social interaction skills during free time activities. Immediate increases in social initiations were observed for only one of the participants, and modest improvements were noted for another. These findings suggest that additional components (e.g., token economy, picture prompts, verbal prompts, and so on) may be necessary when using social stories to enhance social initiations. Although findings for the social stories intervention are promising, further research in this area is necessary to determine who might best be served by this intervention and under what conditions.

IV. Self-Management

Self-management is an effective treatment shown to produce rapid behavior change while increasing independence and decreasing reliance on adult supervision (Koegel et al., 1992; Pierce and Schreibman, 1994). Self-management requires individuals to self-monitor and self-reinforce their own behavior in the absence of a treatment provider (Stahmer and Schreibman, 1992). Typically, individuals record their behavior with a wrist counter, checklist, or with tokens. Rewards are then provided for a predetermined number of responses recorded on the wrist counter or for a certain number of tokens earned. Self-management techniques have been used to teach a variety of skills, including daily living skills (Pierce and Schreibman, 1994), play (Stahmer and Schreibman, 1992), and social skills, as well as to reduce disruptive behavior and to promote generalization across settings (Koegel et al., 1992; Newman et al., 1995; Pierce and Schreibman, 1994).

For example, Macduff et al. (1993) used photographic activity schedules to teach four boys with autism to follow a variety of home living and leisure activities without adult supervision. The participants were between the ages of 9 and 14, each with severe language impairments and a history of disruptive and aggressive behaviors. The participants were highly dependent on verbal prompts (i.e., 1–2/min) to complete these activities and move on to activities in other settings. Their schedules consisted of six photographs, each depicting an activity and secured in a three-ring binder. A multiple baseline design was employed across participants to examine the impact the activity schedule had on remaining on-task and on-schedule.

At the conclusion of the study, all four participants were able to complete a complex chain of home living and leisure activities for at least an hour without adult prompts, suggesting that a transfer of stimulus control from adult prompts to the picture schedule occurred. In addition, generalized responding was evidenced, as the boys were able to complete activities in a variety of sequences and different locations. No resistance to using the schedule occurred, and an anecdotal decrease in disruptive behavior was noted.

The previous study was expanded by training parents to implement picture schedules at home to teach their children to complete home living and leisure tasks (Krantz *et al.*, 1993). Three children between the ages of 6 and 8, all with autism, and their parents participated in the study. The children were already using picture schedules in a clinic setting; thus, the activities chosen were largely identical to the activities they engaged in at the clinic. The home schedules were in place over a 4–5-hour period, from after school until after dinner. A multiple baseline design across participants was selected to measure the effect of the schedules on engagement with the activity, disruptive behaviors (e.g., aggression, property destruction, tantrums), and social initiations.

Increased engagement in tasks and social initiations and decreased disruptive behaviors, occurred for all participants and were maintained for 2–10 months thereafter. However, parent training time was lengthy, ranging from 53 to 90 hours per parent across an average of 20 sessions.

Self-management procedures of teenagers with autism have also been investigated by teaching these teenagers to use a written schedule to independently transition from one activity to another in a classroom setting (Newman *et al.*, 1995). The three participants, ranging in age from 14 to 17 years, functioned between a third and fifth grade level in reading and math. The schedule included the activities and their start times, written in an order on a large sheet of oak tag and posted on the blackboard. To assist with transitions, a digital clock was placed in the front of the classroom near the schedule.

A self-reinforcement component was part of this procedure; it required the students to place seven tokens in their back pockets at the beginning of the day. The participants were then instructed to remove one token at the start of each new activity and put it in their front pockets. At the end of the day, front pocket tokens were exchanged for preferred activities. A multiple baseline design was selected to examine the effect of the schedule on correctly identifying transition times and accuracy of self-reinforcement.

At the conclusion of the study, all three students were able to accurately transition from one activity to another without supervision. However, accuracy of transferring tokens was variable, and students sometimes forget to transfer tokens from one pocket to another. While changes were maintained over a 1-month follow-up period, generalization to other activities in other settings was not assessed.

A self-management procedure has also been combined with video modeling to teach high-functioning children with autism how to give compliments to their peers (Apple *et al.*, 2005). Two experiments were conducted, the first using video modeling alone and second combining video modeling with a self-management procedure. The rules for how and when to give compliments were explained within the 1-min video segments, making additional instruction after watching the video unnecessary. A multiple baseline design across participants was employed. The findings suggested that video modeling was effective for response type compliments (e.g., Oh yeah!) but not for initiated compliments (e.g., Neat airplane!). Subsequently, additional reinforcement contingencies were added, which then produced an increase in initiated compliments. However, this procedure required extensive ongoing supervision from an adult.

Experiment two added a self-management procedure that required two of the participants to use a wrist counter to record initiated compliments and a third participant to record initiations with a checklist. Immediate and independent compliment-giving initiations were observed for all participants with the addition of the self-management procedure, and adult supervision was unnecessary.

V. Conclusions

Although autism is a severe and lifelong disability, skill remediation that targets communication and social behavior can help facilitate independence and may positively affect long-term prognosis. It is often recommended to begin intervention as early as possible, usually in toddler to preschool years, especially in the area of language acquisition. The trend in behavioral research appears to be moving in the direction of more naturalistic teaching methods that can be carried out both at home and school. While some of these interventions are complex and require the presence of an expert, many others, such as video priming and social stories, may be implemented by teachers and parents without much difficulty. Other interventions, including self-management, may help teach children with autism to monitor their behavior in the absence of an adult, thereby facilitating independence and encouraging self-reliance.

References

Apple, A. L., Billingsley, F., and Schwartz, I. S. (2005). Effects of video modeling alone and with self-management on compliment-giving behaviors of children with high functioning ASD. *J. Posit. Behav. Interv.* **7,** 33–46.

Bartman, S., and Freeman, N. (2003). Teaching language to a two-year-old with autism. *J. Dev. Disabil.* **10,** 47–53.

Bondy, A., and Frost, L. (2002). "The picture exchange communication system." Pyramid Educational Products, Newark, DE.

Brownell, M. D. (2002). Musically adapted Social Stories to modify behaviors in students with autism: Four case studies. *J. Music Ther.* **2,** 117–144.

Charlop-Christy, M. H., Carpenter, M., LeBanc, L. A., and Kellet, K. (2002). Using the picture exchange communication system (PECS) with children with autism. Assessment of PECS acquisition, speech, social-communicative behavior, and problem behavior. *J. Appl. Behav. Anal.* **35,** 213–231.

Gray, C. (1998). Social stories 101. *In* "The Morning News," Vol. 10, no. 1, pp. 2–6. Jenison Public Schools, Michigan.

Gray, C. (2000). "The New Social Stories Book." Future Horizons, Arlington, TX.

Koegel, L. K., Koegel, R. L., Hurley, C., and Frea, W. D. (1992). Improving social skills and disruptive behavior in children with autism through self-management. *J. Appl. Behav. Anal.* **25,** 341–353.

Koegel, R. L., and Frea, W. D. (1993). Treatment of social behavior in autism through modification of pivotal social skills. *J. Appl. Behav. Anal.* **26,** 369–377.

Koegel, R. L., Dyer, K., and Bell, L. K. (1987a). The influence of child preferred activities on autistic children's social behavior. *J. Appl. Behav. Anal.* **17,** 243–252.

Koegel, R. L., O'Dell, M. C., and Koegel, L. K. (1987b). A natural language teaching paradigm for teaching nonverbal autistic children. *J. Autism Dev. Disord.* **12,** 207–218.

Krantz, P. J., MacDuff, M. T., and McClannahan, L. E. (1993). Programming participation in family activities for children with autism: Parents' use of photographic activity schedules. *J. Appl. Behav. Anal.* **26,** 137–138.

Kuttler, S., Myles, B. S., and Carlson, J. K. (1998). The use of social stories to reduce precursors to tantrum behavior in a student with autism. *Focus Autism Dev. Disord.* **13,** 176–182.

MacDuff, G. S., Krantz, P. J., and McClannahan, L. E. (1993). Teaching children with autism to use photographic activity schedules: Maintenance and generalization of complex response chains. *J. Appl. Behav. Anal.* **26,** 89–97.

Michael, J. (1985). Two kinds of verbal behavior plus a possible third. *Anal. Verbal Behav.* **3,** 1–4.

Newman, B., Buffington, D. M., O'Grady, M. A., McDonald, M. E., Poulson, C. L., and Hemmes, N. S. (1995). Self-management of schedule following in three teenagers with autism. *Behav. Disord.* **20**(3), 190–196.

Norris, C., and Dattilo, J. (1999). Evaluating effects of a social story intervention on a young girl with autism. *Focus Autism Dev. Disabil.* **14,** 180–186.

Partington, J. W., Sundberg, M. L., Newhouse, L., and Spengler, S. M. (1994). Overcoming an autistic child's failure to acquire a tact repertoire. *J. Appl. Behav. Anal.* **27,** 733–734.

Partington, J. W., and Sundberg, M. L. (1998). "The assessment of basic language and learning skills." Behavior Analysts, Danville, CA.

Pierce, K. L., and Schreibman, L. (1994). Teaching daily living skills to children with autism in unsupervised settings through pictorial self-management. *J. Appl. Behav. Anal.* **27,** 471–481.

Pierce, K. L., and Schreibman, L. (1995). Increasing complex social behaviors in children with autism: Effects of Peer-implemented pivotal response training. *J. Appl. Behav. Anal.* **28,** 285–295.

Pierce, K. L., and Schreibman, L. (1997). Using peer trainers to promote social behavior in autism: Are they effective at enhancing multiple social modalities? *Focus Autism Develop. Disabil.* **12,** 207–218.

Scattone, D., Wilczynski, S. M., Edwards, R. P., and Rabian, B. (2002). Decreasing disruptive behaviors of children with autism using social stories. *J. Autism Dev. Disord.* **32,** 535–542.

Scattone, D., Tingstrom, D., and Wilczynski, S. M. (in press). Increasing appropriate social interactions of children with autism spectrum disorders using social stories (manuscript under review).

Schreibman, L., and Whalen, L. (2000). The use of video priming to reduce disruptive transition behavior in children with autism. *J. Posit. Behav. Interv.* **2,** 3–14.

Stahmer, A. C. (1995). Teaching symbolic play to children with autism using pivotal response training. *J. Autism Dev. Disord.* **25,** 123–141.

Stahmer, A. C., and Schreibman, L. (1992). Teaching children with autism appropriate play in unsupervised environments using a self-management treatment package. *J. Appl. Behav. Anal.* **25,** 447–459.

Sundberg, M. L., and Partington, J. W. (1998). "Teaching language to children with autism and other developmental disabilities." Behavior Analysts, Danville, CA.

Sundberg, C. T., and Sundberg, M. L. (1990). Comparing topography-based verbal behavior with stimulus selection-based verbal behavior. *Anal. Verbal Behav.* **8,** 31–42.

Swaggert, B. L., Gagnon, E., Bock, S. J., Earles, T. L., Quinn, C., Myles, B. S., and Simpson, R. L. (1995). Using social stories to teach social and behavioral skills to children with autism. *Focus Autistic Behav.* **10,** 1–16.

Thiemann, K. S., and Goldstein, H. (2001). Social Stories, written text cues, and video feedback: Effects on social communication of children with autism. *J. Appl. Behav. Anal.* **34,** 425–446.

Tincani, M. (2004). Comparing the picture exchange communication system and sign language training for children with autism. *Focus Autism Dev. Disabil.* **19,** 152–163.

Zanolli, K., Daggett, J., and Adams, T. (1996). Teaching preschool age autistic children to make spontaneous initiations to peers using priming. *J. Autism Dev. Disord.* **26,** 407–422.

CASE REPORTS WITH A CHILD PSYCHIATRIC EXPLORATION OF CATATONIA, AUTISM, AND DELIRIUM

Jan N. M. Schieveld

Department of Psychiatry and Neuropsychology, University Hospital Maastricht
The Netherlands

I. Introduction
II. Case Histories
   A. Case History 1
   B. Case History 2
III. Methods
IV. Results
V. Discussion
VI. Conclusions
VII. Epicrisis
   References

"... I have not a body, I am one."
Rudolph Nurejev

This chapter starts with some remarks on the conceptual history of catatonia, which begins with Kahlbaum and continues with Kraeplin, Bleuler, and Leonhard. The Diagnostic and Statistical Manual, 4th ed., Text Revision, criteria for catatonia and the multicausal origin of the disorder are discussed. So, not only schizophrenia and mood disorders associated with catatonia, which is the primary form, are introduced but also an extensive list of somatic disorders—resulting in secondary catatonia—along with the work of Gelenberg and Wing. Next, two very difficult cases, of boys with autism, catatonia, and one of them with mental retardation as well, are presented. Major textbooks, PubMed, and Medline were used for a select literature search. The results show the main and really relevant but scarce data concerning primary and secondary catatonia. In the discussion the topics are this dearth in knowledge, the concept of catatonia and its similarities with delirium, and the relation catatonia–autism, and where to find the data. The conclusions summarize the main points and end with a gentle reminder, or is it an appeal?

I. Introduction

The noun catatonia is a Greek neologism derived from, *cata* (down, overall) + *teinein* (tense), meaning pervasive (motor) tension, and was coined in 1874 by

Kahlbaum, when he published his monograph *Die Katatonie oder das Spannungsirresein* (Catatonia or the tension madness). Here he described for the first time, in a clear fashion, the combination of alternating motor abnormalities, psychiatric symptoms, and different outcomes. He compared catatonia with the "general paralysis of the insane" (i.e., dementia paralytica), and reported on his findings at autopsy. Then there were always neuropathological changes, and he noticed a close comorbid correlation with tuberculosis. Kraepelin (1899), as well as Bleuler (1911) recognized the clinical description and adopted the name in their textbooks in relation with the schizophrenic psychosis. Leonhard (1979) published in the second half of the last century extensively on the "psychomotor psychosis," distinguishing three types: motility psychosis, periodic catatonia, and catatonic schizophrenia. He was also the first who published extensively on catatonia in childhood (Leonhard, 1960). The Diagnostic and Statistical Manual, 4th ed., Text Revision (DSM-TR-IV) (APA, 2000) defines the essential feature of catatonia as a marked psychomotor disturbance in which the clinical picture is dominated by at least two of the following five psychopathological signs: (1) motor immobility, (2) excess motor activity, (3) extreme negativism or mutism, (4) peculiarities of voluntary movement, and (5) echolalia or echopraxia.

Major textbooks invariably claim that catatonia in adults is uncommon (Sadock and Sadock, 2005). The DSM states that catatonia occurs in at least 5% of adult inpatient samples. However, higher rates have been reported in recent studies. For example, van der Heijden (2004) and van der Heijden *et al.* (2005) studied a large sample ($N = 19,309$) of schizophrenics. Although the diagnosis of catatonic schizophrenia dropped from 7.8% in 1980–1989 to 1.3% in 1990–2001, a possible underdiagnosis of catatonic schizophrenia was found in an independent sample. In a consecutive sample of patients admitted with psychosis, their application of a systematic catatonia rating scale showed that 18% fulfilled criteria for catatonia.

About 25–50% of catatonia cases occur in association with mood disorders, 10% in association with schizophrenia, and the rest in association with other mental disorders, these are the primary forms. However, catatonia can also occur in a general medical condition, especially in various metabolic, neurological, or infectious diseases (Gelenberg, 1976), and these are the secondary forms. Catatonic symptoms can also be a side effect of medication (Gelenberg, 1977; Wing and Shah, 2000b). Besides that, it is notorious in neuroleptic malignant syndrome and in Stauder's lethal catatonia. Children can also suffer from catatonia, especially in a comorbid fashion with autism (Wing and Atwood, 1987). Dhossche and Bouman (1997a) reviewed the literature between 1966 and 1996 and found just 30 published cases of catatonia: 11 with atypical psychosis (also in combination with infantile autism and mental retardation), 10 with general medical condition, 6 with a mood disorder, and 3 with schizophrenia.

Next, two challenging cases of autistic boys with catatonia are presented, after all: one picture tells more than a thousand words. This is followed by a summary of the state of affairs in child psychiatry anno 2005 on catatonia, especially in relation to autism.

II. Case Histories

A. Case History 1

A was diagnosed with autism and severe mental retardation at the age of 2.4 year, and later on he proved to have complete aphasia as well. An extensive pediatric–metabolic–neurological–genetic assessment was completely normal. He has an older brother, now 17-year old, who has been diagnosed with pervasive developmental disorder not otherwise specified (PDD-NOS) and mild mental retardation. His intensive medical assessment was also completely normal. The family history is uneventful, except for coronary heart disease. Both parents are well-informed health professionals.

A always had extensive rituals, but around the age of 8 he became increasingly passive. His development came to a complete standstill, to the exasperation of his parents. That was the reason why he was transferred to an outpatient clinic for an intensive multidisciplinary daily treatment program, from 8.30 a.m. to 4 p.m., based on the learning principles of behavioral therapy. His parents, and sometimes his brother as well, received support, counseling, and mediation therapy. Then a long-term period of waxing and waning of problematic behaviors started.

When he was 9-year old, A developed periods of startled gazing, with anxiety, self-injurious, and extreme varieties of stereotypical behavior (e.g., clinging upside down on a couch when left alone, self-fixating in his coat, etc.). His overall level of functioning deteriorated. After ruling out epilepsy, a working diagnosis of psychosis was made and we started with haloperidol 1.2 mg administered 2 times a day, with some effect—but he never fully returned to his previous level of functioning. Later on, he developed mild extrapyramidal side effects and was switched to risperidone 1 mg/day, with the same success but without these side effects. A was admitted to an inpatient unit for further evaluation and treatment. During the next few years, we tapered off his medication twice, but he quickly became chaotic, very obsessed, and anxious with startled gazing. We never succeeded in stopping this medication. Almost 2 years later, in the fall of 2003, 13-year old and with a weight of 45 kg, he developed full-blown catatonia with immobility/stupor and posturing/catalepsy over the course of a few weeks, without any psychiatric, somatic, or psychosocial cause or precipitant. His score on a catatonia rating scale (Bush *et al.*, 1996a) was 40 and we added lorazepam, slowly increasing to 8 mg/day, 3-3-2, which proved the best dosage schedule. His catatonia scores decreased from 40 to 32 to 24 in July 2004. At that time, he started to speak words for the first time in his life, "mama, papa"! In May 2004, he weighted 60 kg with a length of 1.58 m. Recently there has been a recurrence of catatonic behavior, especially, posturing/catalepsy and walking just as a very stiff shadow at a distance of just 2 cm away from his personal nurse. The dose of lorazepam has been increased to 14 mg/day, decreasing catatonia scores from 35 to 29. His treatment is ongoing. He seems to learn nothing anymore. His quality of life is enhanced, but for how long?

B. Case History 2

B was born a terme and at the end of an uneventful pregnancy. His birth however was complicated by cord strangulation and hypoxia. He was described as a peaceful baby. At age 2, B did not speak any words and was overly quiet and passive. Hearing deficits were suspected by his mother. Audiological testing indicated normal hearing.

At age 3, he was evaluated by a child neurologist and child psychiatrist because of problems with speech development and poor social interactions. Psychiatric observation showed avoidant gaze, limited interest in social interactions, limited emphatic contact, inappropriate smiling, and bizarre postures. His vocabulary was limited to a few words. He also used many jargons and was echolalic. B was greatly fascinated by flickering lights and twirling objects. Neurological investigations were normal, including urinary amino acids, CSF, and a pneumo encephalography! The EEG was slightly abnormal, with mild general slowing. A diagnosis of autistic disorder (AD) was made and he was placed in special education. His adjustment at school and home improved over time. Social functioning remained poor. Insistence on routines and obsessive preoccupations were constantly present throughout this period. Developmental progress was somewhat jerky. At age 5, his speech greatly improved unexpectedly in a short period. He also showed more interests. Cognitive testing was done at age 11 and showed a total IQ (WISC-R) of 103.

When B was 15, he started to complain about command auditory hallucinations. There was also marked a worsening of obsessive–compulsive behaviors, including obsessive slowness. He became literally in slow motion and sometimes even in a complete standstill. His psychomotor retardation alternated with outbursts of aggression and agitation. School performance deteriorated gradually. He was admitted to a child psychiatric hospital for further evaluation and treatment (with Jan Schieveld).

Initial psychiatric assessment showed a disheveled, leptosomic, adolescent with long fingernails. There were alternating episodes of agitation and severe psychomotor retardation, staring in space, waxy flexibility, posturing, and decreased verbal output mainly in dialect. There were no other formal thought disorders. He reported command hallucinations. Obsessions were elaborate and had a delusional quality. There were very extensive wars going on inside his head between different countries (with good and bad ones). He suffered gravely during those battles, and experienced these especially inside his head. He described some of these wars afterwards in his so-called "history books." His mood was constricted. No focal neurological abnormalities were found. The CT of brain and routine laboratory tests were normal. Chromosomal analysis showed 46 XY karyotype without fragile X. Schizophrenia with catatonia and paranoid features, superimposed on AD, was diagnosed.

He was treated with several antipsychotic medications but his condition did not improve. B continued to complain about hearing voices. He was aggressive and difficult to manage in the ward. He ran away from the ward twice, trying to rescue an-indeed-kidnapped boy with whom he had fallen in love. Episodes of catatonic behavior were among the most prominent features throughout the first year of admission. At that point, clozapine was prescribed in monotherapy (400 mg/day). Improvement was seen after a few weeks. B became less aggressive and more social. Catatonic recurrences continued. The frequency of stuporous episodes decreased after lorazepam (2.5 mg three times a day) was added to this medication regimen. A few months later, B was discharged from the hospital. Follow-up shows a young man in his early twenties with autistic symptoms but without catatonic–psychotic recurrences. He attends a day treatment center and lives with his parents. He continues to take clozapine and a small dose of lorazepam (2.5 mg/day) (Dhossche, 1998).

III. Methods

We performed a PubMed and Medline search with the terms: catatonia, catatonia and children, catatonia and autism, catatonia and autism spectrum disorder (ASD), and autism and motor signs. Next, we studied the index entries, catatonia/children/autism, in the most important standard textbooks on: adult (neuro-) psychiatry, child (neuro-) psychiatry, mental retardation, and on autism. And we studied the following books: *Differential Diagnosis in Neuropsychiatry* by Roberts (1984), *Movement Disorders in Neurology and in Neuropsychiatry* by Joseph and Young (1999), *Textbook of Psychosomatic Medicine* by Levenson (2005), and *Catatonia* by Fink and Taylor (2003).

IV. Results

In PubMed and in Medline, there were the following hits, respectively, catatonia 1754 and 1742, catatonia in children 105 and 36 (both dating from 2005 back to 1960), catatonia and autism 33 and 14 (from 2005 back to 1964), catatonia and ASD 52 and 0, autism and motor signs 260 and 2. The publication of Leonhard in 1960 *Über kindliche Katatonien* (On catatonia in children) is very beautifully written (with four pictures of a catatonic child and with an abstract in Cyrillian as well). He gives an overview on catatonia, up until 1960, and the relation with childhood schizophrenia: "Ich habe bisher unter den kindlichen Schizophrenien nur Katatonien gesehen" (I have—up until now—only seen the catatonic forms of childhood schizophrenia). And he presents very lucid and extensive clinical descriptions of four children (with their full names, and their relevant family anamnesis), with the catatonic form of childhood schizophrenia, who had been examined by him. Their genders and age were: 3 boys of 5, 6, and 13 years, and 1 girl of 13 years. And he discusses the differential diagnosis from catatonia with chorea minor, chorea major, and tic disorder; and from the marked impairment of social interaction with schizophrenia and autism. And in 1971, in one of the most famous children's book in the Dutch language, *Pluk van de Petteflet*, a clear description of catatonia was already given: "De directeur deed de kooi open. De arme krullevaar keek angstig en bleef stokstijf staan. Hij moest naar buiten worden gedragen." (The director opened the cage. The poor krullevaar looked anxious and stood like a statue. He had to be carried away.) (Schmidt, 1971).

Epidemiological findings vary greatly, for example, Dhossche and Bouman (1997a) reviewed the world literature and found 29 cases (2 with autism), with gender ratio: 18 males and 11 females, age range was 8–18, with a mean age of 14 years. And they even described, also in 1997, a 17-year-old male adolescent with: Prader–Willi syndrome, mild mental retardation, psychosis, and catatonia. The catatonia was successfully treated with lorazepam 4 mg, and the psychosis with risperidone 6 mg. They tapered off the medication 2 months after discharge uneventfully, and there was no relapse in the 5-year follow-up period (Dhossche and Bouman, 1997b). Cohen *et al.* (1999) reported on a consecutive case series of inpatients: 9 cases (2 with PDD-NOS) with gender ratio: 6 males and 3 females, age range was 13–17 years. They estimated the incidence of catatonia in an inpatient adolescent population as 0.6%. Wing and Shah (2000a) assessed catatonia in a large sample of referred adolescents and young adults (age 15–19 years) with autism and found that 17% met criteria for catatonia. Details can be found in the chapter by Wing. Slooter *et al.* (2005) estimated the incidence of malignant pediatric catatonia, defined as occurring below 18 years, at 0.16 per million.

There was one recent and very lucid conceptual review article named, Catatonia in psychiatric classification: A home of its own (Taylor and Fink, 2003). The standard textbooks, whether on adult (neuro-) psychiatry, child (neuro-) psychiatry, mental retardation or autism, deal scarcely on the subject of catatonia (and autism), with the exception of a nice chapter by Williams in the textbook of Lewis (Williams, 2002). Four contributions stand out: Roberts (1984) with a chapter named Catatonia and stupor in his *Differential Diagnosis in Neuropsychiatry*; Joseph and Young (1999) with a chapter on Catatonia (by Joseph himself) in their *Movement Disorders in Neurology and Neuropsychiatry*; and Masand *et al.* (2005) with their chapter Mania, catatonia and psychosis in James L. Levenson's *Textbook of Psychosomatic Medicine* (2005). They are all instructive in their own right: Roberts from a neuropsychiatric point of view, Joseph with more the neurological vision; and Masand deals nicely with the aspects in relation to general hospital medicine. Fink and Taylor (2003) with *Catatonia* is the overall nec plus ultra.

V. Discussion

There are no clear data on the epidemiology of childhood catatonia, and those that exist vary greatly. That is very worrisome given its great clinical importance in child neuropsychiatry, vide supra, as well as in understanding psychiatry at all. One explanation is the relatively rarity of its occurrence, another one is the lack of knowledge about the issue in us, child psychiatrists. Moreover, there is the lack of a clear concept of catatonia and of reliable rating scales. Again, child psychiatry falls way behind the rest of the flock of psychiatry. Many adult psychiatric disorders have its prodromal phase in child or adolescent psychiatry, and we child psychiatrists are obliged to realize this fact and to act accordingly.

What is catatonia: a symptom, a syndrome, a disease entity, a category, or a dimension in psychopathology? (van der Heijden, 2004). There are great conceptual similarities indeed between catatonia and delirium: both have multiple etiologies, a wide range of clinical manifestations, variable courses, and both have a hyperactive and a hypoactive form (Taylor and Fink, 2003). Delirium is the neuropsychiatric expression of a destabilizing underlying other neuropsychiatric—or internal medicine related—disorder and it is driven by the I WATCH DEATH causes (Wise, 1987), and so is secondary (76% [sic] according to Masand quoting Carroll *et al.* 1994) catatonia as well (Gelenberg, 1976, 1977; Wing and Shah, 2000b). Delirium has a considerable variation in clinical signs in adults and in children (Schieveld and Leentjens, 2005); it has many faces and a kaleidoscopic, fluctuating course; it is a spectrum disorder, and so is catatonia. (A spectrum disorder is a phenomenological set of different signs and symptoms, which are internal coherent, and can fluctuate in the course of time. This can be seen partially or completely in one patient but mostly only in a set of different patients. Other examples of spectrum disorders are, autism and epilepsy.) Both are neuropsychiatric spectrum disorders that overlap to some degree. For example, there are also similarities in the treatment modalities such as with a benzodiazepine (but not in the dosages), or with an antipsychotic, or even with electroconvulsive therapy (ECT) in lethal (delirious?) catatonia. Catatonia and delirium are perhaps both (and sometimes even similar) final common pathways for many complex (child-) neuropsychiatric disorders; they seem to be "reaction types of the brain." There are more famous examples in medicine of this kind (e.g., fever or anemia), which are more often than not a sign of an underlying disease, rather than a separate disease entity in itself. Catatonia is in my opinion a set of different, but coherent, syndromes consisting of subsets of maximal 5 of the core symptoms as recognized by the DSM-IV-TR and again this conceptualization probably holds with delirium as well. (And perhaps there exist no primary forms of catatonia at all, only secondary ones, when one regards affective disorders, schizophrenia, and autism as true medical conditions—brain diseases.)

There are differences as well, delirium is always secondary to a medical condition and catatonia only most of the time. And the brain structures and neural circuitries that are affected in catatonia and delirium may be different; in delirium, it is assumed that there are, especially, at the level of the neurotransmitter systems, general metabolic disturbances; and in catatonia, the disturbances may be more specifically located at the level of the basal ganglia. After all the brain is not one single organ, but a set of different organ systems, with different vulnerabilities (Gualtieri, 1990).

It was Piet Eikelenboom, M.D., Ph.D, professor of adult neuropsychiatry in Amsterdam who concluded, "when psychiatry and neurology were divorced, neurology became mindless, and psychiatry brainless" (Njiokiktjien, 2004). This may explain why there are more publications on autism and motor signs, than autism and catatonia, and why those writings on motor signs are mostly not from the hands of child psychiatrists. Diagnostic criteria of autism show many similarities with items of catatonia rating scales. It seems therefore incomprehensible that so little is known about the relation between these two major conditions (see the low number of hits in medical databases). Previous studies have supported a positive correlation between the severity of autism and the severity of mental retardation. If further studies confirm that catatonia is a common syndrome in autism, it logically follows that the relation between autism and catatonia should be studied intensely in severe mentally retarded people, a long neglected research population. After all: there almost everybody, or nearly no one, suffers from autism (Gualtieri, pg. 147, 1990), so, "ad fontes."

VI. Conclusions

Childhood catatonia, both the primary as well as secondary form, exists in all the combinations known in adult psychiatry: with mood disorders, psychosis, schizophrenia, a general medical condition, as a side effect of medication, with autism, with mental retardation, and even with the lethal variety. It deserves a place in nearly any major child psychiatric differential diagnosis, but that is seldom done. There are many similarities between catatonia and delirium, both seem to be, at least partially, final common pathways (i.e., "reaction types" of the brain), but there are differences as well. The epidemiology is unclear owing to lack of studies. Childhood catatonia, especially the primary form, is perhaps not so rare if one would systematically look for it. So child psychiatrists: remember your medical roots and the human body. Or in other words, take good care of our opening phrase from that great ballet dancer, for more than one reason.

VII. Epicrisis

How can we evaluate the medical histories of these boys?

Boy B, Case History 2, is typical for AD until the onset of psychosis. Hallucinations and delusions in AD are considered rare but have been reported, mostly in high-functioning patients (Kurita, 1999; Petty *et al.*, 1984). Obsessions are always difficult to differentiate from delusions, also in this boy. Auditory hallucinations were prominent and the catatonic symptoms were equally disabling. But most of all impressed me from then (1991) until this day, the very frightening visual and haptic hallucinations *inside* his head (cf. the Pensieve in Rowling, 2000). The whole clinical presentation, its longevity, and nearly complete unresponsiveness to pharmacological treatment are characteristic of schizophrenia. The progression of stereotypical movements, compulsions, withdrawal into full catatonia is remarkable but not uncommon in adult psychiatry, and has been observed by others (Wing and Shah, 2000a; Zaw *et al.*, 1999). In that respect, autistic symptomatology appears indeed as a forme fruste of full catatonia, or is it the other way around!

During the protracted inpatient course the catatonic syndrome seemed refractory to treatment with various typical antipsychotics. It is possible that catatonia was sustained or exacerbated by this medication regimen.

The marked response to clozapine may indicate superior effect in childhood schizophrenia and/or in catatonia (Frazier *et al.*, 1994 quoted by Marriage, 2002) and there is anecdotal evidence that clozapine improves behavioral problems in autistic children as well (Zuddas *et al.*, 1996). The positive response to lorazepam suggests specificity of benzodiazepine treatment in catatonic stupor, in accord with controlled studies in general psychiatric patients (Bush et al., 1996b; Ungvari et al., 1994).

Boy A, Case History 1, lives indeed in a home for severe mentally retarded children. The psychiatric findings, clinical course, and response to medications, documented over several years, force me to conclude—at last—that there are several important comorbid diagnoses to make: autism with almost complete expressive language disorder, mental retardation, schizophrenia, and catatonia. And indeed, regarding both case histories there are some diagnostic and academic questions left to ask: after all, where does autism stop and catatonia begin or, for that matter, catatonic schizophrenia?

Acknowledgments

I thank Kirsten Venrooij, medical secretary, Eric Dumont, M.Sc., educational psychologist, Diana Kroes and her staff for their support and cooperation; Dirk Dhossche, M.D., Ph.D., child psychiatrist for his inspirational e-mails, thoughts, and support; and Karen Gillaerts, M.D., child and adolescent psychiatry resident for her critical reading of the last draft.

References

American Psychiatric Association (APA) (2000). "Diagnostic and Statistical Manual of Mental Disorders," 4th ed. Text Revision (DSM-IV-TR). APA Press, Washington, DC.
Bleuler, E. (1911). "Dementia Praecox Oder Gruppe Der Schizophrenien." Leipzig und Wien, Franz Deuticke.
Bush, G., Fink, M., Petrides, G., Dowling, F., and Francis, A. (1996a). Catatonia: I: Rating scale and standardized examination. *Acta Psychiatr. Scand.* **93,** 129–136.
Bush, G., Fink, M., Petrides, G., Dowling, F., and Francis, A. (1996b). Catatonia: II: Treatment with lorazepam and electroconvulsive therapy. *Acta Psychiatr. Scand.* **93,** 137–143.
Carroll, B. T., Anfinson, T. J., Kennedy, J. C., Yendrek, R., Boutros, M., and Bilon, A. (1994). Catatonic disorder due to general medical conditions. *J. Neuropsychiatry* **6,** 122–133.
Cohen, D., Flament, M., Dubos, P. F., and Basquin, M. (1999). Case series: Catatonic syndrome in young people. *J. Am. Acad. Child Adolesc. Psychiatry* **38**(8), 1040–4046.
Dhossche, D., and Bouman, N. (1997a). Catatonia in children and adolescents. *J. Am. Acad. Child Adolesc. Psychiatry* **36**(7), 870–871.
Dhossche, D., and Bouman, N. (1997b). Catatonia in an adolescent with Prader-Willi Syndrome. *Ann. Clin. Psychiatry* **4,** 247–253.
Dhossche, D. (1998). Catatonia in autistic disorders (brief report). *J. Autism Dev. Disord.* **28,** 329–331.
Fink, M., and Taylor, M. A. (2003). "Catatonia: A Clinician's Guide to Diagnosis and Treatment." Cambridge University Press, Cambridge.
Frazier, J., Gordon, C. T., McKenna, K., Lenane, M. C., Jih, D., and Rapoport, J. L. (1994). An open trial of clozapine with 11 adolescents with childhood onset schizophrenia. *J. Am. Acad. Child Adolesc. Psychiatry* **33,** 658–663.

Gelenberg, A. J. (1976). The catatonic syndrome. *Lancet* **1,** 1339–1341.
Gelenberg, A. J. (1977). Catatonic reactions to high-potency neuroleptic drugs. *Arch. Gen. Psychiatry* **43,** 947–950.
Gualtieri, C. T. (1990). "Neuropsychiatry and Behavioral Pharmacology." Springer-Verlag, New York.
Joseph, A. B., and Young, R. R. (1999). "Movement Disorders in Neurology and Neuropsychiatry," 2nd ed. Blackwell Science, Malden.
Kahlbaum, K. (1874). "Die Katatonie oder das Spannungsirresein, Eine klinische Form psychischer Krankheit." Verlag Von August Hirschwald, Berlin.
Kraepelin, E. (1899). "Psychiatrie. Ein Lehrbuch für Studirende und Aerzte. 6," vollständig umgearbeitete Auflage. Leipzig, Barth.
Kurita, H. (1999). Delusional disorder in a male adolescent with high-functioning PDDNOS. *J. Autism Dev. Disord.* **29,** 419–423.
Leonhard, K. (1960). "On catatonia in children. Psychiatric Neurologic Medical Psychology." Über kindliche Katatonien. Psychiatrie, Neurologie und medizinische Psychologie. *Zeitschrift für Forschung und Praxis*. Mitteilungsorgan der Gesellschaft für Psychiatrie und Neurolgie der Deutsche Demokratische Republik. Jahrgang 12, Januar 1960, Heft I, Seite 1–12.
Leonhard, K. L. (1979). "The Classification of Endogenous Psychoses" (E. Robins, Ed. and R. Berman, trans.), 5th ed. Irvington Publishers, New York.
Levenson, J. L. (2005). "Textbook of Psychosomatic Medicine." American Psychiatric Publishing, Inc., Arlington, VA.
Marriage, K. (2002). Schizophrenia and related psychoses. *In* "Practical Child and Adolescent Psychopharmacology" (S. Kutcher, Ed.), Cambridge University Press, Cambridge.
Masand, P. S., Christopher, E. J., Clary, G. L., Mago, R., Levenson, J. L., and Patkar, A. A. (2005). Mania, catatonia, and psychosis. *In* "Textbook of Psychosomatic Medicine" (J. L. Levenson, Ed.), American Psychiatric Publishing, Inc., Washington, DC.
Njiokiktjien, C. (2004). "Gedragsneurologie van het kind." Suyi Publicaties, Amsterdam.
Petty, L., Ornitz, E., Michelman, J., and Zimmerman, E. (1984). Autistic children who become schizophrenic. *Arch. Gen. Psychiatry* **41,** 129–135.
Roberts, J. K. (1984). Catatonia and Stupor. *In* "Differential diagnosis in Neuropsychiatry." John Wiley and Sons, Chichester.
Rowling, J. K. (2000). "Harry Potter and the Goblet of Fire." Bloomsbury Publishing, London.
Sadock, B. J., and Sadock, V. A. (2005). "Kaplan & Sadock's Comprehensive Textbook of Psychiatry." Lippincott Williams and Wilkins, Philadelphia.
Schieveld, J. N. M., and Leentjens, A. F. G. (2005). Delirium in severely ill young children in the pediatric intensive care unit (PICU). *J. Am. Acad. Child Adolesc. Psychiatry* **44**(4), 392–394.
Schmidt, A. M. G. (1971). "Pluk van de Petteflet," p. 145. Em. Querido's Uitgeverij B.V., Amsterdam.
Slooter, A., Braun, K., Balk, F., van Nieuwenhuizen, O., and van der Hoeven, J. (2005). Electroconvulsive therapy for malignant catatonia in childhood. *Pediatr. Neurol.* **32**(3), 190–192.
Taylor, M. A., and Fink, M. (2003). Catatonia in psychiatric classification: A home of its own. *Am. J. Psychiatry* **160**(7), 1233–1244.
Ungvari, G., Leung, C., Wong, M., and Lau, J. (1994). Benzodiazepines in the treatment of the catatonic syndrome. *Acta Psychiatr. Scand.* **89,** 285–288.
van der Heijden, F. (2004). "Atypicality: Clinical Studies Concerning Atypical Psychosis." Academic Thesis, University of Utrecht, The Netherlands.
van der Heijden, F., Tuinier, S., Arts, N., Hoogendoorn, M., Kahn, R., and Verhoeven, W. (2005). Catatonia: Disappeared or under-diagnosed? *Psychopathology* **15,** 3–8.
Williams, D. T. (2002). Neuropsychiatric signs, symptoms, and syndromes. *In* "Child and Adolescent Psychiatry, A Comprehensive Textbook" (M. Lewis, Ed.), 3rd ed. Lippincott Williams and Wilkins, Philadelphia.

Wing, L., and Atwood, A. (1987). Syndromes of autism and atypical development. *In* "Handbook of Autism and Pervasive Developmental Disorders" (D. L. Cohen and A. M. Donellan, Eds.), pp. 3–19. Silver Spring, VH Winston & Sons, MD.

Wing, L., and Shah, A. (2000a). Catatonia in autistic spectrum disorder. *Br. J. Psychiatry* **176,** 357–362.

Wing, L., and Shah, A. (2000b). Possible causes of catatonia in autistic spectrum disorder. (Reply to Chaplin). *Br. J. Psychiatry* **177,** 180–181.

Wise, M. G. (1987). Delirium. *In* "Textbook of Neuropsychiatry" (R. E. Hales and S. C. Yudofsky, Eds.), The American Psychiatric Press, Washington, DC.

Zaw, F., Bates, G., Murali, V., and Bentham, P. (1999). Catatonia, autism, and ECT. *Dev. Med. Child Neurol.* **41,** 843–845.

Zuddas, A., Ledda, M., Fratta, A., Muglia, P., and Cianchetti, C. (1996). Clinical effects of clozapine on autistic disorder. *Am. J. Psychiatry* **153,** 738 (letter).

ECT AND THE YOUTH: CATATONIA IN CONTEXT

Frank K. M. Zaw*

Senior Clinical Lecturer (Hon.)
Division of Neurosciences, Department of Psychiatry
University of Birmingham, United Kingdom

I. Introduction
II. ECT and the Youth
    A. Extent of the Usage of ECT
    B. Effectiveness of ECT
    C. Deficiencies in the Studies
    D. Current State of Affairs
III. Indications for ECT
    A. General Considerations
    B. Catatonia: A Definite Indication for ECT
    C. Malignant Catatonia
    D. Periodic Catatonia
IV. Pre-ECT Consideration
    A. ECT or no ECT
    B. Preparation for ECT
    C. Mode of Action
V. Administration of ECT
    A. Basic Requirements
    B. Machine, Electrode Placement, and Current
    C. Course of ECT?
    D. Contraindications
VI. Adverse Effects of ECT
VII. Concurrent Medication
VIII. Mortality
IX. Patients' Satisfaction and Attitude to ECT
X. What Follows ECT?
    A. Medication
    B. Maintenance ECT
    C. Repeat Course of ECT
XI. Conclusions
XII. Conflict of Interest
    References

Electroconvulsive therapy (ECT) has been in psychiatric practice for well over half a century, but it continues to incite controversy. However, it is regarded amongst psychiatrists as a safe and effective treatment and at times even a

*Corresponding address: Huntercombe Stafford Hospital, Ivetsey Bank, Wheaton Aston, Staffordshire ST19 9QT, United Kingdom.

lifesaver. It offers a fairly swift but a time-limited response, opening up opportunities for initiation of more longer lasting treatments. The use of ECT in the youth is limited, and as such good studies are few and far between. The recent Practice Parameters by the American Academy of Child and Adolescent Psychiatry, specifically addressing ECT in adolescents, is indeed a welcome addition. Electrocovulsive therapy is as effective in the youth as it is in the adults, and the indications and contraindication are the same. The administration of ECT follows the same general principles in all age groups. One particular indication is of the use in catatonia, a motor syndrome that could occur with affective disorders, schizophrenia or medical conditions, in which it is considered to be extremely effective. The association between catatonia and autism and spectrum disorders has been noted, and in this situation, ECT is considered by some to be effective. Ethical considerations and that of capacity and informed consent are of paramount importance as are the human rights. Working in partnership with the parents/carers all the way is a must. The lack of information leaflets on ECT especially designed for young patients and their parents has to be rectified soon. Registers based on geographical health regions for those below the age of 18 will assist tremendously in epidemiological studies as well as pave the way toward more evidence-based studies that are essential.

I. Introduction

Electroconvulsive therapy (ECT) was introduced as a medical treatment for psychiatric disorders in 1938 by Celleti and Bini (Pallanti, 1999). To the lay public, it denotes something terrible and shocking. It also implies the severity of the condition for which it was required. For ECT to be considered, the clinical situation has to be very serious and calls for an urgent action. Therefore, by the same notion, immediate favorable effect is expected from intervention of this nature. It also indicates that other treatments that have been tried have failed. Indeed a last resort! Or is it really?

The use of the words "electric" as well as "shock" and "convulsive" are understandably quite frightening and emotive. That ECT is frightening, painful, and horrific was dramatized through films like "One Flew Over The Cuckoo's Nest." The black and white film era, when in such films, the scenes of making a corpse come to live through the use of electric current with wires connected to the limbs and the head, and during the process the recipient would shake very vigorously in the whole body, had produced ripples of fear and disgust and a spine chilling experience to most viewers.

The use of ECT remains controversial even in adults. Views on its efficacy and safety vary. There are those who consider it the most effective treatment available in psychiatry and completely safe. The opposing view is that it is ineffective and causes brain damage. Despite its dramatization as a treatment to be avoided, many lives have been claimed to be saved through its use. It is notable that the scientific community including the clinicians directly involved with the use of ECT is positive about ECT, whereas the lay public's view is quite different. The body of knowledge on ECT has been quite limited and unclear. This also adds to the controversy. Historical misuse and inaccurate media portrayal (AACAP, 2004) has led to its reputation being under question.

In those below the age of 18, ECT continues to invite a lot of negative reactions and criticisms, some of which are understandable. Most are based on not very well informed emotive reactions. Baldwin and Jones (1998) addressed the question "Is ECT unsuitable for children and adolescents?" and argued the case against ECT in this age group. There had been those who had called out for the moratorium on the use of ECT on the one hand and those with a more open mind for the "judicious use" on the other. In the middle, there are those who find it very difficult to decide either positively or negatively. Those who are knowledgeable about ECT are far more positive about its continuing use (Ghaziuddin *et al.*, 2001a). The study by Parmar (1983) indicated that more than 70% of child and adolescent psychiatrists had never given ECT to patients. It can therefore be assumed that the majority of child psychiatrists do not look at ECT favorably as a treatment option and appear to consider it as a last resort (Ghaziuddin *et al.*, 2001a). The fact that the pattern of psychiatric disorders in children and adolescents is different to that of the adults also requires consideration here. Over the years, official guidelines on the use of ECT in the UK and the United States of America in children and adolescents have very much been short sections within those for adults (APA, 2001; Freeman, 1995; NICE, 2003). It is apparent that both the lay and the professionals need to be better informed and have a better understanding and experience for the debate to be more balanced. Those interested amongst the lay public should be kept informed about ECT and the information made available in a user-friendly manner. There already exist downloadable leaflets/booklets for adults at respective Web sites by the Royal College of Psychiatrists (RCPsych.), UK, Mental Health Act (MHA) Commission, UK, and American Psychiatric Association (APA). The ECT Guide (2005) by Scottish ECT Audit Network (SEAN) is commendable. There is still a lack of information of this nature addressing children and adolescents, which needs to be rectified soon.

There have been studies on the use of ECT in children and adolescents since 1940s with Heuyer in 1942 and Bender in 1947 reporting on their case series (AACAP, 2004; Rey and Walter, 1997). This was followed mainly by occasional case study and a number of small case series. Thus the literature has remained very limited for a considerable period of time.

It was not until the late 1990s that Rey and Walter (1997) reviewed the literature of over 50 years on ECT in children and adolescents and published their useful findings. They focused on the efficacy of ECT as well as its adverse effects. There were others who contributed in kind (Cohen *et al.*, 1997; Ghaziuddin *et al.*, 1996; Kutcher and Robertson, 1995; Moise and Petrides, 1996; Schneekloth *et al.*, 1993). In many ways the studies that came into the literature in the early and the late 90s have paved the way toward better information and understanding.

In the younger age group, the ethical debate centers around who decides! The issue of whether young people are really competent or have the capacity to decide and give informed consent has been a continuing concern. The European Convention of Human Rights has become crucial within psychiatry. The rights of an individual have to be balanced against the treatment that is proposed.

The recent publication of the Practice Parameter for use of ECT with Adolescents (AACAP, 2004) is a notable landmark in addressing the issues directly for the youth. With such recent additions to the literature, it is hoped that more informed and healthy debate will follow.

II. ECT and the Youth

A. Extent of the Usage of ECT

Electroconvulsive therapy usage in adults in the United States of America was found to be highly variable. This could suggest a lack of consensus about its use (Hermann *et al.*, 1995). There is evidence of a decline in the use of ECT in adults (Eranti and McLoughlin, 2003). Does this mean that there are further or new concerns? In certain States in the United States of America, the use of ECT depends on the age of the patient. For example, ECT is not allowed for those under 12 in California and 14 in Tennessee. In Colorado and in Texas it is 16 (Moise and Petrides, 1996) as it is in the former USSR (Rey and Walter, 1997). In the UK, the figures as of 1991, covering a period of 10 years, were 65 cases under 18 of which those 16 and above constituted some 60% (RCPsych., 1995a). In an Australian survey amongst the 42 patients under the age of 18, who underwent ECT, the youngest were 14-year-old patients (Walter and Rey, 1997). In their review on the use of ECT in children and adolescents, Rey and Walter (1997) found 98 in whom the age was specified; the mean age was 15.4, the youngest was 7, and 5 were younger than 12.

The use in the prepubertal, as has been reported, is very rare. It was for major depressive disorder with pronounced catatonic symptoms, and the response was considered positive (Cizadlo and Wheeton, 1995). There has been a study that reported on a negative response too (Guttmacher and Cretella, 1998).

TABLE I
PATIENTS GIVEN ECT 1977–1996[a]

Country/state/city	Age	Numbers (averaged per year)
California	Under 18	8 per year
New York (Stony Brook)	16–19	22 in 12 years
Edinburgh	17	5 in 10 years
London (Bethlem)	Adolescents	3 in 10 years
Paris (Salpetriere)	15–18	7 in 2 years
Australia (NSW)	14–18	42 in 6 years
USSR	Under 16	0

[a]Adapted from Rey and Walter (1997).

Covering a period 1977–1996, up to 8 patients per year were given ECT with an average of 3.6 patients a year. The details are given in Table I.

Taking various figures into consideration, including the United States of America, of around 500 children and adolescents (aged 11–20) a year receiving ECT (Thompson and Blaine, 1997), roughly one ECT per year, per million population, appeared to be the extent of the use.

However, in the early 1980s in the UK, 31 out of 433 psychiatrists in a survey indicated that they had used ECT in children younger than 16; among the diagnoses were psychotic depression, catatonic stupor, mania, schizophrenia, and anorexia nervosa (Pippard and Ellam, 1981). The number of children and adolescents from the UK reported in the literature as having ECT was however much less than this study discovered, indicating that there were far more, who received ECT, than the numbers reported.

In the United States of America, there has been what appeared to be some revival of interest of the use of ECT in children and adolescents in the context of the diagnosis of catatonia (Dhossche and Stanfill, 2004).

B. EFFECTIVENESS OF ECT

Confirmation of effectiveness in children and adolescents has been noted for a very long time (AACAP, 2004). This has been illustrated as a good response in 2/3 with marked improvement and complete remissions in most. The indications are also considered to be similar to adults. It was however, acknowledged that such a conclusion has to be qualified in the absence of systemic evidence (Rey and Walter, 1997).

Most studies, early and later, confirmed the effectiveness of ECT ranging from 40 to 100% with best results for mania and depression and less for schizophrenia (Rey and Walter, 1997). Most beneficial effects appeared to be seen in mood disorders. Kutcher and Robertson (1995) reported a significantly better outcome in those who had ECT for bipolar disorder as compared to those who did not and claimed that the efficacy in mood disorder was of the order of 50% and above. There were also some complete remissions. However, in schizophrenia the effect is not as much as it is in affective disorders (Cohen et al., 1997; Ghaziuddin et al., 1996; Moise and Petrides, 1996; Rey and Walter, 1997; Strober et al., 1998). The one exception is the report that concluded that adolescents responded poorly to ECT (Guttmacher and Cretella, 1998). There was a patient suffering from schizophrenia and another with Tourette's syndrome in this study.

National Institute of Clinical Excellence (2003) stated clearly that ECT may be of benefit in the rapid control of mania and catatonia. Real ECT has been considered more effective than sham, and bilateral more effective than unilateral.

C. Deficiencies in the Studies

As far as age is concerned, some studies included those who were 18 or older. When the age range can be as wide as from 11 to 20 (Thompson and Blaine, 1987), this contaminates what might appear at first glance to be specific findings for children and adolescents. In such studies, prepubertals are lumped together with the adolescents as well as the adults.

The limitations in the earlier studies included lack of diagnostic specificity. The concept of schizophrenia and autism in the 1940s when compared to the present day is different; from autism as childhood schizophrenia to these two disorders being distinct (Kolvin, 1971). Bertagnoli and Borchardt (1990) in reviewing the studies between 1947 and 1990 noted that there was a lack of diagnostic clarity. Mostly they were of small sample sizes, of retrospective study design, and heterogeneous diagnosis.

The outcome criteria were lacking, and there were other methodological problems as well. The adverse effects were not systematically studied. Studies usually did not contain data from follow-up. Therefore, there has been limited information on the duration of positive effect as well as the longer term side-effects.

Until recently most of the information was extrapolated directly from the adult literature. Thus, the true value of the ECT as a useful treatment in children and adolescents has been open to question and criticism. With the information that has been more specific to this age group (AACAP, 2004; Rey and Walter, 1997; Walter and Rey, 1997) the situation has moved toward a better position. Specially designed studies in children and adolescents, such as that by Kutcher and Robertson (1995), are needed urgently to resolve the pertinent issues (AACAP, 2004) and rectify the deficiencies further.

D. Current State of Affairs

In the UK, the formation of ECT Accreditation Service (ECTAS) (Caird et al., 2004a,b) by the RCPsych. Research Unit followed the NICE Guidelines on ECT (NICE, 2003). The aim is to improve the standards and quality of the administration of ECT, while taking into account the view of the service users as well as the clinicians. The standards set out are applied each year in self and external peer-review. So far, this activity is mainly confined to the adult services and should include the child and adolescent mental health services by linking up with the Quality Network of In-patient Child and Adolescent Psychiatry (QNIC), which is also under the same organization.

In the 2nd ed. of Handbook on ECT (RCPsych., 2005), there is a section "Use of ECT in Children and Adolescents for Treatment of Depression," which is an update on the 1st ed. of the Handbook on ECT (RCPsych., 1995a). College Guidelines on ECT (Scott, 2005) is an update for prescribers, giving the key points contained in the 2nd ed. of Handbook of ECT. There has also been the formation of SEAN (www.sean.org.uk) with audit activities for adults as well as for those under the age of 18.

In the United States of America, The Practice of ECT: Recommendations for treatment, training and privileging (APA, 2001), has only one page dedicated to children and adolescents under special populations giving an extremely summarized account of the review of the recent literature, policies, and procedure. The most recent "Practice parameter for use of electroconvulsive therapy with adolescents" (AACAP, 2004) is comprehensive. It contains a very full review of the literature in children and adolescents, past and recent, the adverse effects together with the technical details and procedure. The consent issue has also been discussed more elaborately. This is indeed the first worthwhile and comprehensive document of its kind dedicated to adolescents. Even within such a dedicated publication, there are still many aspects of ECT that have to be extrapolated from the adult literature.

III. Indications for ECT

A. General Considerations

It should be noted that the indication for ECT in children and adolescents is the same as that for adults (AACAP, 2004; NICE, 2003; RCPsych., 2005). In a paper that examined in some detail the similarities and differences between adolescents and adults in terms of the usage of ECT and its effectiveness, the conclusions are that ECT is as effective for adolescents as it is for adults, 58% achieving remission (Bloch *et al.*, 2001). The authors did, however, note that the diagnosis in the adult population is somewhat different from that of a younger age group with preponderance of "psychotic spectrum" in the adolescents and "affective spectrum" in the adults. The same is also observed in some other studies (Moise and Petrides, 1996; Schneekloth *et al.*, 1993). However, there exists opposing views (Rey and Walter, 1997; Walter and Rey, 1997).

B. Catatonia: A Definite Indication for ECT

We owe to Laszlo Meduna for the knowledge of the favorable effect of convulsive seizures in catatonia. Seizures were initially induced by camphor and later by pentylenetetrazol (Metrazol) before electric current was used in 1938.

Catatonia is a syndrome with marked changes in muscle tone and activity. It could appear as catatonic stupor or excitement. There have been many concepts of catatonia. To cite just a few, catatonia, especially catatonic stupor has been considered secondary to a primary psychiatric disorder, notably schizophrenia (Hamilton, 1984). It has also been considered as a syndrome of motor dysregulation (Fink and Taylor, 2003). Fink and Taylor (1991) called for a separate category for catatonia, distinguishing it as a diagnostic category in its own right. The catatonia syndromes with subtypes like retarded, excited, malignant, and periodic varieties have been described (Fink and Taylor, 2003). These authors included primary akinetic mutism under this heading. There are those who argue for catatonia as a state phenomenon and demonstrated changes in the regional cerebral blood flow, illustrating a patient with catatonia, whose rCRB was corrected with ECT with consequent resolution of the catatonic symptoms (Galynker *et al.*, 1997).

Catatonia appeared only in association with schizophrenia in the International Classification of Diseases (WHO, 1992) even though it has also been noted that this set of symptoms that constitute the diagnosis of catatonia can accompany affective disorder (Abrams and Taylor, 1976; Taylor and Abrams, 1977). The Diagnostic and Statistical Manual (APA, 1994, 2000) classifies catatonia under schizophrenia, catatonic features as a specifier for mood and catatonic disorder due to a medical condition.

As in adults, catatonia is seen in children and adolescents (Dhossche and Bouman, 1997; Rey and Walter, 1997) and usually diagnosed among adolescent psychiatric inpatients with affective disorders and schizophrenia (Zaw, 2004). However, the prevalence is difficult to determine.

Catatonia can now be diagnosed by having only two out of the five criteria (APA, 2000). This could mean that with symptoms such as peculiarities of voluntary movement, which includes stereotype movements and prominent mannerisms or grimacing, and echolalia or echopraxia, the diagnosis can be made. Such symptoms as described are very much a part of the clinical profile of those with autism and autistic spectrum disorder (ASD). Therefore it is possible for those with this neurodevelopmental disorder to qualify for the diagnosis of catatonia more easily.

The link between autism and spectrum disorders on the one hand and catatonia on the other has been an ongoing debate for quite a few years. Wing and Attwood (1987) noted that certain features of catatonia, such as posturing, freezing, abulia, and bradykinesia, can be associated with ASDs. However, Realmuto and August (1991) suggested that despite many similarities, there was no evidence that autism and catatonia co-occur more often than by chance. The more recent studies in autism and spectrum disorders suggested that catatonia is part of the clinical presentation. Wing and Shah (2000) drew conclusions from their work on more than 500 patients and stated that catatonia is a later complication of ASDs. There have been no treatment trials of ECT in their patient population. There have been further studies associating catatonia with autism hypothesizing that certain types of autism may be the earliest expression of catatonia (Dhossche, 2004). There are those who suggested that autism and catatonia are not related and these merely coexist (Zaw *et al.*, 1999). Whereas others claimed that catatonia in autism is a distinct subtype with a particularly poor outcome with ECT (Ghaziuddin *et al.*, 2005). The use of ECT to ameliorate symptoms of catatonia in autism has been put forward by Dhossche and Stanfill (2004) arguing for the shared GABA theory in these two conditions. It is an interesting argument when there has not been any effective treatment for autism. If proven effective, it would indeed be a positive step. It is evident that there are proponents as well as opponents for the notion of the links between catatonia and autism as well as the efficacy of ECT for this patient group. It is to be noted that autism and its spectrum disorders have not been accepted so far as an indication for ECT in any of the published and widely accepted literature (AACAP, 2004; APA, 2001; NICE, 2003; RCPsych., 2005). Specific studies are required to answer pertinent questions directly rather than having to extrapolate from small series or single case studies.

Catatonia has been regarded as one of the main indications for ECT (APA, 2001; NICE, 2003; RCPsych., 2005). In the younger age group with 75% immediate response rate (Rey and Walter, 1997), it has to be considered seriously as we will be extremely hard pressed, if not impossible, to find a treatment procedure as effective within psychiatry or for that matter, medicine as a whole. However, there are definite implications on the use of ECT with the wider concept of catatonia through Diagnostic and Statistical Manual, 4th ed., text revision (DSM–IV-TR) (APA, 2000) as has been described.

Whether catatonia of all etiologies and affiliations respond to ECT equally well is not known. Catatonia in the classical sense, stupor and excitement, as would be seen in affective disorders or schizophrenia appeared very similar. They might even be the same. There have been no distinguishing features offered or proposed in the literature.

One missing aspect of studies is the duration of the effect of ECT in catatonia. This therapy is quite notorious for its short-term effects and the propensity for the relapse of symptoms (Aaronson *et al.*, 1987; Sackeim *et al.*, 1990) and has generally been accepted for the short-term use (NICE, 2003). There has been some evidence in the literature (Rey and Walters, 1997) that the rate of immediate response in catatonia to ECT is higher by one-third than when determined 6 months later. About half of the patients were lost to follow-up among this group and therefore limits the interpretation. Whether catatonic features are more or less likely to relapse than other psychiatric symptomatology is not known and therefore restricts the discussion on the continuation or maintenance therapy with ECT.

C. Malignant Catatonia

Neuroleptic malignant syndrome (NMS), with very similar symptoms to that of malignant catatonia or lethal catatonia responds well to ECT treatment. There are case reports (Chandler, 1991; McKiney and Kellner, 1997; Slooter *et al.*, 2005) suggesting the usefulness of this treatment modality. There have been only limited studies of this disorder in the youth, but ECT has been considered as an established lifesaver in this clinical condition (Fink and Taylor, 2003). The use of ECT in NMS has been considered in the ECT Handbook 2nd ed. (Scott, 2005) as experimental. There does not appear to be any special requirements (Scheftner and Shulman, 1992) except when extensive muscular rigidity is seen as succinylcholine could cause hyperkalaemia and a nondepolarizing relaxant should be considered instead (APA, 2001). The presence of metabolic or cardiovascular instability should be given due attention throughout the course of ECT (APA, 2001).

D. Periodic Catatonia

Gjessing's name has been more or less synonymous with "periodic catatonia," but the notion of periodicity goes back as far as Kraepelin (Gjessing, 1976). Periodic catatonia is defined for the purposes of this chapter as symptoms that are considered as indicative of catatonia, for example, posturing, extreme staring, and other motor symptoms (e.g., waxy flexibility, difficulty in initiating, and maintaining motion), which are episodic, with clear periods in between, lasting hours to days or weeks. In some, the catatonic features recur each day at specified times lasting only for a few hours (Zaw, 2005b). This can either be part of the wider catatonic picture or as a defined episodic phenomenon. Because of the periodic nature, the consideration for ECT could very well be different to the catatonia that did not have such a periodicity. Whether this is a less severe version of catatonia or categorically different is difficult to determine. If the episode of catatonia were to last for a significant period of time, ECT could perhaps be considered.

IV. Pre-ECT Consideration

A. ECT OR NO ECT

Not all of those with catatonia will require ECT. The NICE Guidelines (2003) recommends the use of benzodiazepines most notably Lorazepam, as the first choice. Barbiturates have been claimed to dissolve the catatonic symptoms in some but not in others (McCall *et al.*, 1992). Moreover, there have been claims about certain medication like zolpidem, an alpha$_2$ agonist that have been suggested as a test for catatonia both in the adult population (Thomas *et al.*, 1997) as well as in the younger age group (Zaw and Bates, 1997). In our experience, zolpidem successfully dissolved the catatonic symptoms in three patients. One patient was given ECT following the zolpidem test with total resolution of the catatonic symptoms. In another patient, ECT was considered but the use of zolpidem over 3 days was enough to dissolve the catatonic symptoms completely, thus ECT was no longer required (Zaw, 2005a). In the third patient who showed symptoms of periodic catatonia, the use of zolpidem in a timely way, targeting the symptoms that usually appeared in the evening, by a dosage of zolpidem an hour or two before the symptoms emerged, proved very successful (Zaw, 2005b).

There are no clear guidelines on the use of such medication, including the number of medication to be tried and the duration, or what constitutes a failure. Generally the decision to administer ECT is based on clinical opinion that it was time. Objective criteria are required. Zaw *et al.* (1999) suggested some of the defining characteristics in the patient's physical and mental state that they used. The Practice Parameters (AACAP, 2004) gives a sound principle to be made use of, and this includes: (1) the accuracy of diagnosis; (2) the consideration of the severity and the persistence of symptoms, including the degree of disability, and whether or not it is life–threatening; and (3) the lack of response to other appropriate treatment.

B. Preparation for ECT

1. *Consent, Ethics, and Human Rights*

In UK, for those less than 16 years, parental consent is required in England and Wales but not in Scotland. When consent is not obtained, ECT should not be given unless the patient's life is at risk (from suicide or physical debilitation secondary to depressive illness) and the patient detained under the legislation, for example, MHA (1983) (RCPsych., 1995b).

In catatonia, patients are often mute, bradykinesic, negativistic, and determining their capacity and obtaining informed consent becomes a challenging and difficult issue, especially in children and adolescents. In addition, their understanding and intellectual level of functioning could create additional problems. Moreover, the Human Right issues have to be considered to the fullest extent. One cannot lose sight of the fact that the use of ECT remains controversial, much more so for children and adolescents. Comprehensive clinical assessments are required to determine the level of capacity and allow appropriate discussion with the patient and parents.

Parents must be involved fully in the discussions and their informed consent obtained and documented. Guidelines (AACAP, 2004) should be adhered to. Practical difficulties that might arise should be well thought out and appropriate actions planned well in advance. Legal advice should be sought as appropriate, and the existing mental health legislation referred to.

Information on the benefits of ECT as well as the risks, including possible adverse effects, for example, memory loss, should be given, and discussed fully with sensitivity and without pressure or coercion. Involvement of advocates is to be encouraged. In educating the patient and parents, appropriate teaching aids, which are user friendly, should be made use of, for example, written materials, pictures, and films. Patients, who had had ECT, and their parents can be a most useful source of information and very much in line with the current trend of user involvement.

It is good practice to allow a period of time between obtaining consent and the initiation of treatment. It is crucial to inform that consent can be withdrawn at any time. Informed consent is a process.

Ethics in ECT has been comprehensively dealt with in a monograph by Ottoson and Fink (2004).

2. Second Opinions

The guidelines for those under 18 years produced by the RCPsych. (Freeman, 1995) as well as that by the APA (2001) are commendable. In addition, a further set of guidelines by the AACAP (2004), specifically prepared for the youth, as well as the NICE (2003) guidelines are now available. Two independent opinions should be sought from consultant psychiatrists with experience and expertise in children and adolescents and ECT.

In the UK there is a shortage of consultant child and adolescent psychiatrists who are deemed to have sufficient knowledge of ECT in the age group they specialize in (Parmar, 1983). This could also be true in other countries as well in view of the fact that ECT appeared to be used sparingly for children and adolescents (Ghaziuddin et al., 2001a; Rey and Walter, 1997; Thompson and Blaine, 1987; Walter and Rey, 1997).

One way forward would be to have regional centers where the experience and expertise can be developed in such a way that the appropriately trained expert second opinion consultants can be organized for, as and when required, without much difficulty.

C. Mode of Action

There has been no generally accepted comprehensive theory or hypothesis on the mechanism of action of ECT (NICE, 2003). Some of the theories put forward so far include an alteration in the postsynaptic response to central nervous system (CNS) neurotransmitters (NICE, 2003). There is also a suggestion that regulation of brain derived neurotropic factor, which reverses the damage or protects further damage, is important. (Vaidya, 1998). Others include suggestions that 5HT1A and 5HT3A receptor sensitization through ECT may play an important role (Ishihara and Sasa, 1999). Neuropeptides are also implicated (Mathe, 1999).

Psychological theories of the mechanics of ECT are many and varied. They have developed over the years, some gaining more acceptance than others. A few of these psychological theories are described and reviewed fairly comprehensively by Johnstone (1999).

V. Administration of ECT

A. Basic Requirements

The current view is that ECT should be administered to adolescents only on an inpatient basis (AACAP, 2003). Administration of ECT requires a well-organized team of adequately trained psychiatrist, an anesthetist, and

TABLE II
MEDICATION

	Mainly	Alternatives
Anesthetic agent	Methohexital	Etomidate, Thiopental sodium, Ketamine
Muscle relaxant	Succinylcholine	Atracurium, Mevacurium
Anticholinergic	Atropine	Glycopyrrollate

nursing staff. Each has a different but specific role and function. Thorough assessments and required clinical action, and care before, during, and after each ECT have to be provided adequately and appropriately. Electroconvulsive therapy session for the youth should be arranged separately from adult wards.

Thorough clinical evaluation, including the diagnosis, severity of illness, past treatments, medical and psychotherapeutic, compliance, duration, and response are essential. Physical examination and laboratory investigations are dictated by clinical assessment. Pregnancy test is required for all female patients (AACAP, 2004).

Anesthetic agents, muscle relaxants, and anticholinergic agents are as shown in Table II. Anticholinergic medication is administered immediately before ECT to prevent bradycardia, arrthymia, or occasional ECT-induced cardiac asystole. It also protects the cardiovascular system from vagal discharge in instances of incomplete or missed seizures.

Ventilation with 100% oxygen is required before administering the electrical stimulus.

B. Machine, Electrode Placement, and Current

Electroconvulsive therapy machine should contain a device for recording EEG. Cerebral dominance for language and verbal memory has to be determined before ECT is started. Unilateral electrode placement applied to the nondominant hemisphere should be used and if the response is inadequate to change to bilateral. Bilateral electrode placement may be considered in a severely ill patient with catatonia in which speed of response is critical. Brief pulse instead of sine wave and low dose electricity should be used, but adequate convulsive dosage must be ensured to produce seizures lasting 30–90 s. Electrode placement and dose of electricity strongly influence clinical outcome (AACAP, 2004). No significant difference exists between brief pulse and sine waveform in the adults (The UK ECT Review Group, 2003).

C. COURSE OF ECT?

The number of ECT received by young people varied from 2–21 in those studies reported in the literature between 1952 and 1990 (Bertagnoli and Borchardt, 1990).

The number of ECT to be given cannot be decided accurately in advance. Appropriately comprehensive clinical reviews should take place after each ECT. The patient's overall response is considered after the first 5 or 6 treatments before completing a usual course consisting of 10–12 treatments (AACAP, 2004). When the response is achieved, when there are adverse effects, or if the patient withdrew consent ECT is stopped (NICE, 2003).

As far as the frequency of ECT is concerned, there does not appear to be agreed guidelines. Within the literature, the frequency varied from twice a week to more than twice a week and sometimes only the number of ECT treatments is being mentioned without the frequency. It is said that most centers in the United States of America administer ECT three times a week to adults and adolescents (AACAP, 2004). There does not appear to be any notable difference between administering ECT twice a week, three times a week, and once a week regarding depressive symptoms (The UK ECT Review Group, 2003). However, it appears that the biweekly treatment is most common practice in the UK and endorsed by NICE Guidelines (2003).

D. CONTRAINDICATIONS

There is no absolute contraindication, only relative; this include tumors of the CNS with raised Cerebrospinal fluid (CSF) pressure, active chest infection, and recent myocardial infarction. Medical consultation is recommended in those with concurrent physical illness (AACAP, 2004).

VI. Adverse Effects of ECT

There does not appear to be any study on the impact of ECT on the developing brain. The studies that were reviewed (NICE, 2003) using brain-scanning techniques did not provide any evidence that ECT causes brain damage.

The cognitive effects of ECT are well described (Sackeim, 1992). Adverse effects of ECT include impairment of memory, new learning, and those associated with the use of anesthesia. In addition, tardive and prolonged seizures can also be observed. Headache, nausea, vomiting, muscle aches, confusion, and agitation have also been reported.

Rey and Walter (1997) warned that one cannot be certain that more serious effects did not occur in the younger age group. The NICE Guidelines (2003) stated that the risk with ECT may be greater in children and young people. Most of the studies on the adverse effects of ECT have been done in and extrapolated from the literature in adults. Direct extrapolation could very well be inappropriate as has already been alluded to. The knowledge regarding ECT in adults, do not necessarily translate as being true in the younger age group. The notion of the developing brain in the children and adolescent age group with the possibility of more serious or different side effects has to be borne in mind. There have not been long-term follow-up studies in children and adolescents who had undergone ECT.

Impairment of memory, long and short term, both retrograde, and anterograde amnesia have been implicated with the use of ECT. They are seen more in bilateral and unilateral dominant electrode placements. The effect on the memory does not appear to differ between diagnoses.

Most studies claim that the effect on memory is fairly transient and by 6 months post ECT, the recovery was complete (NICE, 2003). It is noteworthy that some studies have concluded that there were no long-term effects on memory functions of adolescents, who were treated with ECT (Cohen et al., 2000; Ghaziuddin et al., 2001b).

According to The UK ECT Review Group (2003), which looked at the safety of ECT, the tentative conclusions included that the cognitive impairments mostly reflected changes in memory, especially recent memory and more memory impairment is seen in real ECT than in simulated ECT or drug therapy. It also stated that bilateral electrode placement causes more cognitive impairments than the unilateral, and the frequency of ECT is also important, that is, three times a week causing more cognitive impairments than once a week. The dosage of electric current is also important in the sense that high dose causes more impairment than low dosage. However, it also stated that the cognitive function at 6 months is not significantly different on measures of subjective memory impairments, new learning, and remote memory than at base line. Personal memory was no worse in those treated with high dose.

As clearly stated by NICE Guidelines (2003), the cognitive impairment complained of was variable and qualitative studies have indicated that the impairment may be prolonged or permanent. There has been a dearth of literature that looked at the long-term effects.

Global cognitive impairment before treatment, which is likely to be associated to the psychiatric illnesses and the prolonged disorientation in the acute postictal period, which could be associated with electrode placement, and dosage have been found to be associated with persistent retrograde amnesia for autobiographical memory (Sobin et al., 1995).

The illness symptoms of major psychiatric disorders are known to affect cognition. Depressive illness being considered as a neurodegenerative disorder, cognitive deficits could very well be part of the illness process. At the height of the illness, with lack of insight, the lower level of cognitive functioning might not be too apparent initially, and only with the amelioration of symptoms of illness through the use of ECT, this might become more noticeable. There have been pertinent issues raised about the instruments used to determine the cognitive functioning before and after ECT. Routine neuropsychological tests have been felt not to address the type of memory loss reported by the patients, for example, erasing of autobiographical memories and retrograde amnesia. The sensitivity of the instruments and also how it is applied has been brought into question. Obviously, it is possible that these were a direct effect of ECT as has been claimed. The cognitive deficits could also be a coincidental finding. It is also likely that there is a combination of factors, each playing a part in bringing about a measurable effect.

Other adverse effects include prolong and tardive seizures. Lower seizure threshold in children and the association with prolonged seizures and post-ECT seizures should be kept in mind. Prolonged seizures have been described in the younger age group (Ghaziuddin *et al.*, 1996; Guttmacher and Cretella, 1998; Moise and Petrides, 1996). They are those lasting longer than 180 seconds. They are associated with greater ictal confusion and amnesia and inadequate oxygenation with increase hypoxia-related risks. Tardive seizures are rare but potentially serious. They are usually encountered in those with normal EEG and those not receiving seizure-lowering medication. Neurology consult should be sought if there were to be prolonged, recurrent, or tardive seizures.

The risk associated with anesthesia also needs to be considered. It has been stated that adolescents are not at additional risk from ECT or at increased risk of anesthesia-related complications in the immediate recovery period (AACAP, 2004).

Other possible adverse effects include changes in cardiovascular dynamics with the risk of cardiovascular event, status epilepticus, laryngospasm, and peripheral nerve palsy.

There have been a few reports in which ECT had to end prematurely because of substantial side effects, for example, a switch to mania, (Paillere-Martinot *et al.*, 1990), increasing agitation (Slack and Stoudemire, 1989), and NMS following one ECT (Moise and Petrides, 1996).

VII. Concurrent Medication

The AACAP (2004) advises that whenever possible, ECT should be administered without concurrent medication, especially psychotropics. Some medication may either lower the seizure threshold or raise it and might make the side effects worse. A review by Pritchett *et al.* (1993) is helpful. The combination of lithium and ECT may cause acute brain syndrome (Penney *et al.*, 1990). Carbamazepine, raising the seizure threshold, could fail to induce a fit as would a benzodiazepine. Prolonged seizures have been reported with trazodone (Lanes and Ravaris, 1993) and theophylline (Zwillich *et al.*, 1975).

VIII. Mortality

Generally, the safety record of ECT has been most impressive. Low overall mortality has been observed in patients treated with ECT (The UK ECT Review Group, 2003). Adult fatality rate associated with ECT has been cited as 0.2 per 10,000 and anesthesia mortality rate as 1.1 per 10,000. Adolescents have been considered not to be at additional risk (Aubas *et al.*, 1991). No fatalities have been reported in children and adolescents as a direct effect of ECT (Rey and Walter, 1997).

IX. Patients' Satisfaction and Attitude to ECT

Controversy still exists whether ECT is beneficial and the patients are satisfied with the treatment. Since the early 50s, people began to be concerned about the patients' attitudes toward ECT (Fisher *et al.*, 1953). Freeman and Kendall (1980) examined this important topic some 25 years ago. There have been a few studies along the way, and the recent and most notable study in adults was by Rose *et al.* (2003). In all these studies, children and adolescents were not included. Walter *et al.* (1999) however, looked specifically at the attitudes of adolescents toward ECT.

All these surveys concluded that the attitude of the patients receiving ECT was positive. Reported patient satisfaction depends on the methods used to elicit the response. It was pointed out by Rose *et al.* (2003) of the definite bias of these studies being conducted by professionals involved with ECT, meaning that, those who are positive about ECT are bound to get positive results from their surveys. The interval between the treatment and the interview, the number of questions, the complexity of the interview, and the setting of the interview are all very important factors that needed to be taken into consideration.

In a recent study by Johnstone (1999), 20 people, who reported having found ECT upsetting, were interviewed about the experience. Such studies are also very biased and the findings quite limited and limiting.

Even though there are inherent difficulties in conducting a study that is systematic, nonbiased, and accurate, such difficulties need to be overcome through properly conducted and methodologically sound procedures. The claim that over 80% of patients respond well to ECT and the statement "as far as we know, electroconvulsive therapy does not have any long-term effects on your memory or intelligence" (RCPsych., 1995) has now been challenged not only in terms of the stated figures, but also in terms of its methodology arriving at such a conclusion (Rose et al., 2003).

Studies determining the attitude and the level of satisfaction should be independent to observe the objectivity. They should be conducted jointly by clinicians and nonclinicians, who are not involved with ECT in any way, to get the required balance.

Despite the limited use of ECT in the youth, there should be specific studies for this age group.

X. What Follows ECT?

A. Medication

The effect of ECT can be considered as short lived. ECT is by no means an end-all and be-all of the treatment. This therapy is a part of the treatment and should be regarded as treatment during the acute phase of the illness. Maintenance therapy with appropriate medication should be initiated after the last ECT. Once there has been amelioration of the symptom with ECT, the underlying illness has to be treated with appropriate medication. Author's own study on the use of ECT in a young person, who presented with catatonia, is a good example (Zaw et al., 1999).

B. Maintenance ECT

Maintenance ECT is not recommended (NICE, 2003) even for adults.

C. Repeat Course of ECT

This can be considered if the diagnosis is appropriately indicated, when other treatments proved ineffective and in potentially life-threatening condition with previous favorable response (NICE, 2003).

XI. Conclusions

Electroconvulsive therapy is a procedure that attracts special safeguards under Common Law for voluntary patients and under both current and proposed mental health legislation for those receiving compulsory treatment in the UK. The call for a moratorium on the use of ECT in the youth on the one hand as well as the free usage of ECT on the other should be stopped. Injudicious use could very well continue to fuel the controversy that has existed for so many years. The public should be well and appropriately informed. There has been a recent publication on the use of ECT in adolescents (AACAP, 2004), which offers some very useful and practical advice. This is the first specific publication for the younger age group, and this could very well act as an initial force for others to follow.

Ethical considerations and the issue of capacity and informed consent are of paramount importance in children and adolescent when seriously and severely ill as for ECT to be considered. It is extremely complex and requires due attention and action. The full involvement of the parents should happen from the beginning and continued throughout the treatment not just for ECT.

Two independent and appropriate second opinions should be sought before ECT should be decided on.

ECT is an effective treatment for catatonia in all age groups. However, not all of those with the diagnosis of catatonia under DSM–IV-TR (2000) criteria will require ECT. A proportion will respond favorably to appropriate medication like benzodiazepines, lorazepam, and zolpidem. The clinical decision on the use of ECT in catatonia should follow the same principle of diagnosis, severity, and urgency as described in the recent AACAP publication (2004). Clinical urgency could dictate that ECT was used as the first choice. This therapy can be a lifesaver under certain defined conditions, for example, in catatonic stupor and extreme excitement and NMS, when the physical state of the patient is critical. Avoid concurrent medication, especially neuroleptics.

ECT offers a fairly swift but time-limited response even in those where favorable response was seen. ECT is by no means the end-all and the be-all. It is one part of the treatment. Maintenance ECT is NOT recommended.

Cognitive assessment is essential before, at the end of treatment, and 3 and 6 months posttreatment (AACAP Guidelines, 2004). Longer term studies are required.

There certainly are indications that more adolescents and even children could have received ECT without entering the literature base (Pippard and Ellam, 1981). There has to be a mechanism that would capture such valuable information that tends to get lost otherwise. There should be a separate Regional Register that feeds into the National Register of all those below the age of 18, who had received ECT. Formation of child and adolescent section of the regional ECT centers with children and adolescent psychiatrists to develop the experience and expertise over a period of time has become essential. By so doing, appropriate audits as well as well conducted studies, from the pooled data can become possible. In this day and age of evidence-based practice, the literature on the use of ECT in children and adolescents is still quite sparse. There are no information leaflets specifically prepared for children and young people in an age appropriate manner. No youth or child friendly versions exist. This should be rectified immediately by appropriate authoritative bodies, for example, RCPsych., APA, and similar others, with the involvement of appropriate professionals and the lay. User involvement is essential.

Finally, it is most important to make ECT as maximally effective as possible, keep side effects to a minimum and tailor the treatment to the individual need. ECT should only be administered in a service in which all relevant staff keep up-to-date with emerging evidence and have the necessary skills to a high standard. Our patients deserve the very best.

XII. Conflict of Interest

None.

Acknowledgment

I thank Professor Femi Oyebode for his helpful advice. I also thank Dr. Demi Onalaja for information relating to the ECT services in adults in the UK, Dr. Wai Phyo for his assistance with the literature search and Dr. A. Scott for a copy of the section on "Use of ECT in Children and Adolescents for treatment of depression," in ECT Handbook, 2nd ed. (2005).

References

AACAP Official Action (2004). Practice parameter for use of electroconvulsive therapy with adolescents. *J. Am. Acad. Child Adolesc. Psychiatry* **43,** 1521–1539.

Abrams, R., and Taylor, M. A. (1976). Catatonia: A prospective clinical study. *Arch. Gen. Psychiatry* **33,** 579–581.

American Psychiatric Association (1994). "Diagnostic and Statistical Manual (DSM IV)." APA Press, Washington, DC.

American Psychiatric Association (2000). "Diagnostic and Statistical Manual. Text Revision (DSM IV TR)." APA Press, Washington, DC.

American Psychiatric Association (2001). "The Practice of ECT: Recommendations for Treatment, Training and Privileging." APA press, Washington, DC.

Aaronson, T. A., Shukla, S., and Hoff, A. (1987). Continuation therapy after ECT for delusional depression: A naturalistic study of prophylactic treatments and relapse. *Convulsive Ther.* **3,** 251–259.

Aubas, S., Biboulet, P., Daures, J. P., and du Cailar, J. (1991). Incidence and aetiology of cardiac arrest during the preoperative period and in the recovery room. *Annales francaises d' anesthesie et de reanimation* **10,** 436–442.

Baldwin, S., and Jones, Y. (1998). Is electroconvulsive therapy unsuitable for children and adolescents. *Adolescence* **33,** 645–655.

Bertagnoli, M., and Borchardt, C. (1990). A Review of ECT for children and adolescents. *J. Am. Acad. Child Adolesc. Psychiatry* **29,** 302–307.

Bloch, Y., Levcovitch, Y., Bloch, A. M., Medlovic, S., and Ratzoni, G. (2001). Electroconvulsive therapy in adolescents: Similarities to and differences from adults. *J. Am. Acad. Child Adolesc. Psychiatry* **40,** 1332–1336.

Caird, H., Worrall, A., Eds. Revised by Fortune, Z., and Cresswell, J. (2004a). "The ECT Accreditation Service (ECTAS) Standards for Administration of ECT," 2nd ed. Pub. No. CRU033, The Royal College of Psychiatrists' Research Unit, 6th, Floor, 83 Victoria Street, London.

Caird, H., Worrll, A., and Lelliott, P. (2004b). The electroconvulsive therapy accreditation service. *Psychiatric Bull.* **28,** 257–259.

Chandler, J. (1991). Psychogenic catatonia with elevated creatinine kinase and automatic hyperactivity. *Can. J. Psychiatry* **36,** 530–532.

Cizadlo, B., and Wheeton, A. (1995). Case study: ECT treatment of a young girl with catatonia. *J. Am. Child Adolesc. Psychiatry* **34,** 332–335.

Cohen, D., Paillere-Martinot, M. L., and Basquin, M. (1997). Use of electroconvulsive therapy in adolescents. *Convulsive Ther.* **13,** 25–31.

Cohen, D., Taieb, O., Flament, M., Benoit, N., Chevret, S., Corcos, M., Fossati, P., Jeammet, P., Allilaire, J. F., and Basquin, M. (2000). Absence of cognitive impairment at long term follow up in adolescents treated with ECT for severe mood disorders. *Am. J. Psychiatry* **157,** 460–462.

Dhossche, D. (2004). Autism as an early expression of catatonia. *Med. Sci. Monit.* **10,** 31–39.

Dhossche, D., and Bouman, N. H. (1997). Catatonia in children and adolescents. *J. Am. Acad. Child Adolesc. Psychiatry* **36,** 870–871.

Dhossche, D., and Stanfill, S. (2004). Could ECT be effective in autism? *Med. Hypotheses* **63,** 371–376.

ECT in Scotland A Guide to ECT: May 2005 The Latest Evidence. www.sean.org.uk

Eranti, S., and McLoughlin, D. (2003). Electroconvulsive therapy – state of the art. *Br. J. Psychiatry* **182,** 8–9.

Fink, M., and Taylor, M. A. (1991). Catatonia: A separate category for DSM IV. *Integrative Psychiat.* **7,** 2–10.

Fink, M., and Taylor, M. A. (2003). "Catatonia – A Clinician's Guide to Diagnosis and Treatment." University Press, Cambridge.

Fisher, S., Fisher, R., and Hilkevitch, A. (1953). The conscious and unconscious attitudes of psychotic patients towards electric shock treatment. *J. Nerv. Ment. Dis.* **118,** 144–152.

Freeman, C. (1995). "ECT in those under 18 years old." The Royal College of Psychiatrists- The ECT Handbook. The Second Report of the Royal College of Psychiatrists' Special Committee on ECT (Council Report CR 39). Royal College of Psychiatrists, London.

Freeman, C. P. L., and Kendell, R. E. (1980). ECT: Patients' experiences and attitudes. *Br. J. Psychiatry* **137,** 8–16.

Galynker, II., Weiss, J., Ongseng, F., and Finestone, H. (1997). ECT treatment and cerebral perfusion in catatonia. *J. Nucl. Med.* **38,** 251–254.

Ghaziuddin, N., Kaza, M., Ghazi, N., King, C., Walter, G., and Rey, J. M. (2001a). Electroconvulsive therapy for minors: Experiences and attitudes of child psychiatrists and psychologists. *J. ECT* **17,** 109–117.

Ghaziuddin, N., King, C., and Naylor, M. (1996). Electroconvulsive treatment in adolescents with pharmacotherapy refractory depression. *J. Child Adolesc. Psychopharmacol.* **6,** 259–271.

Ghaziuddin, N., Laughrin, D., and Giordani, B. (2001b). Cognitive side effects of ECT in adolescents. *J. Child Adolesc. Psychopharmacol.* **10,** 269–276.

Ghaziuddin, M., Quinlan, P., and Ghaziuddin, N. (2005). Catatonia in autism: A distinct subtype? *J. Intellect. Disabil. Res.* **49,** 102–105.

Gjessing, R. (1976). "Contributions to the Somatology of Periodic Catatonia." Oxford Pergamon Press (cited in Fink, M., and Taylor, M., 2003). "Catatonia: A Clinician's Guide to Diagnosis and Treatment." University Press, Cambridge.

Guttmacher, L. B., and Cretella, H. (1998). Electroconvulsive therapy in one child and three adolescents. *J. Clin. Psychiatry* **49,** 20–23.

Hamilton, M. (1984). Fish's Schizophrenia 3rd ed. Bristol, Wright.

Hermann, R. C., Dorwart, R. A., Hoover, C. W., and Brody, J. (1995). Variation in ECT use in the United States. *Am. J. Psychiatry* **152,** 869–875.

Ishihara, K., and Sasa, M. (1999). Mechanism underlying the therapeutic effects of ECT on Depression. *Jpn. J. Pharmacol.* **80,** 185–189.

Johnstone, L. (1999). Adverse psychological effects of ECT. *J. Ment. Health* **8,** 69–85.

Kolvin, I. (1971). Studies in the childhood psychoses 1. Diagnostic criteria and classification. *Br. J. Psychiatry* **118,** 381–384.

Kutcher, S., and Robertson, H. A. (1995). Electroconvulsive therapy in treatment resistant bipolar youth. *J. Child Adolesc. Psychopharmacol.* **5,** 167–175.

Lanes, T., and Ravaris, C. L. (1993). Prolonged ECT seizure duration in a patient taking Trazodone. *Am. J. Psychiat.* **150,** 525.

Mathe, A. (1999). Neuropeptides and electroconvulsive treatment. *J. ECT* **15,** 60–75.

McCall, W. V., Shelp, F. E., and McDonald (1992). Controlled investigations of the amylobarbital interview for catatonic mutism. *Am. J. Psychiatry* **149,** 202–206.

McKiney, P., and Kellner, C. (1997). Multiple ECT's late in the course of neuroleptic malignant syndrome. *Convulsive Ther.* **13,** 269–273.

Moise, F., and Petrides, G. (1996). ECT in adolescents. *J. Am. Acad. Child Adolesc. Psychiatry* **35,** 312–318.

NICE (National Institute of Clinical Excellence) UK (April 2003). "Guidance on the Use of Electroconvulsive Therapy." Technology Appraisal 59.

Ottoson, J. O., and Fink, M. (2004). "Ethics in Electroconvulsive Therapy." Brunner, Routledge.

Paillere-Martinot, M. I., Zivi, A., and Basquin, M. (1990). Utilisation de l'ECT chez L'adolescent. *Encephale* **16,** 399–404.

Pallanti, S. (1999). Images in Psychiatry – Ugo Cerletti, 1877–1963. *Am. J. Psychiatry* **156,** April.

Parmar, R. (1983). Attitudes of child psychiatrists to ECT. *Psychiatry Bull.* **17,** 12–13.

Penney, J. F., Dinwiddie, S. H., Zorumski, C. F., and Wetzel, R. D. (1990). Concurrent and close administration of lithium and ECT. *Convulsive Ther.* **6,** 139–145.

Pippard, J., and Ellam, L. (1981). "Electroconvulsive Treatment in Great Britain, 1980." Headley Brothers Ltd., London. The Invicta Press.

Pritchett, J. T., Bernstein, H. J., and Kellner, C. H. (1993). Combined ECT and antidepressant therapy. *Convulsive Ther.* **9,** 256–261.

Realmuto, G. M., and August, G. J. (1991). Catatonia in autistic disorder: A sign of co-morbidity or variable expression. *J. Autism Dev. Disord.* **21,** 517–528.

Rey, J., and Walter, G. (1997). Half a century of ECT use in young people. *Am. J. Psychiatry* **154,** 595–602.

Rose, D., Fleishmann, P., Wykes, T., Leese, M., and Bindman, J. (2003). Patients' perspectives on electroconvulsive therapy: Systematic review. *Br. Med. J.* **326,** 7403, 1363–1365.

Sackeim, H. A. (1992). "The cognitive Effects of Electroconvulsive Therapy. Cognitive Disorders: Pathophysiology and Treatment," pp. 183–228. Marcel Dekker, New York.

Sackeim, H. A., Prudic, J., Devanand, D. P., Decina, P., Kerr, B., and Malitz, S. (1990). The impact of medication resistance and continuation pharmacotherapy on relapse following response to electroconvulsive therapy in major depression. *J. Clin. Psychopharmacol.* **10,** 96–104.

Scheftner, W. A., and Shulman, R. B. (1992). Treatment choice in neuroleptic malignant syndrome. *Convulsive Ther.* **8,** 267–279.

Schneekloth, T. D., Rummans, T. A., and Logan, K. M. (1993). Electroconvulsive therapy in adolescents. *Convulsive Ther.* **9,** 159–166.

Scott, A. (2005). College guidelines on electroconvulsive therapy: An update for prescribers. *Adv. Psychiatric Treat.* **11,** 150–156.

Scottish ECT Audit Network (SEAN) www.sean.org.uk.

Slack, T., and Stoudemire, A. (1989). Reconstitution of neuroleptic treatment with Molindone in a patient with a history of Neuroleptic Malignant Syndrome. *Gen. Hosp. Psychiatry* **11,** 365–367.

Slooter, A., Braun, K., Balk, F., van Nieuwenhuizen, O., and van der Hoeven, J. (2005). Electroconvulsive therapy of malignant catatonia in childhood. *Pediatr. Neurol.* **32,** 190–192.

Sobin, C., Sackeim, H. A., Prudic, J., Devanand, D. P., Moody, B. J., and Mcelhiney, M. C. (1995). Predictors of Retrograde Amnesia Following ECT. *Am. J. Psychiatry* **152,** 995–1001.

Strober, M., Rao, U., and DeAntonio, M. (1998). Effects of electroconvulsive therapy in adolescents with severe endogenous depression resistant to pharmacotherapy. *Biol. Psychiatry* **43,** 335–338.

Taylor, M. A., and Abrams, R. (1977). Catatonia: Prevalence and importance in the manic phase of manic depressive illness. *Arch. Gen. Psychiatry* **34,** 1223–1225.

The Royal College of Psychiatrists Special Committee on ECT. The second report of the Royal College of Psychiatrists, ECT Handbook 1st ed. (1995a). Council Report CR 39. Royal College of Psychiatrists London.

The Royal College of Psychiatrists Special Committee on ECT (2005). The third report of the Royal College of Psychiatrists. "ECT Handbook," 2nd ed. Council Report CR 128. Royal College of Psychiatrists, London.

The Royal College of Psychiatrists Fact Sheet on ECT (1995b). The Royal College of Psychiatrists, London.

The UK ECT Review Group (2003). Efficacy and safety of electroconvulsive therapy in depressive disorders: A systematic review and meta-analysis. *The Lancet* **361,** 799–808.

Thomas, P., Rascle, C., Mastain, B., Maron, M., and Vaiva, G. (1997). Test for catatonia with zolpidem. *Lancet* **349,** 702.

Thompson, J. W., and Blaine, J. D. (1997). Use of ECT in the United States in 1975 and 1980. *Am. J. Psychiatry* **144,** 557–562.

Vaidya, D. (1998). Molecular and cellular actions of chronic electroconvulsive seizures. *J. ECT* **14,** 181–193.

Walter, G., and Rey, J. (1997). An epidemiological study of the use of ECT in adolescents. *J. Am. Acad. Child Adolesc. Psychiatry* **36,** 809–815.

Walter, G., Koster, K., and Rey, J. (1999). Electroconvulsive therapy in adolescents: Experience, knowledge, and attitude of recipients. *J. Am. Acad. Child Adolesc. Psychiatry* **38,** 594–599.

Wing, L., and Attwood, A. (1987). "Syndromes of autism and atypical development." *In* "Handbook of Autism and Pervasive Developmental Disorders" (D. Cohen and A. Donellon, Eds.), pp. 3–19. Wiley, New York.

Wing, L., and Shah, A. (2000). Catatonia in autistic spectrum disorders. *Br. J. Psychiatry* **176,** 357–362.

World Health Organisation (WHO) (1992). "1992–ICD–10. International Classification of Diseases – Mental and Behavioural Disorders."

Zaw, F. (2004). "Catatonia in Adolescents," Presented as Poster. The Annual Meeting of the Child and Adolescent Section of the Royal College of Psychiatrists.

Zaw, F. (2005a). "The Use of Zolpidem to Treat Catatonia in Adolescents." (submitting for publication).

Zaw, F. (2005b). "Periodic Catatonia in an Adolescent with Schizophrenia Treated Successfully with Zolpidem." (submitting for publication).

Zaw, F., and Bates, G. (1997). Replication of the Zolpidem test for catatonia in an adolescent. *Lancet* **349,** 1914.

Zaw, F. K., Bates, G. D., Murali, V., and Bentham, P. (1999). Catatonia, autism, and ECT. *Dev. Med. Child Neurol.* **41,** 843–845.

Zwillich, C. W., Sutton, F. D., Jr., Neff, T. A., Cohn, W. M., Matthay, R. A., and Weinberger, M. M. (1975). Theophylline induced seizures in adults: Correlation with serum concentration. *Ann. Intern. Med.* **82,** 784–787.

CATATONIA IN AUTISTIC SPECTRUM DISORDERS: A MEDICAL TREATMENT ALGORITHM

Max Fink,* Michael A. Taylor,[†,‡] and Neera Ghaziuddin[‡]

*School of Medicine, State University of New York, Stony Brook, New York 11794, USA
[†]Rosalind Franklin University of Medicine and Science, North Chicago, Illinois 60064, USA
[‡]University of Michigan School of Medicine, Ann Arbor, Michigan 48109, USA

I. Introduction
II. Catatonia
    A. Catatonia in Autism
III. Treatment Algorithm
IV. Theory
    References

> Diseases desperate grown, by desperate appliance are relieved; or not at all.
> Hamlet, Act IV, Scene 3, Line 9

Autism is a developmental syndrome with an unknown biology and inadequate therapeutics. Assessing the elements of the syndrome for the presence of depression, psychosis, mania, or catatonia, offers opportunities for systematic intervention. Since almost all descriptions of autism highlight the presence of motor symptoms that characterize catatonia, an assessment for this eminently treatable syndrome is recommended for all patients considered to be autistic. A minimum examination includes a catatonia rating scale and for those patients with defined catatonia, a lorazepam test. For those whose catatonia responds to lorazepam, high dose lorazepam therapy is recommended. If this fails, electroconvulsive therapy is recommended. The assessment and treatment of catatonia offers positive medical therapy for the victims of autism and their families.

I. Introduction

Autism is a developmental disorder characterized by social deficits, impaired expressive prosody and language, and restricted repetitive behaviors. The syndrome has received much study and is now officially classified among the psychiatric disorders as a group of syndromes that include pervasive developmental disorder (PDD), Asperger syndrome, Rett syndrome, and disintegrative disorder. The term autistic spectrum disorder (ASD) is increasingly used to describe these conditions. Concurrent medical and psychiatric conditions have been described in persons with PDD, including seizure disorder, mood disorders, and schizophrenia. Some studies find patients with PDD to exhibit catatonia in adolescence and early adulthood (Wing and Gould, 1979).

Stereotyped behaviors, tics, obsessive–compulsive features, selective mutism, staring, rocking, repetitive head banging, grunting, and screaming are the disorder's more prominent symptoms. These motor features are also characteristic of catatonia, and the association of catatonia in PDD is increasingly recognized (Fink and Taylor, 2003).

Present nosology, however, does not define a coherent syndrome for ASD, and as increasing numbers of parents are alerted to the maturational difficulties of their children, the term "autism" is used as a catchall label for those with peculiar developmental histories. The diagnosis rests on an appreciation of the child's developmental history and a checkup of a list of superficial behaviors. Neither a specific diagnostic test nor an effective biological treatment is known, so parents demand social and psychological interventions of relatively unproven merit. Parents even demand (and obtain) biological treatments, as reflected in the demand for intravenous secretin injections for their children before any study was done (Levy and Hyman, 2005).

Although electroconvulsive therapy (ECT) is the most effective treatment for catatonia in all its forms and for depressive illness, the role of ECT for catatonia in autism is not established. Some examples illustrate the limited experience.

Patient W: A 17-year-old boy with history of autism, recurrent depressive illness, and mild intellectual disability was hospitalized with progressive worsening of mood accompanied by generalized slowing of movements (Ghaziuddin *et al.*, 2005). Prior to hospitalization, he was described as a well-adjusted youngster with autism. He would speak spontaneously and was able to participate in simple conversation. In addition to English, he could speak and read Hebrew. After autism was diagnosed at age 4, he received educational, speech, and social-skill interventions throughout his school years.

The decline, marked by mutism and severely reduced food intake, was dated back 3 years when he became progressively slower in movement and had to be prompted to initiate simple activities such as dressing and eating. He would stand or sit alone for prolonged periods undertaking no spontaneous activities. He would spend much time in the bathroom, washing his hands or looking in the mirror. His parents reported repetitive abnormal movements of blinking, eye-rolling, and jerking of his neck.

An acute deterioration was noted when the patient returned from a few week's summer camp. He had lost weight, his speech had become sparse, and his movements and facial expression were further slowed. Over the next 4 months he spoke progressively less and exhibited continuous motor tics, rapid blinking, and head-turning. Laboratory investigations to rule out a general medical or neurologic cause were negative.

A diagnosis of obsessive–compulsive disorder and depressive mood disorder was made, and he was treated with citalopram, risperidone, and ziprasidone, with no improvement. Movements slowed further, described by his parents as "watching a movie in slow motion." Catatonia was diagnosed, and he was treated in hospital with bilateral ECT with dramatic improvement in his movements, speech, appetite, and interpersonal interactions. He became more verbal, his movements were more spontaneous, and he was able to care for his activities of daily living. Antidepressants were prescribed, and the improvement was maintained. His family considered his response to ECT as life-saving.

Patient P: A second experience from the same clinic illustrates the difficulties in using ECT. A 17-year-old boy with normal intelligence and a diagnosis of PDD functioned well until 6 months earlier when he began kneeling for long periods because of what he described as "pain" in the perineal region. His self-care skills declined; he was unable to attend school and urinated in inappropriate places. Finally hospitalized, he exhibited bradykinesia and bradyphrenia. He sat in a kneeling position for long periods while mumbling to himself. Catalepsy and automatic obedience were noted. Extensive laboratory workup, including an EEG and MRI, found no abnormalities. Extensive medication trials, alone and in combinations, of paroxetine, clomipramine, venlafaxine, fluvoxamine, mirtazepine, duloxetine, valproate, ziprasidone, and quetiapine elicited no benefit.

Considering the lack of response and the severity of illness, a course of ECT was recommended. After all state regulations and treatment guidelines were met, the patient was treated with 18 bilateral ECT. During the ECT course, the patient displayed brief periods of remarkable improvement (speaking spontaneously, joking, and laughing). When the course ended, however, he rapidly declined to his pre-ECT condition and was again rigid, exhibiting psychomotor retardation, and described as being "stuck to the wall in a glue-like manner." Post-ECT laboratory workup and EEG were normal. A careful review of his treatment course revealed that ziprasidone (up to 60 mg/day) had been concurrently administered, despite injunctions by the ECT team to stop all antipsychotic medications.

This patient illustrates the difficulty of treating adolescents. Most child and adolescent psychiatrists are unfamiliar with catatonia, its treatment implications, and ECT. When ECT is selected for the treatment of catatonia, all antipsychotic medications are best discontinued to prevent relapse.

Misinterpreting catatonic features for OCD or aspects of Giles de la Tourette's syndrome can defer appropriate management. In the following vignette, catatonia was not recognized until mutism, posturing, staring, and negativism became blatant. Appropriate treatment relieved the signs of catatonia and enabled a return to school and productive work, although socialization and communication skills remained impaired.

Patients A, B: Identical twin boys, born prematurely, developed normally until at age 4 they were noted to have speech and fine motor difficulties, lack of imaginative play, and poor social skills. A Pervasive Developmental Disorder (later Asperger's syndrome) was diagnosed and the boys began the characteristic course of special education (speech, occupational, visual, and cognitive therapies) and psychotherapy. They did well in standardized state academic tests and in athletics, but in 8th grade, B's social problems became overwhelming and a new round of consults ensued, followed by prescriptions of SSRI antidepressants and psychotherapy. His behavior improved.

The next year, motor changes were observed in both boys, more so in B. He would freeze in inaction, posture strangely, and stare. At times, he refused to eat or leave his room. During an interview with a psychiatrist, he offered one-word answers. On other days, he appeared to be well. His behavior was diagnosed as an expression of OCD, and fluvoxamine was prescribed. (At other examinations, Tourette's syndrome was diagnosed.) His brother exhibited similar signs, but to a lesser degree.

That June, B took a full day of examinations, came home, and slept for 16 hours. He passed his examinations, but his behavior worsened. He held his hands in stiff postures, grimacing often, performed circling rituals, and was indecisive and periodically mute. Fluoxetine was prescribed and B became violent and agitated. Psychosis was diagnosed and risperidone prescribed. He became manic and restless, paced endlessly, and exhibited tics. For the next year, he attended a special school.

When B was at age 17, a consultant diagnosed catatonia and a trial of lorazepam was recommended. With this treatment, he became more verbal and interactive, and the signs of catatonia less prominent. As higher dosages were required to sustain the benefit, he became over-sedated. With reduced dosing, the catatonic signs returned.

The brother's behavior in all this time was similar, although less severe, with similar treatments and similar benefits.

At age 19, a second consultant examined both boys, finding them slow in movement, staring and posturing, responding sparsely to questions. Both admitted to being sad and unhappy at school, with few friends, spending most of the time in their rooms. History were consistent with the diagnosis of catatonia in a depressive mood disorder and ECT was recommended.

A course of bilateral ECT at two treatments a week was begun for B. After nine treatments, he was more verbal, related better to others, was no longer staring, posturing, or negativistic, and he appeared happy. He received additional weekly treatments (total of 19) and was able to attend vocational school, work part-time, and take part in community outreach programs. The signs of autism, depression, and catatonia improved.

After B's response, the family agreed to the brother's treatment. He, also, showed a resolution of the catatonic signs, odd behaviors, and he became fully verbal. He, also, attends a technical school, works part-time, and volunteers in community groups.

The parents report that both boys are better integrated in their community and the family. However, there was no change in their underlying diagnosis of pervasive developmental disorder.

II. Catatonia

Catatonia is a well-defined motor syndrome first delineated by Karl Kahlbaum in 1874. Kraepelin precipitously endorsed the syndrome as a form of dementia praecox, tying catatonia to schizophrenia. This association persisted in Diagnostic and Statistical Manual, 3rd ed. (DSM-III) as catatonia and was recognized only as a subtype of schizophrenia. The DSM-IV continues this association and only perfunctorily recognizes catatonia associated with general medical conditions and mood disorder. This limitation persists despite the well-documented association of catatonia in patients with manic and depressive mood disorders, toxic and infectious conditions, and epileptic states. Catatonia is also specifically recognized as a potentially toxic response to all antipsychotic agents in the neuroleptic malignant syndrome (NMS) and to selective serotonin reuptake inhibitor (SSRI) antidepressants in the toxic serotonin syndrome. Many medications elicit catatonia (Fink and Taylor, 2003).

Simplified diagnostic criteria for catatonia have been proposed (Table I).

The presence of two or more signs for 24 hours or longer is considered sufficient for the diagnosis. The features need not include mutism or immobility as many catatonic patients, particularly those with mania, are ambulatory and speak.

The recognition of catatonia is facilitated by the use of rating scales, which define the features of stereotypy, mutism, negativism, posturing, staring, and

TABLE I
Diagnostic Criteria for Catatonia (Fink and Taylor, 2003)

A. Immobility, mutism, or stupor of at least 1 hour duration, associated with at least one of the following: catalepsy, automatic obedience, or posturing, observed or elicited on two or more occasions
B. In the absence of immobility, mutism, or stupor, at least two of the following, which can be observed or elicited on two or more occasions: stereotypy, echophenomena, catalepsy, automatic obedience, posturing or negativism, ambitendency

echopraxia, among others. For a discussion of the merits and application of catatonia rating scales (CRSs) see Fink and Taylor (2003).

A. CATATONIA IN AUTISM

These reports highlight the association between catatonia and autism, although they do not clarify the nature of the association. Explanations vary from autism as a manifestation of the catatonia syndrome, catatonia as a complication of autism, and catatonia as a cause of deterioration in patients with autism.

Catatonia is well defined in adults, and diagnostic criteria are detailed (Fink and Taylor, 2003). Yet, specific criteria for catatonia in children or adolescents are lacking. The studies that describe catatonic features in young patients, however, suggest that the criteria applied to adults are applicable to children and teenagers, and the presentation of catatonia is consistent across all age groups (Hare and Malone, 2004; Takaoka and Takata, 2003; Wing and Shah, 2000).

Several published reports suggest an association between catatonia and autism, with dramatic improvements when autistic patients with catatonia are treated for the syndrome of catatonia. A description of three autistic patients reported that they shared the symptoms of mutism, echopraxia/echolalia, and stereotypy with catatonia (Realmuto and August, 1991). The authors concluded that catatonia in autism may be a variant of the autistic condition.

The clinical presentation, course, and treatment response of a 14-year-old boy with catatonic stupor and with a preexisting diagnosis of autism reported that the patient displayed mutism, akinesia, motor rigidity, waxy flexibility, posturing (including psychological pillow), facial grimacing, and other involuntary movements of his upper extremities (Zaw *et al.*, 1999). Intravenous injection of sodium amobarbital failed to resolve the motor symptoms. A course of ECT, however, elicited sustained relief of catatonic stupor without a change in the symptoms of autism.

The largest study of the association of catatonia and autism is the report by Wing and Shah (2000). Using a semi-structured diagnostic schedule to gather information from parents and care-givers of 504 patients referred to a clinic for ASDs, they reported that 17% of patients aged 15 or older had features of catatonia. These patients were more likely than the comparison group to have had, before the onset of the change in behavior, impaired language, and passivity in social interaction of skill and behavior. An earlier study reported 6.5% of autistic children developing catatonia by their late teens (Gillberg and Steffenburg, 1987).

Dhossche (2004) described catatonic symptoms to be common in autism and implicated abnormal gamma-aminobutyric acid (GABA) function in both disorders. Additionally, neuroimaging studies showed small cerebellar structures in both disorders. Hypothesized susceptibility genes for autism and catatonia on the long arm of chromosome 15 are being evaluated (Dhossche, 2004).

In an 18-year-old boy with ASD that worsened over 4 years with progressive stereotypy, rocking, posturing, and motor rigidity until it required hospital care, the behaviors were interpreted as psychological difficulties in execution, not in planning (Hare and Malone, 2004). Behavioral training relieved the symptoms and allowed the patient to carry on more normal daily motor activities.

III. Treatment Algorithm

Stereotypy and abnormal motor features are so prominent in identifying autism that an assessment for catatonia is essential in every examination of a child or adolescent for developmental disorder. When found, the symptoms of catatonia are treatable.

The first effective treatment for catatonia was described by Bleckwenn (1930) in the administration of intravenous sodium amobarbital in high doses. For the next half-century, barbiturates were widely used. With concerns about the toxicity of these agents, benzodiazepines were developed as suitable replacements.

The first patients successfully treated with convulsive therapy were suffering with catatonia (Meduna, 1985). In the succeeding decades, convulsive therapy was found to be remarkably effective in nonmalignant and malignant catatonia. ECT is a definitive treatment for catatonia and is prescribed when benzodiazepines fail (Table II).

The intravenous administration of amobarbital, lorazepam, or diazepam relieves catatonia in more than half the patients. The challenge test is simple to perform. The symptoms are rated in a CRS. An intravenous line is established. A syringe containing 2–4 mg lorazepam in 2 ml (or larger volumes of solution) is prepared and 1 mg injected. In the next 2–5 min, the symptoms of catatonia are again rated. If no change is observed, the second 1 mg lorazepam is injected, and the assessment repeated. For patients who show a reduction in CRS scores of 50% or more, treatment with increasing doses of lorazepam are prescribed.

TABLE II
Catatonia Treatment Algorithm (Fink and Taylor, 2003)

1. Score Catatonia Rating Scale
2. Lorazepam Challenge Test
    3a. If positive, treat with oral lorazepam, 6–24 mg/day
    3b. With rapid improvement, continue lorazepam maintenance for 6 months
    3c. If catatonia does not resolve within a week, treat with bilateral ECT
4. If lorazepam challenge test is negative, treat with bilateral ECT

Dosages as high as 24 mg/day have been necessary for effective relief, as illustrated in the following patient.

Patient Y: A 14-year-old girl with developmental difficulties associated with massive brain cysts was hospitalized on a child neurology service because of mutism, rigidity, failure to eat or drink, incontinence, fever, and tachycardia. Well until the year before, she had developed a "depressive-like" condition that was treated first with antidepressant medications and then, after a period of increased activity, impulsivity, and irritability, with haloperidol. As manic behavior abated, mutism and rigidity emerged. An extensive neurologic investigation was unhelpful, and a neuropsychiatric assessment was requested.

When first seen, the girl was seated in her room, intravenous line and nasogastric tube in place. A nurse was seen to be encouraging the girl, who was hunched over a writing pad, pen in hand, poised to write. Other than a few scratches on the pad, however, she had written nothing and had been in the posture for some time. Her only responses were a brief whisper, but she understood instructions. Gegenhalten and waxy flexibility could be elicited. About 2 mg of lorazepam were administered through the IV line, and over the next 2 min the girl straightened her position, lifted her head, turned to the nurse, smiled, looked at her pad, and began to write. Lorazepam throughout the day was prescribed. The family refused ECT. Dosages of lorazepam were increased over a 3-week period. An effective dosage of 24 mg/day was eventually required to permit her to go home and return to school.

If lorazepam therapy is effective and the underlying cause of the catatonia has resolved, a 6-month maintenance lorazepam course at the dose that resolved the catatonia is recommended as otherwise relapse is likely, as observed in patient Z.

Patient Z: A 17-year-old boy was hospitalized after he ceased to speak, eat, and drink. He spent hours standing in the middle of his room, not responding to his parents. Catatonia was diagnosed, and lorazepam therapy was begun. Dosages were increased, and after several weeks, at an oral dose of 14 mg/day, catatonia resolved.

Because he was from another state, discharge was planned for the following Thursday and lorazepam was to be tapered. On Friday it was reduced to 12 mg/day. On Monday he had relapsed, and lorazepam was increased to 14 mg/day. By Wednesday catatonia had again resolved, and he was discharged the next day to successfully remain on 14 mg/day for the next 6 months.

An additional description of an autistic young woman who developed a manic syndrome with catatonic features that successfully responded to ECT is cited by Fink (1999).

If the patient has no response to the lorazepam challenge or fails to respond to high doses of lorazepam, ECT is the treatment of choice and often life-saving. To deny ECT at this point can lead to permanent disability or death. While improvement has followed with the usual schedules of three treatments a week, daily treatment for 3–5 days is often necessary, particularly in the more malignant conditions. The efficacy of bilateral electrode placement is better documented than is unilateral electrode placement. Continuation ECT is typically necessary. While the guidelines for frequency and number of treatments are poorly defined, the present suggestion is that weekly and biweekly treatments may be needed for 3–6 months to ensure a stabilized response.

The consideration of ECT in children and adolescents is fraught with difficulties. These are not engendered by the treatment itself or its risks, for the treatments have extremely low morbidity and almost nonexistent mortality rates. Fears that ECT will interfere with brain development is stated as reasons for its nonuse. Yet, follow-up studies of children and adolescents treated with ECT show no signs of brain damage or impairment of the anticipated educational development. Children successfully treated with ECT return to normal schooling and perform as well as their peers (Cohen et al., 1997; Rey and Walter, 1997).

ECT use is inhibited by the treatment's stigmatization where psychodynamic and social concepts have long dominated the thinking and care of children. One consequence is legislative restrictions in several states that limit the ages in which ECT may be used (Ottosson and Fink, 2004). In addition to the hurdle of educating parents on the utility and safety of the treatment to assure their consent, the practice guidelines require the concurrence of two specialists in pediatric psychiatry before ECT can be administered (APA, 2001). Such a requirement inhibits the proper use of ECT since the number of qualified pediatric psychiatrists is small (Ghaziuddin et al., 2001).

The appearance of catatonic signs or of psychosis is a frequent justification for the use of antipsychotic agents. But the incidence of NMS increases in patients with incipient signs of catatonia. The relief afforded catatonia by antipsychotic drugs is small, and the risk for worsening is sufficiently strong to preclude their use (Fink and Taylor, 2003). Antipsychotic agents should not be used in patients with autism who exhibit the signs of catatonia.

IV. Theory

The association of catatonia and ASD and the efficacy of ECT for persons so afflicted are consistent with the understanding of the pathophysiology of the two conditions and the effects of ECT. Disorders of movement are ubiquitous in ASD, and stereotypic movements—hand flapping, tics, compulsions, and other repetitive movements that are characteristic of catatonia—are commonly described (Bodfish et al., 2000; Brasic, 1999; Green et al., 2002; Gritti et al., 2003; Kano et al., 2004; Miliyerni et al., 2002). Catatonia is the paradigm of a movement dysregulation disorder (Taylor and Fink, 2003).

The brain structures considered to be abnormal in ASD, particularly the frontal circuitry, are also perturbed in catatonia and can be seen as directly eliciting the abnormal movements common to both conditions (Acosta and Pearl, 2004; Carper and Courchesne, 2005; Sears et al., 1999). The brain structural abnormalities associated with ASD emerge progressively and appear to disrupt the functional connections of the frontal circuitry with other brain regions, including the limbic system (Bauman and Kemper, 2005; Courchesne and Pierce, 2005).

Catatonia is most commonly associated with mood disorder, and as many as half of adult catatonic patients will simultaneously have a depressive illness or mania (Fink and Taylor, 2003). Autistic persons are also at greater risk for mood disorder, and mood disorder is prevalent in the relatives of sufferers (DeLong, 2004).

Autistic persons are also at greater risk for epilepsy. Up to one-third of sufferers experience overt seizures and as many as 75% will have epileptiform patterns on the EEG (Danielsson et al., 2005; Kagan-Kushnir et al., 2005; Tuchman and Rapin, 2002). The association between catatonia and seizure disorder in younger patients is strong (Fink and Taylor, 2003). Persons with ASD are thus likely to develop catatonia for three reasons: the sharing of a motor dysregulation process, the comorbidities of mood disorder, and epilepsy.

Barbiturates, benzodiazepines, and ECT are effective in relieving catatonia and seizure disorder, and their anticonvulsant activity is one explanation for this common effect. Each intervention raises seizure thresholds. Carbamazepine is another anticonvulsant that has been used successfully to relieve catatonia. A neurohumoral theory of the role of GABA has been formulated in catatonia. A similar formulation has been proposed for the catatonia in autism (Dhossche, 2004).

ECT is also effective in relieving mood disorder without catatonia. Thus, an autistic person with a depressive or a manic mood disorder has an ECT-responsive syndrome. The two common underlying processes eliciting catatonia in such an age group are themselves ECT-responsive conditions.

References

Acosta, M. T., and Pearl, P. L. (2004). Imaging data in autism: From structure to malfunction. *Semin. Pediatr. Neurol.* **11,** 205–213.

American Psychiatric Association APA 2001. "*Electroconvulsive Therapy: Treatment, Training and Privileging,*" 2nd ed. APA Press, Washington DC.

Bauman, M. L., and Kemper, T. L. (2005). Neuroanatomic observations of the brain in autism: A review and future directions. *Int. J. Dev. Neurosci.* **23,** 183–187.

Bleckwenn, W. J. (1930). The production of sleep and rest in psychotic cases. *Arch. Neurol. Psychiatry* **24,** 365–372.
Bodfish, J. W., Symons, F. J., Parker, D. E., and Lewis, M. H. (2000). Varieties of repetitive behavior in autism: Comparisons to mental retardation. *J. Autism. Dev. Disord.* **30,** 237–243.
Brasic, J. R. (1999). Movements in autistic disorder. *Med. Hypotheses* **53,** 48–49.
Carper, R. A., and Courchesne, E. (2005). Localized enlargement of the frontal cortex in early autism. *Biol. Psychiatry* **57,** 126–133.
Cohen, D., Dubos, P. F., Hervé, M., and Basquin, M. (1997). Épisode catatoniques à l'adolescence: Observations cliniques. *Neuropsychiatr. Enfance Adolesc.* **45,** 597–604.
Courchesne, E., and Pierce, K. (2005). Why the frontal cortex in autism might be talking only to itself: Local over-connectivity but long-distance disconnection. *Curr. Opin. Neurobiol.* **15,** 225–230.
Danielsson, S., Gillberg, I. C., Billstedt, E., Gillberg, C., and Olsson, I. (2005). Epilepsy in young adults with autism: A prospective population-based follow-up study of 120 individuals diagnosed in childhood. *Epilepsia* **46,** 918–923.
DeLong, R. (2004). Autism and familial major mood disorder: Are they related? *J. Neuropsychiatry Clin. Neurosci.* **16,** 199–213.
Dhossche, D. (2004). Autism as early expression of catatonia. *Med. Sci. Monit.* **10,** RA31–RA39.
Fink, M. (1999). *Electroshock: Restoring the Mind,* pp. 47–49. Oxford University Press, New York.
Fink, M., and Taylor, M. A. (2003). *Catatonia: A Clinician's Guide to Diagnosis and Treatment.* Cambridge University Press, Cambridge, United Kingdom.
Ghaziuddin, M., Quinlan, P., and Ghaziuddin, N. (2005). Catatonia in autism: A distinct subtype? *J. Intellect. Disabil. Res.* **49,** 102–105.
Ghaziuddin, N., Kaza, M., Ghazi, N., King, C., Walter, G., and Rey, J. (2001). The use of electroconvulsive therapy in minors: Experience, attitude of child psychiatrists & psychologist. *J. ECT* **17,** 109–117.
Gillberg, C., and Steffenburg, S. (1987). Outcomes and prognostic function in infantile autism and similar conditions. *J. Autism Devel. Dis.* **17,** 273–287.
Green, D., Baird, G., Barnett, A. L., Henderson, L., Huber, J., and Henderson, S. E. (2002). The severity of motor impairment in Asperger's syndrome: A comparison with specific developmental disorder of motor function. *J. Child Psychol. Psychiatry* **43,** 655–668.
Gritti, A., Bove, D., DiSarno, A. M., D'Addio, A. A., Chippara, S., and Bove, R. M. (2003). Stereotyped movements in a group of autistic children. *Funct. Neurol.* **18,** 89–94.
Hare, D. J., and Malone, C. (2004). Catatonia and autistic spectrum disorders. *Autism* **8,** 183–195.
Kagan-Kushnir, T., Roberts, S. W., and Snead, O. C., IIIrd (2005). Screening electroencephalograms in autism spectrum disorders: Evidence-based guideline. *J. Child Neurol.* **20,** 197–206.
Kano, Y., Ohta, M., Nagai, Y., Pauls, D. L., and Leckman, J. F. (2004). Obsessive-compulsive symptoms in parents of Tourette syndrome probands and autism spectrum disorder probands. *Psychiatry Clin. Neurosci.* **58,** 348–352.
Levy, S. E., and Hyman, S. L. (2005). Novel treatments for autistic spectrum disorders. *Ment. Retard. Dev. Disabil. Res. Rev.* **11,** 131–142.
Meduna, L. (1985). Autobiography. *Convulsive Ther.* **1,** 43–57, 121–138.
Miliyerni, R., Bravaccio, C., Falco, C., Fico, C., and Palermo, M. T. (2002). Repetitive behaviors in autistic disorder. *Eur. Child Adolesc. Psychiatry* **11,** 210–218.
Ottosson, J.-O., and Fink, M. (2004). *Ethics in Electroconvulsive Therapy.* Brunner-Routledge, New York.
Realmuto, G. M., and August, G. J. (1991). Catatonia in autistic disorder: A sign of comorbidity or variable expression? *J. Autism Dev. Disord.* **21,** 517–528.
Rey, J. M., and Walter, G. (1997). Half a century of ECT use in young people. *Am. J. Psychiatry* **154,** 595–602.

Sears, L. L., Vest, C., Mohamed, S., Bailey, J., Ranson, B. J., and Piven, J. (1999). An MRI study of the basal ganglia in autism. *Prog. Neuropsychopharmacol Biol. Psychiatry* **23,** 613–624.

Takaoka, K., and Takata, T. (2003). Catatonia in childhood and adolescence. *Psychiatry Clin. Neurosci.* **57,** 129–137.

Taylor, M. A., and Fink, M. (2003). Catatonia in psychiatric classification: A home of its own. *Am. J. Psychiatry* **160,** 1233–1241.

Tuchman, R., and Rapin, I. (2002). Epilepsy in autism. *Lancet Neurol.* **1,** 352–358.

Wing, L., and Gould, J. (1979). Severe impairments of social interaction and associated abnormalities in children: Epidemiology and classification. *J. Autism Child. Schizophr.* **9,** 11–29.

Wing, L., and Shah, A. (2000). Catatonia in autistic spectrum disorders. *Br. J. Psychiatry* **176,** 357–362.

Zaw, F. K., Bates, G. D., Murali, V., and Bentham, P. (1999). Catatonia, autism, and ECT. *Dev. Med. Child Neurol.* **41,** 843–845.

PSYCHOLOGICAL APPROACHES TO CHRONIC CATATONIA-LIKE DETERIORATION IN AUTISM SPECTRUM DISORDERS

Amitta Shah* and Lorna Wing[†]

*Leading Edge Psychology, Purley CR8 2EA, United Kingdom
[†]Centre for Social and Communication Disorders, Bromley, Kent BR2 9HT, United Kingdom

I. Introduction
II. Effects of Stress
III. Effect of Medical Treatments
IV. Problems with Assessing the Effects of Medical Treatments
V. Psychological Methods of Intervention
VI. General Principles of Psychological Treatment and Management
   A. Initial Assessment
   B. Dealing with Stress Factors
   C. Conceptualization of the Nature of Catatonia-Like Deterioration
   D. Use of Prompts as External Stimuli/Signals
   E. Maintaining and Increasing Activity
   F. Structure and Routine
VII. Management of Specific Problems
   A. Incontinence
   B. Eating Problems
   C. Speech and Communication Problems
   D. Difficulty with Walking
   E. Standing Still or Adopting Fixed Postures
   F. Catatonia-Like Excitement
VIII. Implications for Services and Staffing Levels
IX. Conclusions
   Appendix A
   Individual Case History
   References

The psychological dysfunctions that may underlie catatonia-like deterioration in autism spectrum disorders are discussed. Clinical observation suggests that an important factor is ongoing stress. The evidence for this from research and clinical observation is considered. The lack of evidence concerning the most appropriate medical treatments is discussed. A psychological approach designed for individual needs by relevant professionals and applied by parents and/or caregivers is described. This can be helpful whether or not medical treatments are used. It involves detailed holistic assessment of the individual and their circumstances to highlight possible precipitating stress factors in view of their underlying autism and cognitive/psychological functioning. The overall aim of this approach is to restructure the individual's lifestyle, environment and resolve cognitive/psychological factors to reduce the stress. An eclectic approach is used to find individual strategies in order to provide external goals and stimulation to increase motivation and keep the person engaged and active in meaningful and enjoyable pursuits. The approach describes ways of using verbal and physical prompts as external stimuli to overcome the movement difficulties and emphasizes maintaining a predictable structure and routine for each day. The importance of educating caregivers and service providers to understand the catatonia-like behavior is emphasized. Advice is given on management of specific problems such as incontinence, freezing in postures, eating problems, and episodes of excitement.

> Whether a particular disorder is precipitated or relieved by psychological factors has no bearing on whether a neurological or psychological paradigm is more appropriate for understanding it.
>
> Rogers, 1992

I. Introduction

As described in Chapter 2 by Wing and Shah, in a small minority of people with autism spectrum disorders (ASDs), some catatonia-like and some parkinsonism-like features become very marked or appear for the first time and are severe enough to interfere with the activities of everyday life. This usually happens in adolescence or adult life, though can rarely be seen in childhood. In most cases the onset is gradual and the presence of all the features of classic stupor appears to be rare (Wing and Shah, 2000). This will be referred to here as "catatonia-like deterioration," to avoid the premature conclusion that it is the same as catatonia. A prevalence of 17% was found among people aged 15 years or above who were referred to a diagnostic service (Wing and Shah, 2000).

Because of our special interest in catatonia-like conditions in autistic disorders (ADs), the present authors have seen a large number of individuals with these clinical pictures. The problems of movement, speech, and behavior in catatonia-like deterioration in ADs are similar to those found in acute and chronic catatonic states. But, in our experience, typical catatonic stupor is rare, and we have not seen waxy flexibility in anyone referred to us. The onset of the deterioration in those we have seen was characteristically slow. However, individual cases of autism and catatonic stupor have been reported in the literature (Dhossche, 1998; Ghaziuddin *et al.*, 2005; Realmuto and August, 1991; Zaw *et al.*, 1999).

The precise neuropathology of ASDs is as yet unknown, and there are no known treatments that can cure the characteristic impairments of social interaction, communication, and imagination. At the time of writing, no systematic studies of the causes, nature, and treatment of catatonia-like conditions in ADs have been published. The few papers in the literature that possibly have some bearing on this condition will be discussed, but the results they describe tend to make confusion worse confounded.

II. Effects of Stress

Our clinical observations suggest that ongoing stressful experiences are a major precipitating factor in many individuals who develop catatonia-like deterioration. We have identified various diverse stress factors. These include:

1. External factors, such as being in unstructured environments and programs, which are not autism-friendly. The loss of routine, structure, and meaningful occupation that occurs after leaving school is a common factor. In adolescence, the pressure of major examinations becomes unbearable for some individuals. In adults, an increase in social, adaptive, and independence demands leads to stress and noncoping and eventually giving up. Significant life events, such as bereavement, family break up, house move, and so on may also have a severe effect on some individuals.

2. Psychological factors such as experience of conflict, pressure, and confusion. This can be due to diverse factors such as not having a diagnosis or explanation for one's difficulties or alternatively not accepting a diagnosis or an accumulation of failure experiences. In high functioning people with Asperger Syndrome (AS), an awareness of their limitations and differences from peers/siblings can cause psychological pressure and stress. In some cases, the conflict between parental expectation and the individual's capacity and lack of motivation adds to the pressure.

3. Biological factors, such as illness, pain, and hormonal changes (e.g., during puberty), are also implicated in some cases.

The mechanisms through which the experience of stress leads to catatonia-like deterioration are completely unknown. At a neuropsychological level, it can be speculated that in vulnerable autistic individuals, functions and systems that are weak are further weakened. Catatonia-like deterioration is most commonly seen in individuals who have shown the "passive" type of social impairment and functioning (Wing and Shah, 2000). Individuals in this "passive" subgroup have ongoing difficulties of motivation, volition, and initiating activity and interaction. Uta Frith's theory of central coherence deficit and frontal lobe executive dysfunction in ASDs leading to difficulty in controlling and regulating behavior offers a possible explanation. (Frith, 2003; Hill, 2004). It is possible that individuals in the "passive" subgroup have a weaker system of central control of actions. Ongoing stress may further disrupt this function and give rise to the characteristic problems of initiating and ceasing actions and speech, poor control of behavior, and increased dependence on external prompts.

These speculations give no help in understanding those cases of catatonia-like deterioration that follow a course in which slowness, movement difficulties, and muteness are very evident on some days, but are not seen at all on other days. In instances when the individual is able to "miraculously" overcome the inability to initiate an action or movement for a specific sudden stimulus, it can be hypothesized that the power/shock of the external stimulus jolts the weak central control system into temporary "action." One such example relates to a young man described in Wing and Shah (2000) who was permanently confined to a wheelchair due to catatonia-like deterioration. On one occasion, when his elderly father stumbled and nearly fell, the young man was able to leap from his wheelchair, save his father from falling, but then returned to immobility.

At a psycho-physiological or neurological level, there is no research yet that has identified any specific mechanisms relating stress and catatonia-like deterioration in autistic individuals. The following discussion considers studies and hypotheses that have some bearing on the subject.

Dhossche and Rout (Chapter 4) have offered a biochemical theory involving abnormal gamma-aminobutyric acid (GABA) function that may link autistic symptoms and catatonia. Adaptive and maladaptive stress responses are regulated by GABAergic processes in the hypothalamus (Kovacs et al., 2004). Hypothalamic abnormalities may be present in autism and catatonia.

There is an interesting case study by Loos and Miller (2004) of the link between stress and "shut-down" in a child with high functioning autism. Their description of the "shut-down" state has a lot of overlap with catatonia-like deterioration. These authors have suggested that chronic "stress instability" can lead to "shut-down" episodes and cause nervous system damage and developmental impairments with the autism spectrum. These authors have suggested that a corticotropin releasing factor (CRF) that is important in stress functions may be unstable in some individuals within the autism spectrum.

The subjective experience of anxiety and physiological correlates of anxiety may have some relevance to the mechanisms relating stress and catatonia.

It has been suggested, in relation to typical catatonic stupor, that extremely high anxiety is an important factor. Northoff (2002) investigated retrospectively the subjective experiences of patients who had had episodes of classic acute catatonic stupor. The dominating emotion reported was anxiety that was overwhelming in its intensity. Northoff contrasted these retrospective subjective experiences with those of people with parkinsonism. He noted that the people who had had catatonic episodes seemed unaware that they had had problems with movements, whereas those with parkinsonism were fully aware of this aspect of their condition. People with ASDs, especially those with catatonia-like deterioration, have difficulty verbalizing their feelings. However, one young man in the study by Wing and Shah (2000) who was described in Chapter 2 and another man seen subsequently were able to give accounts of their experiences during catatonia-like deterioration. Both described clearly that they wanted to move, tried to move, but their bodies would not obey. They did not describe feelings of anxiety.

A different picture was found in another young man assessed by Shah. He had high functioning AS and developed catatonia-like deterioration after leaving school. During the psychological assessment it was noticed that this young man showed no signs of movement disorder associated with catatonia when he was focusing on verbal and nonverbal cognitive tests, which he enjoyed and found extremely easy. However, when the assessment required him to express and verbalize his emotions and feelings, he seemed to become very anxious, confused, and immediately showed movement problems such as freezing, slowness in action, and difficulties in talking fluently. It was as if the inability to express his emotions and feelings caused intense anxiety and stress, which had a direct effect on his movements and speech. This young man, despite having above average verbal and nonverbal intelligence was unable to verbalize any aspect of his catatonia-like deterioration or anxiety.

There are, specifically for this group, no published studies of the physiological measures that have been found to be associated with feelings of anxiety (Lader and Wing, 1966). However, with hindsight, it now appears likely that some, perhaps many people, who in the past were diagnosed as having catatonic schizophrenia, in reality had undiagnosed ASDs (Nylander and Gillberg, 2001; Wing and Potter, 2002; Wing and Shah, 2000). People with ADs, who have catatonia-like deterioration, would easily meet the criteria for "catatonic schizophrenia" in Diagnostic and Statistical Manual, 4th ed. (DSM-IV) and International Classification Diseases, 10th revision (ICD-10). Ungvari *et al.* (1999) found that adults diagnosed as having chronic catatonic schizophrenia did not report experiencing anxiety. If they did have ADs, it is unlikely that they would have reported their feelings. However, Venables and Wing (1962) examined skin potential and 2-flash threshold in a group of adults living in institutions and diagnosed as having various types of schizophrenia. They found that the most socially withdrawn, mute or almost mute participants, some diagnosed as having catatonic schizophrenia, had the highest levels of skin potential and lowest 2-flash thresholds of any of the subgroups. The authors suggested that these measures indicated events in the brain, referred to in the paper as "high levels of arousal," which led to marked withdrawal from the environment (that is the poor speech, slowness, and unresponsiveness of chronic catatonia-like conditions). The authors also noted that, in people with "chronic schizophrenia," increased arousal had been found to impair the normal selectivity of perception, making them especially vulnerable to events in the environment. This observation is interesting in the light of the oversensitivity to sensory input that is so characteristic of people with ADs and subjective reports of "switching off" when overloaded with sensory input (e.g., Williams, 2003).

This study left many questions unanswered. Were the brain events associated with the physiological measures and the speech, motor, and social problems a response to environmental pressures? Were the brain events a response to the distress caused by the speech, motor, and social problems? Were the brain events indicated by the physiological measures directly or indirectly the cause of the catatonia-like clinical picture or not related at all? Most important of all, what were the neurological, biochemical, and psychological pathways that led from environmental (or internal, e.g., hormonal) stress to catatonia-like conditions in vulnerable people? Further systematic research is needed to answer questions concerning individual differences in tolerance of stress and experiencing and verbalizing anxiety.

III. Effect of Medical Treatments

Northoff (2002) noted that 60–80% of people, who had had acute catatonic stupor and reported severe anxiety, responded rapidly to high doses of lorazepam, but those who had not been anxious were not helped by this medication (Northoff *et al.*, 1998). Ungvari *et al.* (1999) in the study mentioned earlier carried out a double-blind placebo-controlled crossover trial of lorazepam in a group of people diagnosed as having chronic catatonic schizophrenia. They found that lorazepam in doses of 6 mg/day over 6 weeks was ineffective. As mentioned earlier, these researchers had also found that the participants in the trial did not report subjective anxiety.

The few studies that have been published on the treatment of catatonia-like states in ADs have reported individual cases with acute catatonic stupor that responded to medication and/or electroconvulsive therapy (ECT) (Dhossche, 1998; Ghaziuddin *et al.*, 2005; Realmuto and August, 1991; Zaw *et al.*, 1999). In contrast, the paper by Brasic *et al.* (2000) described the clinical course over 7 years of chronic catatonia-like deterioration in a young man with pervasive developmental disorder. After the first 6 years, he received 25 treatments with ECT with no therapeutic effect. Whether there were any adverse effects was not reported. Among the participants in the study by Wing and Shah (2000), only two had had ECT. One individual had had 3 courses. The first gave 6 months remission of the catatonia-like symptoms, the second gave 3 months remission, and the third was ineffective. The ASD was not affected. One other person had one course of ECT with no effect. We have never seen, in chronic catatonia-like states in ASDs, the dramatic recoveries reported following the administration of lorazepam and/or ECT for typical catatonic stupor (Fink and Taylor, 2003; Northoff, 2002).

IV. Problems with Assessing the Effects of Medical Treatments

It has been pointed out that there are major problems in assessing the effects of medical interventions that have been used for catatonia in people with ADs (Dhossche, personal communication).

First, none of the people in our study (Wing and Shah, 2000), any seen since, or those in the Ungvari *et al.* (1999) study had been given the high doses of lorazepam that are recommended by specialists in the field of catatonia. Also, it is not known whether the administration of ECT in the two cases described earlier followed the modern guidelines for treatment of catatonia (Fink and Taylor, 2003). There are no reports in the literature of clinical trials in which these guidelines for treating typical catatonia have been used in catatonia-like deterioration in ADs. Until research of this kind has been carried out, the questions remain unanswered. But the present authors' clinical experience does not give rise to any optimism concerning the value of the available medical treatments in ASDs.

Second, the effect on response to treatment of the chronicity of the catatonia-like deterioration in ASDs, even when accepted treatment guidelines are followed, is unknown. Chronic "catatonic schizophrenia" is a notoriously treatment-refractory condition (although there are a few case reports describing unexpected and significant gains in function in such patients). Acute-onset catatonia on the other hand, even in its most severe forms, is reported to have a good prognosis, at least at short-term follow-up (Fink and Taylor, 2003). It is possible that early recognition and prompt, adequate treatment may be essential to avoid chronicity. This important issue warrants systematic study in people with catatonia with and without ASDs. Current data do not allow any definite conclusions as it is unlikely that people with ADs with chronic catatonia-like deterioration have received treatment reported to be optimal at the onset of the catatonia or later stages. However, it must be emphasized that catatonia-like features are part of the clinical picture of ASDs from early childhood onwards. Catatonia-like deterioration appears against this background. This raises the question of when and how to recognize the onset of the deterioration. Also in our current state of knowledge, there is no certainty about the relationship between slow-onset catatonia-like deterioration in ASDs and acute catatonia.

The mechanism of action of ECT is unknown. ECT has been effective in individual cases reported in the literature mentioned earlier of children or adolescents with ADs who had typical catatonic stupor (but not for two individuals, known to the present authors, who were not in stupor but had catatonia-like deterioration). Dhossche and Rout (Chapter 4) have offered a biochemical theory involving abnormal GABA function that may link autistic symptoms, catatonia, and the mechanisms of action of ECT.

V. Psychological Methods of Intervention

It appears from the few studies of individuals in print that catatonic stupor in autism may respond to lorazepam in appropriate doses and/or ECT. It is important to emphasize that these treatments should be used if the catatonia is life threatening.

However, for chronic catatonia-like deterioration without stupor, for at least some individuals, even the most appropriate medical treatments may not work or produce only partial remissions. As things are at the time of writing, very little help is available for people with ADs and catatonia-like deterioration. Faced with seemingly intractable problems in the person for whom they are caring, many parents and caregivers have turned to us, the authors, for help because of our special interest in the field. Over the past 10 years, we have discussed ideas on psychological interventions with professionals, care staff in residential homes, and parents who have practical experience of catatonia-like deterioration. We have observed the results of different approaches and collected feed back from those involved.

It is important to point out that medical and psychological approaches are not incompatible. People with catatonia-like deterioration have to be helped and supported throughout the day (and sometimes the night) while awaiting medical treatment, the treatment to take effect, and if it proves ineffective. The psychological approach is appropriate in all these situations. Furthermore, having an understanding and a clear plan to follow give parents and caregivers the confidence that allows them to be calm and supportive, instead of experiencing the anxiety and despair of not knowing what to do. The cognitive reconceptualization of the problem as neuropsychological which is not due to the individual being lazy, willful, stubborn, or manipulative, gives rise to renewed hope and a constructive approach to appropriate interventions. This in itself reduces stress for all concerned, including the affected individual, and has positive benefits.

We have found that the most effective approach is to adapt and extend the principles that are known to be effective in understanding and supporting individuals with ASDs. The approach for each individual depends on the premorbid pattern of autistic impairments and how these have been affected by the catatonia-like deterioration and the precipitating stress factors. The relative rarity of this catatonia-like condition and the resulting wide geographical spread of cases known to us have made it difficult to conduct field trials and carry out a scientific evaluation of the results. From the information we have collected, the psychological techniques of management described later appear in practice to be helpful in improving quality of life for some, if not all, affected individuals. In some cases, they have resulted in complete remission of the deterioration.

It is also true that for some individuals with catatonia-like deterioration, who are of high ability and/or have always been rigidly stubborn and noncompliant, the psychological approaches are difficult to apply and may not help. In our experience, such people may also adamantly refuse to take medication. The presence of a severe comorbid psychiatric condition can also make treatment difficult. The psychological approach appears to be most likely to be helpful in people who had, prior to the deterioration, tended to be passive in their social interactions (Wing, 2005; Wing and Gould, 1979). Predictors of favorable (and nonfavorable) response to the proposed psychological approach should be assessed in future studies.

VI. General Principles of Psychological Treatment and Management

The general principles of treatment and management are discussed in later sections. However, clinicians must always be aware that the program must be adapted to the needs of each individual, on the basis of the clinical picture. It must also be modified in the light of the individual's progress.

A. Initial Assessment

Our clinical experience suggests that early recognition of the onset of catatonia-like deterioration is as important for psychological as for medical treatment. Once a pattern of disability has become chronic, it appears that it is more difficult for any type of intervention to have an effect, though this must not deter clinicians, parents, and caregivers from giving the most appropriate treatments available.

For some people, the diagnosis of an ASD has been made before the appearance of catatonic features. In others, the presence of an AD has not previously been identified. In either case, assessment should include a detailed developmental history, current clinical picture, and psychological examination of skills and disabilities. It is essential to collect information in a systematic form from informants, preferably parents if available. One instrument that can be used is the National Autistic Society's Diagnostic Interview for Social and Communication Disorders (DISCO). This is a semistructured interview that covers areas of development as well as behavior found in ASDs (Wing *et al.*, 2002). It elicits information for the diagnosis of autism and related conditions and catatonia-like deterioration. It also identifies the degree to which the catatonia-like deterioration affects the person in everyday life. This information, together with that obtained from detailed psychological assessment, observation, and other available sources of information, can be used for identifying individual stress factors and needs, and devising an individual program of treatment and management, and as a baseline for its evaluation. Diagnostic criteria indicating a specific associated psychiatric condition should be identified, if present.

The psychological assessment should include appropriate tests of cognitive ability. Tests, such as the Wechsler (1997) Adult Intelligence Scale, Wechsler (1992) Intelligence Scale for Children, and Leiter (1990, 1997) International Performance Scale (a nonverbal cognitive test), are very useful for obtaining quantitative estimates of cognitive level and pattern of abilities, and qualitative assessment of the individual's approach and particular motor difficulties (Shah and Holmes, 1985). It is important to ascertain whether the severe movement difficulties are masking high cognitive abilities. These can be obscured though they are not impaired by catatonia-like deterioration. The information about cognitive level and profile is essential for providing advice about the appropriate level of cognitive stimulation and for selecting structured activities likely to motivate the individual concerned.

The motor components of catatonia-like deterioration can appear in a wide variety of forms. Clinical investigation should cover other identifiable causes of motor problems, any of which could occur in a person who also has an AD. Thorough medical and dental examinations are also needed in case there is a physical condition causing chronic distress. People with ADs, who have poor language, have great difficulty in talking about or showing physical symptoms.

As emphasized by Fink and Taylor (2003), long term administration of neuroleptic drugs can lead to motor disorders including tardive dyskinesia. Continuous mouth and tongue movements, as well as catatonia-like and parkinsonian-like features are described as characteristic of this condition. However, the same phenomena, including continual mouth and tongue movements, can also occur in catatonia-like deterioration in a person with an AD who has never received neuroleptics. Careful consideration should be given to past and present medications as possible causative agents of catatonia-like motor problems.

B. Dealing with Stress Factors

A fundamental aim of the psychological approach for catatonia-like deterioration in ASDs is to identify and reduce the stress being experienced by the person concerned. This will involve restructuring the individual's lifestyle, environment, daily program, and resolving cognitive/psychological sources of stress. Professionals need to draw from a variety of psychological/behavior therapies and use the most appropriate eclectic approach to achieve change for the individual. A related important aim is to provide external goals and stimulation to increase motivation and meaningful, enjoyable activities for the individual. The programs must be adapted to the individual concerned and may involve change at various levels (individual, carer, systems, programs, residential, day services and staffing level, and intensity of support). Hare and Malone (2004) described a detailed program for helping one young man overcome his extreme difficulty in ascending and descending stairs. They emphasized the importance of planning such a program specifically for each individual, rather than having a standardized approach to everyone with "autistic catatonia," to use the terminology they preferred.

C. Conceptualization of the Nature of Catatonia-Like Deterioration

Everyone concerned with the affected person needs to understand that having catatonia-like deterioration is a neurological complication that can occur in ASDs (Wing and Shah, 2000). The effects on movement are not under voluntary control. The person is not being deliberately manipulative, stubborn, willful, obstructive, or lazy. The condition must cause severe distress and frustration to those affected, and they need a sensitive, sympathetic, and understanding approach. Caregivers must aim to be calm, gentle but confident, and positive in their approach. Psychological techniques of reframing information, cognitive restructuring, and shared communication models can be used to inform, motivate, and influence parents, care providers and service providers in finding individual proactive strategies and solutions.

D. Use of Prompts as External Stimuli/Signals

People with catatonia-like deterioration often need to be given verbal or physical prompts to move on or complete an action or a specific activity. The amount, level, and type of prompting necessary will vary for different people, situations, and actions for the same person. The aim of the prompt is to enable the person to carry out movements and actions as smooth as possible. For people with severe catatonia-like deterioration, prompts may not be enough, and they may, at least initially, need substantial physical assistance.

Verbal prompts can vary from quietly calling the person's name to giving specific instructions for each action required. Each prompt may need to be repeated several times before the person is able to respond, and the person will require time to respond.

Physical prompts should start with a minimal light touch. This may be enough to enable the person to begin a movement or action. If not, the physical prompt should be extended so that it guides the movement in the right direction. As the person makes progress, the type and level of prompt can be adjusted and reduced.

Staff members and parents, who care for the person, may find it difficult to decide whether he or she should be assisted by prompts or left to carry out actions in his or her own time. Caregivers may feel that there is a danger of making the person dependent on prompts and thereby reducing the capacity of the person for independent action. Sometimes, staff members feel uneasy about prompting or assisting because this seems to be encroaching on the person's right to privacy and dignity. To resolve this dilemma, it is necessary to understand the behavior in the context of catatonia-like deterioration and the possible long-term negative effects on independence and functioning if the condition progresses. Prompts are necessary to enable individuals with catatonia-like deterioration to overcome the difficulties in the central control of voluntary movement and gradually to regain their independence. If left to their own devices and expected to learn the "hard" way, the person concerned may become increasingly unable to initiate movements and gradually less active and less independent. The inability to initiate and complete movements causes frustration, resulting in stress and tension that, in turn, is likely to exacerbate the difficulty with voluntary movements.

E. Maintaining and Increasing Activity

The effects of catatonia-like deterioration can be reduced substantially by keeping the person active, mobile, and stimulated without putting pressure on them, and by reducing additional demands

Physical activities involving rhythmic, repetitive movements are particularly beneficial. Such activities may include walking, swimming, trampolining, cycling, roller-skating, or roller-blading, ice-skating, dancing, music, and movement. Any other activities that the person seems to find meaningful and enjoyable and is able to carry out relatively easily should be included. Activities that require excessive physical effort and those that the person finds very difficult should be avoided. It may be useful for the person to join in specific physical activities with a small group of people. The momentum of the group is often helpful in enabling the person to begin and continue the activity. They may still need one-to-one support, prompt, and guidance through the activity.

The aim is to keep the individual active goal-oriented and stimulated. It may be necessary to have a one-to-one carer most of the time who can encourage, prompt, and direct throughout the day. Such intensive support may be required in the initial treatment phase and reduced gradually as progress is made.

As well as having problems initiating actions, people with catatonia-like deterioration may find it difficult to stop a repetitive action once started. Examples are continual brushing of hair, repetitively putting a cup to the lips after the drink has been finished, even continuing to ride a bicycle round a track to the point of exhaustion. Again, gentle prompting, verbal or physical, may be necessary to help the person to stop.

F. Structure and Routine

A structured plan of activities of the kind suggested in the earlier section and a predictable routine are necessary for people with catatonia-like deterioration to develop the habit of participation. Habitual actions are much easier for such people, in contrast to new or sporadic activities that are hard, even impossible for them to start. Unpredictability, ad hoc plans, and uncertainty will increase stress and may be the trigger for episodes of freezing or excitement. Principles of the TEACCH approach are useful in devising structured programs, organizing the physical environment, making expectations clear and explicit, and using visual plans for daily schedules (Mesibov et al., 1994).

VII. Management of Specific Problems

A. Incontinence

People with catatonia-like deterioration may show signs of incontinence. This is usually related to the difficulty of getting up from a chair or bed and reaching the toilet in time or the inability to ask to go to the toilet. Such incontinence is puzzling and distressing for a person who previously was fully independent in this respect. This needs to be understood and managed in a discrete and sensitive way. Simple methods, such as regular, frequent physical and/or verbal prompts to go to the toilet and giving enough time to use the toilet, are useful. The person may need assistance with clothes and then be given a verbal prompt to use the toilet. Depending on the severity of catatonia-like deterioration, the person may need physical assistance with personal hygiene.

People with severe catatonia-like deterioration who have difficulty using the toilet independently will benefit from a regular toilet use program. The use of the toilet should be linked to the daily routine by slotting it in at appropriate times, such as immediately on waking, at the beginning and end of each period of activity, before or after each meal time, and before going to bed. This will work only if there is a consistent daily routine that is preplanned and adhered to as strictly as possible. As the person makes progress, the amount of prompting and physical assistance should be adjusted accordingly.

B. Eating Problems

Some people with catatonia-like deterioration develop severe difficulties with eating. The complex motor coordination required for eating with a knife and fork or with a spoon seems to trigger particularly severe difficulties in initiating and completing movements, and ritualistic and repetitive behavior. There are also difficulties with the movements of lips, jaws, and tongue required to take the food off the fork or spoon and to chew and swallow it.

The eating difficulties are often misinterpreted as "playing up," being deliberately slow, or having a poor appetite. Parents and caregivers may be advised by professionals involved not to be concerned, to leave the person to eat if they want to and when they want to. The result may be severe loss of weight and an exacerbation of the other aspects of catatonia-like deterioration.

The problems can be reduced by using verbal and physical prompts and making the process of feeding as easy for the person as possible. Depending on the extent of the difficulties, any of the following methods may have to be used:

1. Using a spoon instead of a fork and knife
2. Adjusting the type and consistency of food so that it is easily scooped onto a spoon. Some people may need liquidized food, which can be fed from a cup
3. Verbal prompts required for each action or each mouthful
4. Physical prompts that may range from touching the elbow lightly to giving hand on hand support and guiding the person's movement in the direction required
5. If prompts are not sufficient, the person may have to be fed
6. If the person is having difficulty opening the mouth, a light touch on the cheek, or touching the lips with the loaded spoon may be effective.

C. Speech and Communication Problems

The planning and execution of the movements required for speech may be as difficult for a person with catatonia-like deterioration as any other motor activity. It is important not to put pressure on them to answer questions or to talk. Instead, others should talk to the person, focusing on the current activity. It is more important to relate to the person through physical activities than through verbal discussions.

Having to verbalize or indicate choices is particularly difficult for people with ASDs, and this is exacerbated by the motor problems of catatonia-like features. Caregivers may need to make the decisions and choices for the person on the basis of their knowledge of the person, their interests, and their likes and dislikes. Gentle suggestion and encouragement without asking directly what the person wants to do or whether or not they want to do something, is the most helpful approach. Visual communication systems and nonverbal personal communicators may be useful for some individuals whose speech is severely affected by catatonia-like deterioration.

D. Difficulty with Walking

Usually, a person with catatonia-like deterioration will be able to walk without stopping if he or she is holding on to or linking with the carer's arm. If the person stops suddenly, a light physical prompt on the back or a verbal prompt will help them to start again. Sometimes, walking as part of a group enables the person to walk at a steady pace without stopping. Some individuals have been helped by being encouraged to take the dog for a walk and hold on to the lead. This provides a continuous external stimulus and goal and can be very effective in keeping a person with catatonia-like deterioration moving at a steady pace.

E. Standing Still or Adopting Fixed Postures

Fixed postures of any kind will be extremely tiring for the muscles, but the person will not be able to express this, or do anything about it. The person who appears to be standing still and staring in space is probably unable to initiate movement needed for a particular activity. A verbal prompt or a light touch may enable them to move. They should then be involved immediately in a different activity.

Some individuals hold up one or both arms and may continue to do this for long periods if left alone. If verbal prompting to lower the arms does not work, physical assistance may be needed.

F. Catatonia-Like Excitement

Episodes of uncontrollable, frenzied, and inappropriate behavior may occur. These may, wrongly, be interpreted as outbursts of "challenging" behavior. This misconception can lead to a fruitless search for "triggers" and "communicative functions" and the application of behavioral methods to discourage the inappropriate behaviors. In the clinical experience of the authors, this has not proved to be a useful approach for the management of catatonia-like excitement. Even more unhelpful is telling the person to stop behaving in that way, to stop being silly, or asking them why they are doing this.

If these episodes are short lived and do not affect the safety of the person, other people, or the environment, the best strategy may be not to intervene, though supervising to ensure safety and give support when the incident is finished.

If these episodes are more severe and longer lasting, intervention at different levels may help the person to calm down more quickly. Verbal suggestions to do something different, mild physical restraint, or simply physically leading the person into a different environment and prompting them to sit down may work. Different strategies are needed to suit different people. When the episode is over, the person should be reassured and encouraged to carry on with the normal routine.

VIII. Implications for Services and Staffing Levels

People with ASDs, who develop catatonia-like deterioration, need an intensive program of help in the right environment, as outlined earlier. Even those who show some response to medical treatment are liable to recurrence if they are in a setting that is not suited to their needs. They require 24-hour care in a setting in which a structured, organized daily program can be planned and carried out consistently. A high staff ratio and sufficient trained staffs are essential. There must also be full access to stimulating physical and occupational activities, of an appropriate kind, in suitable environments.

In the initial stages, the person will require a full time, one-to-one carer for the waking hours in order to carry out the program described. Ideally, each person should have one or two key caregivers who can build up a rapport and a relationship with the person concerned.

IX. Conclusions

As was noted in Chapter 2 by Wing and Shah, the few studies that are relevant or possibly relevant to the unraveling of the nature of catatonia-like deterioration in ASDs generate more questions than answers. The relationships among typical acute catatonic stupor, chronic catatonia, catatonia-like deterioration, and parkinsonism remain unsolved. The relevance of measures of autonomic activity is still to be explored. Stress seems to have a role in the causation but the neurological, biochemical, and psychological mechanisms through which it operates have not been elucidated.

The place of medical treatments also remains a problem. The only published evidence comes from single case studies of individuals with ADs who develop catatonic stupor. We have seen many different medications tried, mostly with no or only short-lived beneficial effects. Our limited experience of ECT in these conditions (and, for Wing, considerable experience of its use for "catatonic schizophrenia" in the old institutions) gives no cause for optimism. Detailed clinical research is called for, carried out with caution, humanity, and respect for each individual concerned. Until clear results are available, the present authors will remain pessimistic about the medical approach, though appropriate medical treatment has to be tried in life-threatening conditions. The psychological methods of helping described in this chapter are safe, humane, person centered, and do lead to improvement in varying degrees in many people affected by catatonia-like deterioration. The suggestions in the Blueprint for assessment and treatment (Section IV) are fully in line with the cautious approach recommended in this chapter.

Many mysteries remain to be solved in this field. It is to be hoped that this book will be a stimulus to research to find the solutions.

Appendix A

Individual Case History

Bernard was 23 at the time of referral. He had attended a special school. He had always been quiet and passive and from 13 to 15 years, had received treatment for social withdrawal and selective mutism. The team treating him had concluded that Bernard was socially withdrawn and selectively mute due to his mother's over involvement with him and his resentment of this, though this was never evident in his overt behavior. The main part of this treatment consisted of psychological therapies aimed at enabling Bernard to express his repressed anger toward his mother. This approach was unsuccessful. At the age of 16, Bernard was diagnosed as having autism.

From 16 to 19 years, he attended a school for children with autism. The report from the school described Bernard as being unable to communicate normally and being able only to whisper in a soft voice when asked a question, provided the questioner was prepared to wait a long time for a response. He was also described as being extremely slow in his movements, often becoming fixed in a stationary position for long periods. After leaving the education system, Bernard was not offered any further training or community service. His mother could see that he was distressed by the absence of any structure to his day. She tried to refer him to the local community team for people with learning disabilities. However the referral was rejected on the grounds that Bernard had a recorded nonverbal IQ in the low average range. His mother had to work hard to care for him at home without any support. Bernard gradually deteriorated in his ability to carry out any movements or activities without help or prompting. During a 5-month period of eating very little he lost a great deal of weight. He was then referred to a mental health psychiatrist. A diagnosis of depression was made, and Bernard was prescribed antidepressants. At a review after a year, the psychiatrist was struck by the deterioration in Bernard's condition and referred him to us.

At the point of our involvement, Bernard, who used to be fully independent in all aspects of self-care, had become totally dependent on his mother. The morning routine of getting up, washing, and dressing was taking up to 5 hours. Bernard was only able to finger-feed bite size pieces of dry foods. However, he frequently became "frozen," and each small meal was lasting for hours. He was severely underweight and still losing more weight. Although previously he had been able to travel independently, he could not now travel on his own as he was unable to get up from his seat on the bus to alight.

Bernard hardly talked at all. When asked questions, he took a long time to respond and even then was able to answer only in monosyllables. He whispered the answers very softly. He seemed unable to lift his head to make eye contact or to interact. He had a fixed expression on his face, and was not able to acknowledge people in any way.

Bernard was spending most of his waking time on a sofa in the living room, in a fixed stooped posture with his head bowed and his arms hanging by his sides. He tried to respond to instructions, but each movement seemed agonizingly difficult. Before each action, his body twitched and jerked, he blinked continuously, and made repetitive movements with his mouth.

A comprehensive assessment was carried out. This included a detailed investigation of developmental history and past patterns of behavior, detailed observations of the current behaviors, responses and movement difficulties, and psychometric assessment using the Leiter International Performance Scale (a nonverbal test). A diagnosis of an ASD and severe catatonia-like deterioration was made. As described earlier, Bernard was showing a full range of catatonia-like phenomena. Stress factors identified included not coping in mainstream school, unsuitable therapies focusing on eliciting emotions, and lack of structure and occupation.

After the diagnosis, a case was made for the local services for people with learning difficulty to be involved and provide support to the family. An intervention plan based on the principles of reducing stress, providing appropriate environment, program, and gradually increasing motivation and activities as outlined earlier was put into practice. This included training of caregivers to understand that Bernard's difficulties were due to catatonia-like deterioration. One-to-one support was provided. He was helped with physical and verbal prompting and a gradual increase in his activity level. Slowly Bernard started to become less rigid in his postures and movements. He was able to complete more and more movements with prompts, which were gradually reduced. He was able to feed himself a range of foods with minimal physical and verbal prompts. He began to respond more quickly. He was able to walk upright and the jerks, twitches, and repetitive mouth movements decreased and then stopped completely. As he progressed, the local health authority was persuaded to find a placement at a small specialist day centre, which provided a highly structured program of activities, one-to-one support and a calm quiet atmosphere. Bernard has continued to make progress and is regaining his mobility, speech, and independence.

References

Brasic, J. R., Zagzag, D., Kowalik, S., Prichep, L., John, E. R., Barnett, J. Y., Bronson, B., Nadrich, R. H., Cancro, R., Buchsbaum, M., and Braithwaite, C. (2000). Clinical manifestations of progressive catatonia. *German J. Psychiatry* **3,** 13–24.

Dhossche, D. (1998). Brief report: Catatonia in autistic disorders. *J. Autism Dev. Disord.* **28,** 329–331.

Fink, M., and Taylor, M. A. (2003). "Catatonia: A Clinician's Guide to Diagnosis and Treatment." Cambridge University Press, Cambridge.

Frith, U. (2003). "Autism: Explaining the Enigma," 2nd ed. Blackwell, Oxford.

Ghaziuddin, M., Quinlan, P., and Ghaziuddin, N. (2005). Catatonia in autism: A distinct subtype? *J. Intellec. Disabil. Res.* **49,** 102–105.

Hare, D. J., and Malone, C. (2004). Catatonia and autistic spectrum disorders. *Autism* **8,** 183–195.

Hill, L. H. (2004). Executive dysfunction in autism. *Trends Cogn. Sci.* **8,** 26–32.

Kovacs, K. J., Miklos, I. H., and Bali, B. (2004). GABAergic mechanisms constraining the activity of the hypothalamo-pituitary-adrenocortical axis. *Ann. NY Acad. Sci.* **1018,** 466–476.

Lader, M. H., and Wing, L. (1966). "Physiological Measures, Sedative Drugs, and Morbid Anxiety." Oxford University Press, London.

Leiter, R. G. (1990). "International Performance Scale (1948 Revision)." Stoelting Co., Chicago.

Leiter, R. G. (1997). "International Performance Scale Revision." Stoelting Co., Chicago.

Loos, H. G., and Miller, I. M. (2004). "Shutdown States and Stress Instability in Autism." Available on-line at: www.cuewave.com/tau/SI-SDin.Autism.pdf.

Mesibov, G., Schopler, E., and Hearsey, K. (1994). Structured teaching. *In* "Behavioural Issues in Autism" (E. Schopler and G. Mesibov, Eds.), Plenum, New York.

Northoff, G. (2002). What catatonia can tell us about "top-down modulation": A neuropsychiatric hypothesis. *Behav. Brain Sci.* **25,** 578–604.

Northoff, G., Krill, W., Gille, B., Russ, M., Eckert, J., Bogerts, B., and Pflug, B. (1998). Major differences in subjective experience of akinetic states in catatonic and parkinsonian patients. *Cogn. Neuropsychiatry* **3,** 161–178.

Nylander, L., and Gillberg, C. (2001). Screening for autism spectrum disorders in adult psychiatric out-patients: A preliminary report. *Acta Psychiatr. Scand.* **103,** 428–434.

Realmuto, G., and August, G. (1991). Catatonia in autistic disorder; A sign of co-morbidity or variable expression? *J. Autism Dev. Disord.* **21,** 517–528.

Rogers, D. (1992). "Motor Disorder in Psychiatry: Towards a Neurological Psychiatry." Wiley, Chichester.

Shah, A., and Holmes, N. G. (1985). The use of the Leiter International Performance Scale with autistic children: A research note. *J. autism Dev. Disord.* **15**(2), 195–205.

Ungvari, G. S., Chiu, H. F. K., Chow, L. Y., Lau, B. S. T., and Tang, W. K. (1999). Lorazepam for chronic catatonia: A randomised, double-blind, placebo controlled cross-over study. *Psychopharmacology* **142,** 393–398.

Venables, P. H., and Wing, J. K. (1962). Level of arousal and the subclassification of schizophrenia. *Arch. Gen. Psychiatry* **7,** 114–119.

Wechsler, D. (1992). "Wechsler Intelligence Scale for Children," 3rd ed. UK. The Psychological Corporation.

Wechsler, D. (1997). "Wechsler Adult Intelligence Scale," 3rd ed. The Psychological Corporation.

Williams, D. (2003). "Exposure Anxiety: The Invisible Cage." Jessica Kingsley, London.

Wing, L. (2005). Problems of categorical classification systems. *In* "Handbook of Autism and Pervasive Developmental Disorders" (F. R. Volkmar, R. Paul, A. Klin, and D. Cohen, Eds.), Wiley, Hoboken, New Jersey.

Wing, L., and Gould, J. (1979). Severe impairments of social interaction and associated abnormalities in children: Epidemiology and classification. *J. Autism Child. Schizophr.* **9,** 11–29.

Wing, L., and Potter, D. (2002). The epidemiology of autistic spectrum disorders: Is the prevalence rising? *Men. Retard. Dev. Disabil. Res. Rev.* **8,** 151–161.

Wing, L., and Shah, A. (2000). Catatonia in autistic spectrum disorders. *Br. J. Psychiatry* **176,** 357–362.

Wing, L., Leekam, S. R., Libby, S. J., Gould, J., and Larcombe, M. (2002). The diagnostic interview for social and communication disorders: Background, inter-rater reliability and clinical use. *J. Child Psychol. Pychiatry* **43,** 307–325.

Zaw, F. K. M., Bates, G. D. I., Murali, V., and Bentham, P. (1999). Catatonia, autism and ECT. *Dev. Med. Child Neurol.* **41,** 843–845.

SECTION V
BLUEPRINTS

BLUEPRINTS FOR THE ASSESSMENT, TREATMENT, AND FUTURE STUDY OF CATATONIA IN AUTISM SPECTRUM DISORDERS

Dirk Marcel Dhossche,* Amitta Shah,† and Lorna Wing‡

*Department of Psychiatry and Human Behavior, University of Mississippi Medical Center
Jackson, Mississippi 39216, USA
†Leading Edge Psychology, Purley CR8 2EA, United Kingdom
‡Centre for Social and Communication Disorders, Bromley, Kent BR2 9HT, United Kingdom

I. Introduction
II. Blueprint for the Assessment of Catatonia in ASDs
   A. Catatonia-Like Deterioration
   B. Medical/Neurological Evaluation
   C. Search for Culprit Medications or Other Substances
   D. Diagnostic Catatonia Criteria for ASDs
   E. Assessment of Severity into Mild, Moderate, and Severe
III. Blueprint for the Treatment of Catatonia in ASDs
   A. Treatment Modalities for Catatonia in ASDs
   B. Ethical and Legal Issues
   C. Comments
IV. Blueprint for the Future Study of Catatonia in ASDs
   A. Blueprint for Future Studies in the Clinical Arena
   B. Blueprint for Future Studies in the Preclinical Arena
   References

> Medicine's ground state is uncertainty. And wisdom – for both patients and doctors – is defined by how one copes with it.
>
> Atul Gawande, M. D. (2002)

The blueprints for the assessment, treatment, and future study of catatonia in autism spectrum disorders (ASDs), which are submitted in this chapter aim to increase early recognition and treatment of catatonia in ASDs, show the urgency of controlled treatment trials, and increase collaborative and interdisciplinary research into the co-occurrence of these two enigmatic disorders.

Catatonia should be assessed in any patient with ASDs when there is an obvious and marked deterioration in movement, pattern of activities, self-care, and practical skills, compared with previous levels, through a comprehensive diagnostic evaluation of medical and psychiatric symptoms. A formal diagnosis should be ascertained using ASD specific criteria for catatonia that takes into account baseline symptoms like muteness, echophenomena, stereotypy, negativism, or other psychomotor abnormalities. Any underlying medical and neurological conditions should be treated, and culprit medications or other substances that may cause catatonia should be eliminated.

Separate treatment blueprints are presented for mild, moderate, and severe catatonia, featuring combinations of a psychological approach developed by Shah and Wing and medical treatments that have shown efficacy in catatonia: lorazepam challenge, lorazepam trial, lorazepam continuation, and bilateral electroconvulsive therapy (ECT). These treatment modalities in themselves are well established. Side effects and complications are known and manageable. Legal, ethical, and practice guidelines governing all treatment aspects should be followed. The treatment blueprints should be viewed as best estimates pending future controlled studies.

The blueprint for the future study of catatonia in ASDs describes promising clinical and preclinical research avenues. Longitudinal studies need to assess the possible effect of early recognition and adequate treatment of catatonia in ASDs in order to avoid the impairment associated with chronicity. Effects of current and new anticatatonic treatments should be examined in experimental models of autism and catatonia. Finally, the role of gamma-aminobutyric acid (GABA) dysfunction in autism, catatonia, and abnormal stress responses in these disorders should be further assessed.

I. Introduction

Autism and catatonia are important psychopathological dimensions with deep historical roots. Their overlap has been noted, but the nature of this overlap remains unresolved. Some people with autism spectrum disorders (ASDs) develop catatonia of varying levels of severity, similar to people with affective disorders, various medical and neurological disorders, and schizophrenia. However, the literature gives no guidance on how to treat catatonia in people with ASDs. It is also unknown if the type of catatonia that develops in ASDs is caused by the same psychological or biological factors as in other major neuropsychiatric disorders. In fact, the knowledge base on catatonia itself is quite limited. Moreover, old and new catatonia studies often concern adult patients with chronic conditions. It remains an open question—how many study subjects in these studies have underlying ASDs? There are no controlled studies of catatonia in ASDs.

However, this lack of understanding should not deter from offering interim guidelines on assessment and treatment of catatonia in ASDs, as it is likely that patients and their families will benefit if catatonia is recognized early and treated appropriately. Our blueprints for assessment and treatment rely heavily on the wealth of clinical experience that has been gathered in the field of catatonia but have been modified to accommodate specifics of ASDs.

There are indications that the emergence of catatonia in ASDs is not benign and often becomes chronic and difficult to treat like many other chronic conditions. Perhaps, early recognition and prompt treatment can alleviate these problems as the current literature emphasizes that early and adequate treatment of catatonia is needed for a favorable prognosis (Fink and Taylor, 2003). Ohta, Kano, and Nagai (see Chapter 3) found evidence that ASD cases with sudden onset of catatonia remitted at a higher rate than cases with a gradual onset, although the number of cases did not allow robust statistical conclusions. Treatments in this observational study were uncontrolled, precluding any conclusions about the effects of early treatment on subsequent course. Only follow-up of patients treated with different standardized treatment protocols will inform on this important issue.

Some authors in this book have much experience with catatonia in general psychiatry, some with catatonia-like problems in ASDs, and others with both. These blueprints should be viewed as best estimates reflecting the authors' collective experience while awaiting confirmation through systematic studies. Expert guidelines in an area with a deficient knowledge base are inherently limited. In the worst-case scenario, guidelines may have a stifling effect on progress and enforce the *status quo*. However, our blueprints are unlikely to have this effect because catatonia is poorly recognized in people with ASDs. Families and clinicians of these patients are often unaware of new developments in the catatonia field.

Autism spectrum disorder and catatonia are conditions that are on the borderline and overlap with so many clinical specialties. Psychiatrists, neurologists, pediatricians, psychologists, and other child specialists often have their own criteria and terminology that are never compared across disciplines. Like with the Tower of Babel, synergy that may advance the field of autism and catatonia is lost due to confusing diagnostic terminology among specialties. As awareness and interest are raised, collaborative and interdisciplinary studies may become feasible. If nothing else, we hope that our blueprints increase early recognition and treatment of catatonia in ASDs, show the urgency of controlled treatment trials, and stimulate preclinical and clinical research into the co-occurrence of these two enigmatic conditions.

II. Blueprint for the Assessment of Catatonia in ASDs

> [Instead] what is going to happen – and this will be entirely salutary – will be a revival of interest in psychopathology. This will happen as surely as the sun will rise because, of the great research themes in psychiatry's 200-year history, psychopathology is the one that offers

the greatest promise, while having the lowest startup cost: it requires merely a return to the bedside.

Edward Shorter, Ph.D.
In "Where is psychiatry heading? The future of psychiatry lies in revisiting the past." CrossCurrents, winter, 2003/2004, University of Toronto

A blueprint for the assessment of catatonia in ASDs is presented in Fig. 1.

It is likely that the different components of the assessment procedure of catatonia in ASDs will fade into each other in clinical practice. For example, establishing the presence of a diagnosis of catatonia will probably go hand-in-hand with estimating the level of severity of functional impairment due to catatonia. In the algorithm, a formal diagnosis is established first, and then a level of severity is determined. This sequence risks being somewhat artificial but is maintained for the sake of didactic clarity.

A. CATATONIA-LIKE DETERIORATION

Catatonia-like deterioration should be considered in any patient when there is an obvious and marked deterioration in movement, pattern of activities, self-care, and practical skills, compared with previous levels (see Chapter 2 by Wing and Shah). More specifically, common features of catatonia-like deterioration are slowness and difficulty in initiating movements unless prompted, odd gait, odd stiff posture, freezing during actions, difficulty crossing lines (e.g., pavement cracks), inability to cease actions, marked reduction in the amount of speech, or complete mutism. Other behaviors include impulsive acts, bizarre behavior, sleeping during the day but awake at night, incontinence, and excited phases.

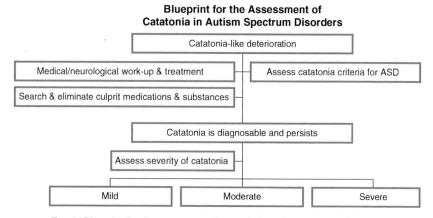

FIG. 1. Blueprint for the assessment of catatonia in autism spectrum disorders.

Catatonia-like deterioration reportedly starts in adolescence and young adulthood. For example, the majority of patients had onset of catatonia between 15 and 20 years of age in the sample of Wing and Shah (2000). Five individuals had brief episodes of slowness and freezing before age 10. In the study of Ohta, Kano, and Nagai (Chapter 3), the average age at onset of catatonia in ASDs was 19 years (SD 6, age range 15–23). Although catatonia has been reported in prepubertal children (Dhossche and Bouman, 1997), its occurrence seems to be rare in this age group. In Chapter 4, Dhossche and Rout speculate that regressive phases in the development of autistic children may be related to catatonia and that young age may modify typical symptoms of catatonia into atypical presentations. They also discuss research strategies to test this hypothesis. However, the present treatment blueprints are only applicable to patients with clearly definable catatonia. Current treatments in children with autism consist mainly of behavioral interventions that are reviewed by Scattone and Knight in Chapter 11. Although the youngest age at which full catatonia in ASDs can first develop is unknown, there seems to be a markedly increased risk in adolescence and young adulthood.

B. Medical/Neurological Evaluation

Possible catatonia-like deterioration in patients with ASDs should prompt a thorough clinical assessment. Physical examination and laboratory investigations (including a pregnancy test for all female patients) are dictated by clinical assessment.

Infectious, metabolic, endocrinological, neurological, and autoimmune diseases have been associated with catatonia, and must therefore be ruled out. For an in-depth review, we refer to two recent publications on this subject (Carroll and Goforth, 2004; Fink and Taylor, 2003).

C. Search for Culprit Medications or Other Substances

All prescribed medications should be evaluated for their potential to induce catatonic symptoms since many medical and psychiatric medications can cause catatonia or catatonia-like conditions (Fink and Taylor, 2003; Lopez-Canino and Francis, 2004). Antipsychotic agents should be discontinued as they are contraindicated in patients with ASDs who exhibit the signs of catatonia because of the reported increased incidence of malignant catatonia or neuroleptic malignant syndrome (NMS) in patients with incipient signs of catatonia. When catatonia is resolved, antipsychotics may be useful for select indications, but reemergence of catatonic symptoms should prompt discontinuation.

Illicit drugs (PCP, mescaline, psilocybin, cocaine, opiates, and opiods), disulfiram, steroids, antibiotic agents (ciprofloxacin), and bupropion have also been associated with the emergence of catatonia in case reports. Withdrawal of benzodiazepines, gabapentin, and dopaminergic drugs, especially if done rapidly, has precipitated catatonia in some patients (Fink and Taylor, 2003).

One author (L.W.) treated a teenage girl with typical autism who developed slowness, freezing during activities, and near muteness when given a monoamine oxidase inhibitor (MAOI) to treat possible underlying depression. When the MAOI was discontinued, the freezing and muteness stopped. Therefore, the use of antidepressant medications, including MAOIs and serotonin selective reuptake inhibitors (Leo, 1996), should also be scrutinized in patients who develop catatonia. As with antipsychotics, antidepressants or any other type of psychiatric medication may be indicated for select indications, when the catatonia is resolved, but reemergence of catatonic symptoms should prompt discontinuation.

D. Diagnostic Catatonia Criteria for ASDs

Different criteria for diagnosis of catatonia are used in clinical practice and research. It is unknown how these different criteria compare and which set of criteria is best, as there is no gold standard. The criteria of Fink and Taylor (2003) (see Table I in Chapter 14 by Fink, Taylor, and Ghaziuddin) are attractive because of their simplicity, clinical relevance, and conformity with Diagnostic and Statistical Manual of Mental Disorder, 4th ed. (DSM-IV). However, their criteria are less applicable to mute patients or the majority of autistic patients with baseline symptoms like echophenomena, stereotypy, negativism, or other psychomotor abnormalities (see Chapter 6 by Stoppelbein, Greening, and Kakooza). Therefore, diagnostic criteria specific for catatonia in ASDs are proposed (see Table I) in which drastically decreased speech replaces mutism, and the duration of symptoms in criteria A and B is longer. The scope of

TABLE I
Diagnostic Catatonia Criteria for ASDs

Criterion A
  Immobility, drastically decreased speech, or stupor of at least 1 day duration, associated with at least one of the following: catalepsy, automatic obedience, or posturing
Criterion B
  In the absence of immobility, drastically decreased speech, or stupor, a marked increase from baseline, for at least 1 week, of at least two of the following: slowness of movement or speech, difficulty in initiating movements, or speech unless prompted, freezing during actions, difficulty crossing lines, inability to cease actions, stereotypy, echophenomena, catalepsy, automatic obedience, posturing, negativism, or ambitendency

symptoms for criterion B is broadened to include overall increases in slowness and other abnormalities of movement and speech.

E. ASSESSMENT OF SEVERITY INTO MILD, MODERATE, AND SEVERE

Once a diagnosis of catatonia is made, the severity of the condition should be determined by assessing the degree to which activities of daily living, occupational activities, and physiological necessities (eating, drinking, and excretion) are affected.

The level of impairment should dictate the need for services and staffing levels accordingly. Mild represents slight impairment in social and vocational activities without hampering efficiency as a whole. Patients with moderate impairment have more pronounced deficits in all areas but remain ambulatory and do not require acute medical services to sustain feeding or vital functions. Patients with acute stupor, immobility for most of the day, in a bedridden state, and in need for parental feedings have severe catatonia that constitutes a medical emergency. Patients with features of malignant catatonia (fever, altered consciousness, stupor, and autonomic instability as evidenced by lability of blood pressure, tachycardia, vasoconstriction, and diaphoresis) also fall in this category.

The use of currently available catatonia rating scales is not endorsed, as none have been tested in autistic populations. It is also unknown if scores on any of these scales adequately pertain to level of severity or other parameters that are crucial for clinical decision-making. We refer to other authors (e.g., Fink and Taylor, 2003; Mortimer, 2004, and so on) for detailed discussions of the relative merits pending future studies using these measures with people with ASDs.

III. Blueprint for the Treatment of Catatonia in ASDs

In Figs. 2, 3, and 4, blueprints are presented for the treatment of mild, moderate, and severe levels of catatonia in people with ASDs, once the diagnosis of catatonia is confirmed and persists despite adequate treatment of any underlying medical conditions and elimination of "culprit" medications or other substances.

A. TREATMENT MODALITIES FOR CATATONIA IN ASDs

1. *Shah–Wing Approach*

This psychological approach is described in Chapter 15 by Shah and Wing. The treatment involves identifying and reducing stress factors, keeping the person active in pursuits that he or she enjoys, using verbal or gentle physical prompts to

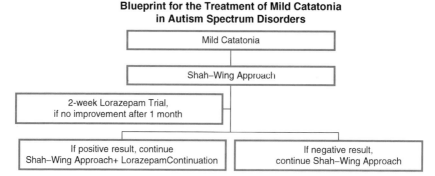

FIG. 2. Blueprint for the treatment of mild catatonia in autism spectrum disorders.

overcome the movement difficulties, and maintaining a predictable structure and routine for each day. The importance of educating caretakers to understand catatonia-like behavior and to realize that this is not under the control of the patient is paramount. Management techniques for specific problems, such as incontinence, freezing in postures, eating problems, and episodes of excitement, can be found in Chapter 15.

This approach by its nature does not carry the risk of physical side effects of medical interventions. Side effects, if any, should be minimal. A limiting factor of full implementation of the program may be the need for increased staffing levels and specialized training of staff as well as easy access to therapeutic physical activities.

2. *Lorazepam Challenge Test*

This option is included primarily for the treatment of severe catatonia in which there is the need for rapid and aggressive intervention. However, in individual cases and particular situations, the lorazepam challenge tests may still be applied, particularly when the onset of moderate catatonia is acute.

The intravenous administration of amobarbital, lorazepam, or diazepam relieves catatonia in more than half the patients as described in Chapter 14 by Fink, Taylor, and Ghaziuddin. Specifically, an intravenous line is established, and a syringe containing 2–4 mg lorazepam in 2 ml (or larger volumes) of solution is prepared, and 1 mg is injected. In the next two to five minutes, any changes are noted. If no change is observed, the second 1 mg lorazepam is injected, and the assessment is repeated.

Lorazepam treatment in any mode of administration (intravenous, intramuscular, per os) has a wide margin of safety. The most common side effect is sedation that may require special attention in people with chronic obstructive

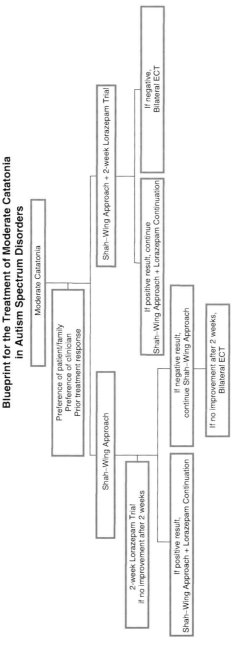

FIG. 3. Blueprint for the treatment of moderate catatonia in autism spectrum disorders.

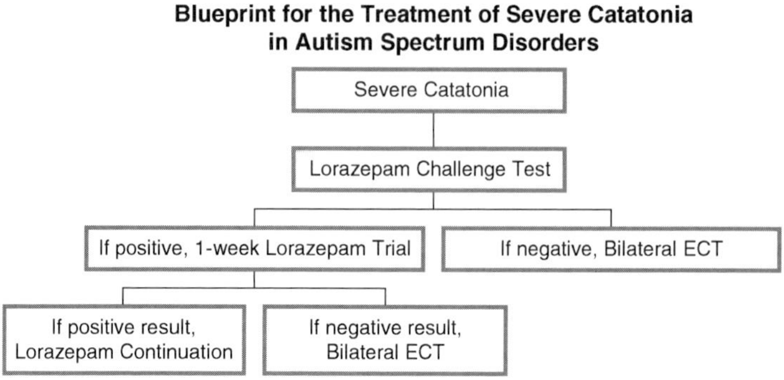

Blueprint for the Treatment of Severe Catatonia in Autism Spectrum Disorders

```
                    ┌─────────────────┐
                    │ Severe Catatonia│
                    └────────┬────────┘
                             │
                   ┌─────────┴──────────┐
                   │ Lorazepam Challenge Test │
                   └─────────┬──────────┘
              ┌──────────────┴──────────────┐
   ┌──────────┴──────────┐          ┌───────┴────────────┐
   │If positive, 1-week  │          │ If negative,       │
   │Lorazepam Trial      │          │ Bilateral ECT      │
   └──────────┬──────────┘          └────────────────────┘
      ┌──────┴───────┐
┌─────┴────────┐ ┌───┴──────────────┐
│If positive   │ │If negative result,│
│result,       │ │Bilateral ECT     │
│Lorazepam     │ │                  │
│Continuation  │ │                  │
└──────────────┘ └──────────────────┘
```

FIG. 4. Blueprint for the treatment of severe catatonia in autism spectrum disorders.

pulmonary disease, obstructive sleep apnea, or obesity. As with any medication, appropriate safeguards should be in place for the occurrence of allergic reactions.

3. *Lorazepam Trial*

Treatment with increasing doses of lorazepam as high as 24 mg/day is recommended. Improvement should be observed at least after 1 week in severe cases before considering bilateral electroconvulsive therapy (ECT). In mild and moderate cases, a 2-week trial is recommended.

4. *Lorazepam Continuation Treatment*

Once the efficacy of lorazepam is demonstrated, a 6–12 months continuation phase starts (see Chapter 14 by Fink, Taylor, and Ghaziuddin). Although the usual dose range of 6–24 mg/day is considered "high" in conventional psychopharmacology, higher doses are reported to be helpful in catatonia. A slow taper is recommended when discontinuing lorazepam after the continuation phase to avoid withdrawal symptoms due to tolerance.

5. *Bilateral ECT*

Electroconvlsive therapy is indicated in severe catatonia when the lorazepam challenge test fails or increased dosages do not bring rapid relieve (see Chapter 14 by Fink, Taylor, and Ghaziuddin). In such circumstances, the ECT may be life saving. It is also indicated in moderate cases.

Technical issues of ECT are paramount to ensure an adequate trial. The efficacy of bilateral (bitemporal or bifrontal) electrode placement is better documented than unilateral placement. The use of bilateral placement is recommended. All psychiatric medications should be stopped prior to initiation of

ECT as well as any other nonpsychiatric medications, if possible (see Chapter 13 by Zaw).

The relief of catatonia often seems to require more frequent seizures than does the relief of major depression. The UK standard practice of two seizures a week is effective for major depression, but this may not be so for catatonia. In severe or malignant catatonia, daily treatment for 3–5 days may be needed. The number of sessions that will be needed before substantial improvement or remission occurs, cannot be predicted. It seems reasonable to assess the patient's overall response after the first 5 or 6 treatments and then again after 10 or 12 treatments.

The cognitive and noncognitive side effects of ECT are well-known. For a detailed description and their management, we refer to Chapter 13 by Zaw for findings in adolescents, which generally also apply to older populations.

Our guidelines do not specifically list options for continuation treatment after an effective ECT course but may include continuation lorazepam treatment or continuation ECT, in addition to the Shah–Wing Approach. Although the guidelines for frequency and number of ECT treatments are poorly defined, weekly and bi-weekly ECT treatments may be needed for up to 6 months to ensure a stabilized response (see Chapter 14 by Fink, Taylor, and Ghaziuddin). One author (D.D.) has treated a 9-year-old boy with acute, malignant catatonia, without autism, with two courses of inpatient ECT (seven and nine treatment sessions), followed by 5 months of continuation ECT (once a week during the first 2 months, followed by biweekly sessions), with the result of complete resolution of catatonia and return to baseline functioning. However, continuation or maintenance of ECT is not recommended even for adults, according to recent British guidelines (National Institute of Clinical Excellence, UK, Guidance on the use of ECT, Technology Appraisal 59, 2003), so court approval may need to be pursued if warranted by clinical considerations.

B. ETHICAL AND LEGAL ISSUES

All treatments should comply with local and national stipulations and laws. This is particularly important for ECT given the level of scrutiny at various levels (see Chapter 13 by Zaw). Electroconvulsive therapy practice guidelines are currently available in USA (Ghaziuddin *et al.*, 2004), UK (National Institute of Clinical Excellence, UK, April 2003, Guidance on the use of electroconvulsive therapy, Technology Appraisal 59), and other countries.

Ethical issues are paramount and involve education of patients and their families and informed consent, particularly when the patient is a child or adolescent (see Chapter 13 by Zaw). For an in-depth discussion of ethical issues involved in ECT, we refer to a recent lucid exposition on this subject (Ottosson and Fink, 2004).

C. Comments

The general treatment approach in catatonia in ASDs is the psychological treatment outlined by Shah and Wing in Chapter 15, that is, the Shah–Wing Approach. The person concerned has to be cared for throughout the day (and night) whether or not medical treatments are in place. The psychological approach is based on good practice in ASDs and extends to cases in which medical interventions do not work.

Treatment options for each severity level are dictated by the expected risk/benefit ratio. For example, ECT is not recommended in mild cases but may be considered in moderate cases as a last resort. Severe catatonia is considered a life-threatening condition and warrants rapid and aggressive ECT treatment constituting lorazepam challenge test, lorazepam trial, and bilateral (see Chapter 14 by Fink, Taylor, and Ghaziuddin). As the level of severity changes, treatment recommendations should be altered accordingly.

The recommended periods to continue an intervention before the next step should be considered are somewhat arbitrary. For example, in mild catatonia, the Shah–Wing Approach should be implemented for 1 month, without effect, before considering a lorazepam trial. Deviations from these periods can be expected in individual cases, especially when the severity level of catatonia fluctuates.

An important limitation is that little guidance can be given on the duration that medical treatments should be continued for maintenance or prophylaxis. From the catatonia literature, continuation treatment for lorazepam is at least 6–12 months (Fink and Taylor, 2003). An important endpoint would be the remission or substantial reduction of catatonia for at least 3–6 months before a successful treatment is stopped (in the case of ECT) or slowly tapered (in the case of lorazepam). These important questions can only be answered in future studies that follow catatonic patients, with and without ASDs, for longer periods. Such studies could also inform on the potential benefits of early and adequate treatment in reducing chronic impairment.

Another limitation is that the treatment recommendations may not apply to chronic catatonia that has become less responsive to medical intervention. As Shah and Wing point out in Chapter 15, in treatment-refractory cases, the psychological approach should still be used as the preferred standard of care.

Another limitation concerns possible poor acceptance of lorazepam treatment and ECT in patients with ASDs and catatonia. There are no treatment studies in catatonia that satisfy currently accepted methodological standards of evidence-based practice. This deplorable situation undoubtedly reflects, in part, the practical and ethical difficulties of conducting randomized studies in severe and life-threatening forms of catatonia. It may also illustrate the backlog of psychiatric research compared with other branches of medicine. On a darker note, the treatments that have shown beneficial effects in catatonia, that is, behavioral-psychological treatment, lorazepam treatment, and ECT are not novel and have virtually little potential for commercial exploitation. Other classes of psychiatric medication, including the newer antidepressive and antipsychotic preparations, cannot be endorsed for the treatment of catatonia because of shown but unquantified risks of worsening the condition. In addition, ECT was, is, and, in all likelihood, will continue to be fraught with controversy and stigma. The public as well as some in the medical community remain reluctant or aversive toward the idea and practice of electrical induction of seizures for therapeutic aims.

Benzodiazepines, especially when prescribed in high dosages as recommended for catatonia (see Chapter 14 by Fink, Taylor, and Ghaziuddin), are also frowned upon by many owing to associated risks of addiction and dependence. In all likelihood, the overall risk for benzodiazepine addiction is negligible in most people with ASDs compared with the risk in people without developmental disorders who suffer from anxiety and depression, especially when co-morbid with substance use disorders. Substance seeking behavior is usually associated with immediate behavioral effects of the substance. However, catatonic patients do not report euphoria or any other pleasant effects of benzodiazepine treatment. There are no known cases of benzodiazepine addiction in people with catatonia (without ASDs) (personal communication of Max Fink).

Acute treatment of catatonia with lorazepam is unlikely to induce tolerance, the pharmacological mechanism underlying dependence, within the course of a few weeks. During the continuation or maintenance phase of the treatment, tolerance may develop, but it has not been examined yet if this occurs for the anticatatonic effect of lorazepam or other benzodiazepines. The need for increasing dosages in time to sustain a benefit has not been reported so far in catatonia in patients without ASDs (personal communication of Max Fink). We recommend that after the recommended 6–12 months lorazepam continuation phase, the dose be slowly tapered as the best strategy to avoid withdrawal symptoms and early relapse.

The experience that higher dosages of benzodiazepines are needed to obtain anticatatonic effects compared to anxiolytic or sedative effects is of theoretical importance (see GABA theory of catatonia in Chapter 4 by Dhossche and Rout). It seems that catatonic patients are able to metabolize large doses of lorazepam (and barbiturates) without the usual sedative effects (Fink and Taylor, 2003). The tolerance of catatonic patients to ECT is a parallel example. Seizures induced daily are necessary in some patients to achieve a benefit, particularly in acute and severe catatonia. In autism, there are anecdotal reports that some patients have a high-tolerance for the usual dosages of sedatives, for example, when they are premedicated for minor surgery or imaging procedures. There have also been reports of negative effects of benzodiazepines on behavioral problems in children with autism (Marrosu et al., 1987) and on overall functioning in autistic youngsters with seizure disorders (Gillberg, 1991) but not in autistic patients with catatonia (although, the experience has been limited).

Some patients with ASDs may respond abnormally to benzodiazepines because of preexisting gamma-aminobutyric acid (GABA) dysfunction or abnormalities in the GABA-benzodiazepine receptor complex (Dhossche *et al.*, 2002; Garreau *et al.*, 1993; see Chapter 4 by Dhossche and Rout). The need to further examine GABA dysfunction in autism and catatonia is emphasized in Chapter 4 by Dhossche and Rout (see also the Blueprint for the Future Study of Catatonia in ASDs). However, this should not discourage from prescribing lorazepam treatment in patients with autism once the diagnosis of catatonia is made. The wide margin of safety of lorazepam and other benzodiazepines should alleviate immediate concerns, but any adverse reactions or complications including tolerance, drug-seeking behavior, or withdrawal, when continuing the treatment for longer periods, should be noted and managed carefully.

In conclusion, despite poor acceptance and lack of rigorous scientific studies, we believe that the available evidence behooves us to include benzodiazepines and ECT as important and unavoidable treatment options for catatonia in any meaningful attempt to ameliorate the immediate and possibly long-term prognosis, regardless of underlying autism or any other disorders, as potential benefits seem to outweigh real and perceived risks.

Our treatment blueprints are contrary to the practice in some residential centers of prescribing large doses of newly developed psychotropic medications, usually antipsychotics but also anticonvulsants and antidepressants, for unspecified behavioral or psychiatric disturbances in people with ASDs and other developmental disorders. This approach, emulating treatment of bipolar and schizophrenic conditions, lacks any scientific basis in ASDs and other developmental disorders.

The interventions featured in our blueprints are labor-intensive, albeit in different ways for the medical and nonmedical modalities. Optimal psychological and behavioral interventions require adequate staffing levels and training, while the medical interventions require specialist resources. Full implementation may require considerable administrative and medical investment, which is a difficult feat in times of economic constraint and scarce research funding for clinical studies.

IV. Blueprint for the Future Study of Catatonia in ASDs

It is only when we do something different that we can expect to achieve above-average success. As an inventor, I have found that it is impossible to determine in advance where the best innovative ideas will come from, or how they will originate.

E. W. "Al" Thrasher (1920–2005)
Benefactor of the Thrasher Research Fund, Salt Lake City, Utah

The most urgent research challenge is the lack of effective treatments in many people with ASDs. The Interagency Autism Coordinating Committee Autism Research Matrix (April 2004, http://www.nimh.nih.gov/autismiacc/congappr commrep.pdf), that is endorsed by the US National Institutes of Health, lists the following goals for achievement within 7–10 years: (1) evidence that 25% of cases of ASDs can be prevented from symptom expression through early identification and early treatment, (2) methods that allow 90% of individuals with ASDs to develop speech, (3) the identification of genetic and nongenetic causes of ASDs (and possible interactions between them), and (4) the development of efficacious drug treatments that target core symptoms of ASDs. Three of the four goals concern existing or new treatments. Some leaders in the field informally predict that new drug treatments for ASDs will be found by serendipity, similarly to the discovery of antipsychotic and antidepressant medications in the 1950s and 1960s. The secretin saga of the last few years offers a salient example, albeit of a false start. However, the history of psychiatry shows that the discovery of antipsychotic and antidepressant compounds has not led to major discoveries of etiological factors involved in psychosis and affective disorders. Perhaps, we should be more prepared to accept that effective treatments might be discovered unexpectedly even if the causes of ASDs remain as elusive as before.

Catatonia has been almost completely absent in autism research and vice versa. However, catatonia in ASDs warrants special attention, as the condition is associated with considerable impairment. However, there is still a lack of awareness of both ASDs and catatonia hampering case finding and the design of good studies in this field. The assessment of catatonia among the myriad of autistic symptoms requires specialized training and clinical acumen.

As noted previously, catatonia research faces another challenge, as it is an indication, at least in its most severe forms, for ECT, an established treatment fraught with controversy and stigma. Current practice is still struggling to give ECT its proper place among other treatments for severe psychiatric disorders, as first-line treatment, last resort, or somewhere in between. Although ECT was widely embraced at first, the advent of antidepressive and antipsychotic medications and subsequent speculations about the biochemical nature of depression and psychosis may have subjected ECT to the tomato effect—the effect that occurs when an efficacious treatment for a certain disease is ignored or rejected because it does not "make sense" in the light of accepted theories of disease mechanism and drug action (Goodwin and Goodwin, 1984). The lack of modern studies of ECT for catatonia that meet acceptable scientific standard highlights the need for such studies that will surely benefit catatonic patients, with and without ASDs, and their treating physicians alike.

A. Blueprint for Future Studies in the Clinical Arena

1. There is an urgent need for treatment trials in autistic populations with catatonia, according to level of severity (see Treatment Blueprints for mild, moderate, and severe catatonia) and at all stages of the condition (acute, fluctuating, chronic phases, initial and continuation treatment, and prophylactic interventions). Response predictors should also be assessed. It is expected that chronic catatonic states in ASDs will be less responsive to treatment than acute or new onset catatonia in ASDs, similar to long-standing catatonia that is associated with other disorders.

2. Studies are warranted into the psychological and biological correlates of catatonia in people with and without ASDs. There is some evidence that stress, increased anxiety, and arousal accompany the development of catatonia in people with (see Chapter 15 by Shah and Wing) and without (see Chapter 8 by Cohen) ASDs. These observations beg for follow-up studies, using modern techniques, along the lines of the earlier studies of Gjessing (1974) in the fifties and of other investigators (Venables and Wing, 1962) (see Chapter 9 by Dhossche, Carroll, and Carroll).

3. The reliability and validity of existing catatonia rating scales need further testing in autistic populations of different ages (see Chapter 3 by Ohta, Kano, and Nagai, Chapter 5 by de Raeymaecker, and Chapter 6 by Stoppelbein, Greening, and Kakooza).

4. Childhood catatonia is poorly studied and probably unrecognized. Catatonia should be studied systematically in children, adolescents, and young adults with psychiatric, neurological, and developmental disorders, especially ASDs and Prader–Willi syndrome (see Chapter 7 by Verhoeven and Tuinier, Chapter 8 by Cohen, and Chapter 12 by Schieveld).

5. The symptom overlap between autistic regression in childhood disintegrative disorders and catatonia should be further assessed (see Chapter 4 by Dhossche and Rout).

6. The presence of catatonic symptoms should be assessed during regression in children with autism (see Chapter 4 by Dhossche and Rout).

7. Catatonia seems to be an important phenotype for genetic studies. Leonhard's periodic catatonia and its corresponding DSM categories should be further assessed (see Chapter 1 by Neumärker).

8. Family and genetic studies of bipolar, schizophrenia, and other psychotic disorders should ascertain catatonic subtypes and analyze data for catatonic subtypes separately (see Chapter 10 by Chagnon).

9. Family and genetic studies using head-to-head comparisons of patients with ASDs and catatonia should be conducted in order to substantiate common biological, neuropsychological, or genetic risk factors (see Chapter 10 by Chagnon).

B. Blueprint for Future Studies in the Preclinical Arena

There are currently very few definitive findings on the biological factors involved in ASDs and catatonia. The etiology and pathophysiology of ASDs, catatonia, and most other neuropsychiatric conditions remain clouded in mystery. Valid animal models of ASDs have not been developed, although there are a few promising models (see Chapter 4 by Dhossche and Rout). There are only a few putative animal models of catatonia (Kanes, 2004).

There is some evidence that GABA dysfunction is present in both autism and catatonia. Abnormal GABA function could be induced by abnormal central stress responses, by genetic predisposition, or a combination of both these and other factors. Preclinical studies should assess the role of GABA in adaptive changes due to anxiety, hormonal changes at the onset of puberty, and other types of psychological or somatic stress, especially in hypothalamic circuits (Kovacs et al., 2004). Findings should be relevant to advance (and test) GABA hypotheses of autism and catatonia (Dhossche, 2004; see also Chapter 4 by Dhossche and Rout).

There is a considerable amount of literature on the effects of electroconvulsive seizures (ECS)—the experimental analog of ECT. However, animal studies of ECT have not been particularly useful to advance human studies or clinical practice (Abrams, 2002). Arguably, this is because current animal models of depression, psychosis, or any other major psychiatric disorders have limited validity and, therefore, remain largely irrelevant for ECT research. However, it is possible that genes and molecules that are involved in ASDs and catatonia will be discovered as research advances. The effects of any promising experimental compound or procedure, including ECS, should be examined in suitable animal models with genetic or molecular defects that are the same or similar to the defects in the human disorder (Dhossche and Stanfill, 2004; see also Chapter 4 by Dhossche and Rout). One can only hope that the promise of the neuroscientific and genetic revolution does not remain a pipe dream for autism and catatonia research and for those afflicted by these disorders and their families.

Acknowledgment

Research support from the Thrasher Research Fund, Salt Lake City, Utah, is gratefully acknowledged.

References

Abrams, R. (2002). "Electroconvulsive Therapy," 4th ed. Oxford University Press, New York.

Carroll, B., and Goforth, H. (2004). Medical catatonia. In "Catatonia From Psychopathology to Neurobiology" (S. Caroff, S. Mann, A. Francis, and G. L. Fricchione, Eds.), American Psychiatric Publishing, Inc. Washington, DC.

Dhossche, D., and Bouman, N. (1997). Catatonia in children and adolescents (letter). *J. Am. Acad. Child Adolesc. Psychiatry.* **36,** 870–871.

Dhossche, D., Applegate, H., Abraham, A., Maertens, P., Bencsath, A., Bland, L., and Martinez, J. (2002). Elevated plasma GABA levels in autistic youngsters: Stimulus for a GABA hypothesis of autism. *Med. Sci. Monitor* **8,** PR1–PR6.

Dhossche, D. (2004). Autism as early expression of catatonia. *Med. Sci. Monitor* **10,** RA31–RA39.

Dhossche, D., and Stanfill, S. (2004). Could ECT be effective in autism? *Med. Hypotheses* **63,** 371–376.

Fink, M., and Taylor, M. (2003). "Catatonia. A Clinician's Guide to Diagnosis and Treatment." Cambridge University Press, Cambridge.

Garreau, B., Herry, D., Zilbovicius, M., Samson, Y., Guerin, P., and Lelord, G. (1993). Theoretical aspects of the study of benzodiazepine receptors in infantile autism. *Acta Paedopsychiatr.* **56,** 133–138.

Gawande, A. (2002). "Complications: A Surgeon's Notes on an Imperfect Science." Metropolitan Books/Henry Holt & Co, New York.

Ghaziuddin, N., Kutcher, S., Knapp, P., Bernet, W., Arnold, V., Beitchman, J., Benson, R. S., Bukstein, O., Kinlan, J., McClellan, J., Rue, D., Shaw, J. A., and Stock, S. (2004). Practice parameter for use of electroconvulsive therapy with adolescents. *J. Am. Acad. Child. Adolesc. Psychiatry.* **43,** 1521–1539.

Gillberg, C. (1991). The treatment of epilepsy in autism. *J. Autism and Dev. Disord.* **21,** 61–77.

Gjessing, L. R. (1974). A review of periodic catatonia. *Biol. Psychiat.* **8,** 23–45.

Goodwin, J., and Goodwin, J. (1984). The tomato effect. Rejection of highly efficacious therapies. *JAMA* **251,** 2387–2390.

Kanes, S. (2004). Animal models. In "Catatonia. From Psychopathology to Neurobiology" (S. Caroff, S. Mann, A. Francis, and G. L. Fricchione, Eds.), American Psychiatric Publishing, Inc. Washington, DC.

Kovacs, K. J., Miklos, I. H., and Bali, B. (2004). GABAergic mechanisms constraining the activity of the hypothalamo-pituitary-adrenocortical axis. *Ann. NY Acad. Sci.* **1018,** 466–476.

Leo, R. (1996). Movement disorders associated with serotonin selective reuptake inhibitors. *J. Clin. Psychiat.* **57,** 449–454.

Lopez-Canino, A., and Francis, A. (2004). Drug-induced catatonia. In "Catatonia. From Psychopathology to Neurobiology" (S. Caroff, S. Mann, A. Francis, and G. L. Fricchione, Eds.), American Psychiatric Publishing, Inc. Washington, DC.

Marrosu, F., Marrosu, G., Rachel, M., and Bigglio, G. (1987). Paradoxical reactions elicited by diazepam in children with classic autism. *Funct. Neurol.* **2,** 355–361.

Mortimer, A. M. (2004). Standardized Instruments. In "Catatonia. From Psychopathology to Neurobiology" (S. Caroff, S. Mann, A. Francis, and G. L. Fricchione, Eds.), American Psychiatric Publishing, Inc., Washington, DC.

Ottosson, J.-O., and Fink, M. (2004). "Ethics of Electroconvulsive Therapy." Hove: Brunner-Routledge, New York.

Venables, P. H., and Wing, J. K. (1962). Level of arousal and the subclassification of schizophrenia. *Arch. Gen. Psychiatry* **7,** 114–119.

Wing, L., and Shah, A. (2000). Catatonia in autistic spectrum disorders. *Brit. J. Psychiatry* **176,** 357–362.

INDEX

A

ABLLS. *See* Assessment of basic language and learning skills
Acute brain syndrome, 224
Acute deterioration
 in autistic patients, 234–235
AD. *See* Autistic disorders
ADI-R. *See* Autism Diagnostic Interview-Revised
Adjustment disorders, 42
Adolescent psychiatry, 4
Adult psychiatry, 203
Affective disorder, 166
Agitation, 199
Akinesia, 23, 50, 137
Akinetic catatonic patients, 138, 142
Amandatine, 137
Amineptine, 142
Amisulpride, 141–142
Amobarbital, 274
Anesthesia, risk associated with, 223–224
Angelman syndrome, 65, 167, 171
Anticholinergic medication, 49
Antidepressant medications, 272, 280–281
Antipsychotic agents, 236
 use of, 240
Antipsychotics, 42, 45–46, 50–52, 124, 140–141, 154. *See also* Carbamazepine
Apathy syndrome, 92
ASD. *See* Autism spectrum disorders
Asperger syndrome (AsD), 12, 33, 167, 247
Assessment of basic language and learning skills (ABLLS), 184–185
Attention deficit/hyperactivity disorder (ADHD) child, 88
Audiological testing, 198
Augmentative communication system, 183
 use of signs as, 182
Autism
 Asperger syndrome, 234
 catatonia in, 233
 catatonic regression in, 59–60
 cerebellum in, role of, 158–159
 clinical manifestations of, 156
 definition of, 233–234
 diagnosis of, 197, 234
 disintegrative disorder, 234
 GABA models of, 69–70
 and Motor Theory of language, 152–153
 pathophysiological theory of, 156
 pervasive developmental disorder (PDD), 234
 Rett syndrome, 234
 similarities between catatonia and, 157–159
 symptoms of children with, 156
 treatment algorithm for, 238
 types of, 113
Autism Diagnostic Interview-Revised (ADI-R), 108, 114
Autism spectrum disorders, catatonia in, 104–110
 assessment of
  catatonia-like deterioration, 270–271
  diagnostic catatonia criteria, 272–273
  medical or neurological evaluation, 271
  medication, scrutiny of, 271–272
  severity condition, assessment of, 273
 future studies of
  in clinical field, 282
  in preclinical field, 283
 treatment modalities of
  bilateral ECT, 268, 276–277
  lorazepam challenge test, 268, 274–276, 278
  lorazepam continuation treatment, 268, 276
  lorazepam trial, 268, 276, 278
  Shah-Wing approach, 268, 273–274, 277
 treatment of
  with benzodiazepines, 279–280
  ethical and issues in, 277
  general approach, 278
  limitations in, 278
  with lorazepam, 279–280

Autism spectrum disorders, catatonia in (cont.)
  options for severity level, 278
  recommended periods of intervention, 278
Autism spectrum disorders (ASD), 215, 234, 240–241, 245–246
  association of GABA receptors and, 171
  catatonia-like deterioration
  catatonia-like features and, relationship of, 35
    prevalence of, 36–37
  catatonia-like features, case studies of
    comparative frequency of, 30–32
    deterioration, 32–34
    frequency of, 25–29
  catatonic features, manifestations of, 23–24
  clinical features of, 23
  diagnosis of, in children, 182–183
  genomic scans for
    methods of, 166–167
    results of, 167–173
  patients, 153, 156
    symptoms of, 157
  patients, effect of
    age, 34–35
    level of ability, 34–35
  patients with catatonia, case study of
    limitations of study, 53
    method, 43–44
    presentation of cases, 44–49
    result interpretation, 51–52
    results, 49–51
    subjects, 43
    treatment suggestions, 52
  risk factors of, in offspring, 166
  stereotypies in, 110–115
  susceptibility regions for, on chromosome 15, 165, 171–172
  Uta Frith's theory in, 247–248
Autistic catatonia, symptoms of, 107
Autistic children, high-functioning. *See also* Self-management
  video modeling for, 190
Autistic disorders (AD), 253–252. *See also* Autism spectrum disorders
  in catatonia, 246
  diagnosis of, 198
Autistic hebephrenia, 13
Autistic impairments, premorbid pattern of, 252–253
Autistic patients, baseline symptoms of, 272

Autistic regression, 55–56
  early, 57–58
  late, 58–59
  role of GABA. *See* Gamma-aminobutyric acid, function
  symptoms comparison with catatonic regression, 60–61
  treatment implications, 62
Autistic spectrum disorders. *See* Autism spectrum disorders
Autistic symptomatology, 203
Autosomal chromosomes, 165, 167

B

Barbiturates, 66
Bayley Motor Scale, 96
Behavioral interventions
  language
    assessment of basic language and learning skills, 182–185
    sign language and picture exchange communication system, 182–185
  self-management, 189–191
  social skills
    pivotal response training (PRT), 185–186
    priming, 186–187
    social stories, 187–189
Behavioral therapy
  for catatonic treatment, 140
  learning principles of, 197
Behavioral treatments
  private clinics, 181–182
  school systems, 181–182
  use of video technology, computers, and naturalistic procedures in, 182
Benzodiazepines, 52, 156, 202, 226
  -GABA receptor complexes, abnormal, 66, 280
  high dosage, effect of, 279
  withdrawal effects of, 272
BFCRS. *See* Bush-Francis Catatonia Rating Scale
Bilateral ECT, 276–278
Bilateral electrode placement, 220, 222
Bipolar affective disorder, 124–126
Bipolar disorder (BP), 166
Bleuler autism, 11
Bradykinesia, 235
Bradyphrenia, 235

INDEX

Brain abnormalities, 172
Brain structural abnormalities, 240
Bromazepam (BZP), 48
Bupropion, 272
Bush-Francis Catatonia Rating Scale (BFCRS), 108–109, 135

C

Calcium-regulated potassium channel (*KCNN3*) gene, 167
Carbamazepine, 66, 124
CARS. *See* Childhood Autism Rating Scale
Catalepsy, 23
Cataleptoid insanity, 22
Catatonia, 218
  adherence to delusional ideas resulting in, 139–140
  adult psychiatry, 203
  age-of-onset of, 49, 51, 61
  akinetic patients with, 138, 142
  in ASD, 42. *See also* Catatonia in autism spectrum disorders
  in autism
    association between, 236–237
    cerebellum in, role of, 158–159
  childhood. *See* Catatonia in young people
  concluding remarks on, 145
  correlation with tuberculosis, 196
  course of, 50–51
  criteria for, 195
  and delirium
    differences, 202
    similarities, 201
  early, 5
  ECT for, 66, 234, 238
  and family history, 50
  feature of, as marked psychomotor disturbance, 196
  history of, 195–197
  hyperanxious or hyperemotional states resulting in, 142–143
  indication for ECT, in youth with, 208–211, 214–216
  malignant, 216, 273, 277
  medication, effect of, 50
  multicausal origin, 195–197
  neuroanatomic model of, 137–138
  neurochemical etiologies for medical, 155
  neurologic conditions associated with, 155
  periodic, 216–217
  phenomenology of, in young people, 132–135
  preceding conditions, 50
  and psychiatric complication, 50
  psychiatric nosography of, in children and adults, 135–136
  psychopathological models of, 138–139
    clinical and heuristic validity, 143
    limitations of, 144–145
  rating scales (CRS), 236
  recognition of, 236
  resistance to delusional thinking or conviction resulting in, 140–141
  in schizophrenia
    genomic scans of, 165–173
  severities of, 44, 49–50
  as syndrome of motor dysregulation, 214
  treated with convulsive therapy, 238
  treatment suggestions, 52–53
Catatonia in autism spectrum disorders (ASD)
  assessment of, 273
    BFCRS, 108–109
    catatonia-like deterioration, 270–271
    diagnostic catatonia criteria, 272–273
    DISCO, 109
    medical or neurological evaluation, 271
    medication, scrutiny of, 271–272
    severity condition, assessment of, 273
  characteristic symptoms, 106–107
  epidemiology, 105–106
  etiology, 108
  future studies of
    in clinical field, 282
    in preclinical field, 283
  risk factors of, 107–108
  treatment modalities
    bilateral ECT, 268, 276–277
    lorazepam challenge test, 268, 274–276, 278
    lorazepam continuation treatment, 268, 276
    lorazepam trial, 268, 276, 278
    Shah-Wing approach, 268, 273–274, 277
  treatment of, 109–110
    with benzodiazepines, 279–280
    ethical and legal issues in, 277
    general approach, 278
    limitations in, 278
    with lorazepam, 279–280
    options for severity level, 278
    recommended periods, of intervention, 278

Catatonia in SZ,
  genomic scans of, 165
  methods of, 166–167
  results of, 167–173
Catatonia in young people
  phenomenology of, 132–135
  prevalence of, 132–133
  symptoms of, 133–137, 144
Catatonia-like deterioration, 107
  in ASDs, 245, 248, 262
  effects of medical treatment, 250–251
  electroconvulsive therapy (ECT) in, 250
  implications for services and staffing levels
    in, 260
  management of specific problems
    catatonia-like excitement, 259–260
    difficulty with walking, 259
    eating problems, 258
    incontinence, 257
    speech and communication problems,
      258–259
    standing still, 259
  at neuropsychological level, 247
  prevalence rate for severe, 105
  psychological methods of intervention, 252
  psychological treatment and management,
    general principles of
    conceptualization of nature of, 255
    dealing with stress factors, 254–255
    initial assessment, 253–254
    maintaining and increasing activity,
      256–257
    structure and routine, 257
    use of prompt as external stimuli/signals,
      255–256
  stress factors in
    biological, 247
    external, 247
    psychological, 247
Catatonia-like features in ASD, case studies of
  catatonia-like deterioration, 32–34
  comparative frequency of, 30–32
  frequency of, 25–29
Catatonic idiocy, 5
Catatonic motor dysfunction, 138
Catatonic regression
  characteristics of, 56
  early-onset, 55
  role of GABA. *See* Gamma-aminobutyric acid,
    function
  symptoms comparison with autistic regression,
    60–61
  treatment implications, 62
Catatonic schizophrenia, 127, 135–136,
  160–161
  case studies of, 140–141
  diagnosis of, 196
  prevalence rate of, 127
Catatonic stupor, 5
  benzodiazepine treatment in, 204
Catatonic symptom(s), 8
  distribution, factor analysis of, 135
Catatonic syndrome. *See specific* Catatonia
Catatonic treatment, behavioral therapy for, 140
Catechol-*O*-methyltransferase gene, 166
CDD. *See* Childhood disintegrative disorder
Central nervous system (CNS)
  abnormalities, 127
  motor structures, 120
Cerebellar lesions, effect of, 159
Cerebellar Purkinje cells, 158–159
Cerebellum
  neuroanatomical abnormalities of, 159
  role of, in catatonia and autism, 158
Cerebral dysfunction, 155
Cerebral edema, symptoms of, 140
Cerebral impairment, 135
Cerebrospinal fluid (CSF) levels, 68
Childhood Autism Rating Scale (CARS), 108,
  113–114
Childhood disintegrative disorder (CDD), 56,
  58–61
Childhood schizophrenia, 22
  catatonic form of, 200
Child neuropsychiatry, clinical importance
  in, 201
Child psychiatry, 4, 10
Children, developmental psychopathology and
  motor impairment in
  ADHD of the combined type, 94
  developmental coordination disorder, 96–97
  dissociated motor development, 94–95
  critical clinical signs in, 94
  disturbed motor and behavior patterns, 92–93
  of low birth weight, 95–96
  paranatal complications
    apathy syndrome, 92–93
    babies of breech presentation, 92
    choreiform syndrome, 93
    hyperexcitability syndrome, 92

INDEX

PDD and body ego problems
  communication impairments, 97
  purposeful hand skills in Rett's disorder, loss of, 97
  repetitive and stereotyped patterns of behaviors, 97
  social interaction, impairment in, 97–98
  subtle forms of PDD NOS, 98
prenatal complications
  apathy syndrome, 92–93
  babies of breech presentation, 92
  choreiform syndrome, 93
  hyperexcitability syndrome, 92
  stereotypic movement disorder, 98–99
Children, low birth weight, motor development in
  previous prematurity syndrome, 95
  respiratory distress syndrome, 96
  transient dystonia syndrome, 95–96
Choreiform movements, behavior problems, in children, 93
Choreiform syndrome, 93
Chromosomal analysis, 199
Chromosomal susceptibility regions, 165–167, 171–173
Chronic catatonic schizophrenia, 251
Ciprofloxacin, 272
Citalopram, 235
Clozapine in monotherapy, 199
Cocaine. *See* Illicit drugs
Cognitive deficits, 223
Cognitive model of voluntary human movements, 138–139
Cognitive stimulation, level of, 254
Cognitive testing, 198
Complex behaviors, use of PRT to teach, 185
Consciousness, quiet state of, 85
Conversation skills, 186
Convulsive therapy, in catatonia, 238
Cortical glutamate-glutamine levels, 69
Corticotropin releasing factor (CRF), 248

D

DCR. *See* Diagnostic Criteria for Research
Deferred imitation, 86
Delirium
  definition of, 201
  differences with catatonia, 202
  similarities with catatonia, 201

Delta-like 4 *Drosophila* (*DLL4*) gene, 165, 172
Dementia hebephrenica's praecox.
  See Hebephrenia
Dementia infantilis, 6
Dementia praecocissima, 6
Dementia praecox, 7, 9, 13, 132
Depressive illness, 223
Depressive mood disorders, 42
  diagnosis of, 235
Desmuslin gene (*DMN*), 172
Deterioration, acute
  in autistic patients, 234–235
Diagnostic and Statistical Manual (DSM), 154
Diagnostic and Statistical Manual for Mental Disorder, 4th edition (DSM-IV), 6, 12–13, 23, 96–98, 132, 272
  catatonic symptoms of, 134
  text revision (TR), 42–43, 49, 51
Diagnostic Assessment for the Severely Handicapped II (DASH-II), measurement of movement disorders by, 114
Diagnostic Criteria for Research, 43
Diagnostic Interview for Social and Communication Disorders (DISCO), 25–26, 30, 32, 34, 109
Diazepam, 274
DISCO. *See* Diagnostic Interview for Social and Communication Disorders
Disruptive transition behavior. *See also* Priming
  use of videotaped instruction to reduce, 186
Disulfiram, 272
DLL4. See Delta-like *Drosophila* gene
DMN. See Desmuslin gene
Dopamine (DA), 108, 111–112
Dopaminergic drugs, withdrawal effects of, 272
Down's syndrome, 30, 36, 57
Drug intoxications and withdrawals, 155
DSM-IV. *See* Diagnostic and Statistical Manual for Mental Disorder, 4th edition
Dystrobevin binding protein 1 (DTNBP1), 172

E

E6-AP ubiquitin-protein ligase (*UBE3A*) gene, 171
Eating problems in catatonic patients, recommendations for, 258
ECS. *See* Electroconvulsive seizures
ECT. *See* Electroconvulsive therapy
Ego development, of child, 83, 89

Electroconvulsive seizures, 62, 283
  in GABA models of autism, 69–70
  and neurogenesis, 71–72
  on Purkinje cell survival, effect of, 70–71
Electroconvulsive therapy (ECT), 141–142, 156, 208–210, 218, 226
  action of, 68–69
  administration of, 219–220
    anticholinergic medication, 220
    ventilation, 220
  adverse effects of
    impairment of memory, 221–223
    new learning, 221–223
  benefits of, 218
  bilateral, 276–278
  in catatonia-like deterioration, 250, 252
  in childhood psychoses, 63–64
    limitation in diagnostic criteria, 64
  contradictions, 221
  course of, 221
  dosage of electric in, 222
  effectiveness of, 211
  extent of use of
    in adults, 210
    in prepubertal, 210–211
  in GABA functioning, role of, 66
  indications for
    general considerations, 213–214
  machine, electrode placement, and current in, 220
  maintenance, 225
  medication
    anticholinergic, 220
  mode of action of, 219
  mortality
    adult fatality rate, 224
  preparation for
    consent, 218
    ethics, 218
    human rights, 218
  psychological theories of mechanics of, 219
  recommendation of, 235
  in relieving mood disorder without catatonia, 241
  repeat course of, 225
  usage, effect of, 222
  used in catatonic patients, 234–235, 237–241
Endorphins, 111
Epileptic seizures, 48, 50

Executive functions, 160–161
Expected risk/benefit ratio, 278

F

Factor analysis of catatonic symptom distribution, 135
Fidgety pattern movement, 84
Fragile X. *See* Down's syndrome
Free-play activities, 185–186
Functional communication
  use of pictures to promote, 183–184
  use of sign language to promote, 182–183

G

GABA. *See* Gamma-aminobutyric acid
GABA$_A$-receptor, 127, 137
  subunits genes, 65
Gabapentin, withdrawal effects of, 272
GAD. *See* Glutamic acid decarboxylase
Gamma-aminobutyric acid (GABA), 108
  dysfunction, 280, 283
  ergic anticonvulsant, 127
  function, 128, 237, 251
    in autistic regression, 64–66
    in catatonic regression, 66–68
    ECT, action of, 66, 68–69
    role of, in catatonia, 241
  mimetic compounds, 119
  models, of autism, 69–70
  receptor genes, 165
    B3 (*GABRB3*), 171
    G3, 169
    5 (*GABRA5*), 171
  theory of catatonia, general criteria for, 67
Gene(s)
  calcium-regulated potassium channel, 167
  catechol-*O*-methyltransferase, 166
  cholinergic receptor nicotinic alpha polypeptide 7, 165, 172
  delta-like 4 *Drosophila*, 162, 165
  desmuslin, 172
  E6-AP ubiquitin-protein ligase, 171
  GABA receptor, 165, 171
  NMDA receptor-regulated 2, 172
  serotonin transporter, 166
  zinc transporter, 165, 172
Genetic Location Database maps, 165, 167
Genetic maps, 167

INDEX

291

Genetic predisposition, 108
Genome scans, 165
 with chromosome 15, 165
 concluding remarks on, 173
 methods of
  for ASD, 166–167
  for catatonia in SZ, 166–167
 results of
  for ASD, 167–173
  for catatonia in SZ, 167–173
Glutamic acid decarboxylase (GAD), 66, 68
Growth hormone (GH) deficiency, 120

H

Hallucinations, 139–141, 199
Haloperidol, 47, 197
Hebephrenia, 5–6, 13
Heller's dementia. *See* Childhood disintegrative disorder
Heller syndrome, 6
Heterozygotes, 70
5-HIAA. *See* 5-hydroxyindoleacetic acid
5-HT. *See* Serotonin, effect of
Human motility, psychoanalysis view of
 mobility and psychoanalytic therapy
  clinical consequence, 88
 motor achievements and their psychodynamic significance
  ego development, 89
  object permanence, 89
  school age child, transition to, 91
  socialization process, 91
  sphincter control, 91
  upright locomotion, 90–91
 young orthopedic patients, impact of immobilization on, 88–89
5-hydroxyindoleacetic acid, 68
Hyperanxious experiences, 142–143
Hyperexcitability syndrome, 92, 94
Hyperkinesia, 48
Hypoperfusion, 156
Hypothalamic dysfunction, 120–121

I

ICD-10. *See* International Classification of Diseases, 10th revision
Illicit drugs, 272
Infantile autism, 11

Infantile catatonia, 5
 differentiated into groups, cases of, 9
Infantile schizophrenia, 10
Interagency Autism Coordinating Committee
 Autism Research Matrix, goals of, 281
International Classification of Diseases,
 10th revision (ICD-10), 6, 12, 23
International Classification of Diseases (ICD), 154
IQ, 21, 25–27, 29, 31–35, 41, 43–46, 48

K

Kanner syndrome, 12
KCNN3. *See* Calcium-regulated potassium channel gene
Kiddie-PANSS, 10
Kluver-Bucy syndrome, 156

L

Landau-Kleffner syndrome (LKS), 58
Language
 acquisition, association of psychomotor function and, in autistic children, 152
 assessment of basic language and learning skills, 184–185
 sign language and picture exchange communication system, 182–184
Learning disability, catatonia-like features in, 36
"Le plaisir d'être cause," 85
Lethal catatonia, electroconvulsive therapy (ECT) in, 202
Lithium, 224
 in PWS treatment, 124–125
Locomotion anxiety, 90
Lorazepam, 137, 217, 226, 239, 250
 for catatonia, 198, 200
 challenge test, 268, 275–276, 278
 side effect of, 274
 continuation treatment, dose range in, 268, 276
 trial, 268, 276, 278

M

Malignant catatonia, 216, 273, 277
Marshfield maps, 165, 167
Memantine. *See* Amandatine
Memory, impairment of, 222

Mental disorders, 194. *See also* Autism syndrome disorders; Bipolar disorder
  major, 3
  autism, 3
  catatonia, 3
  pathological changes, 13
Mental health-care services, 4
Mental retardation, diagnosis of, 197
Mescaline. *See* Illicit drugs
N-methyl-D-aspartate (NMDA) receptor, 137
Modified Rogers Scale, 109
Monoamine oxidase inhibitor (MAOI), 272
Mood stabilizers, 125
Mother-child interaction, 83–85, 90–92, 99
Motivation
  enhancement of, 185
  to learn complex skills, 185
Motor behaviors, abnormal, in babies
  apathy syndrome, 92–93
  born in breech presentation, 92
  hyperexcitability syndrome, 92
Motor development, steps in
  motor acts development
    clinical significance, 86
    gestalt perception and technical insight, 87
    Oedipal phase, 86
    rapprochement phase, 86
    precise movement, 84
    milestones of this phase, 85
    spontaneous movements, 84
Motor symptoms, 140
Motor system, 139
Motor Theory of language, and autism, 152–153
Movement dysregulation disorder, 240
Multiple baseline design
  to assess the level of social initiations, 186
  examine effects of video priming, 187
  to measure
    disruptive behaviors, 190
    social initiations, 190
  used to assess changes in social interaction skills, 188–189
Multisystem disorders. *See* Prader-Willi syndrome

N

NARG2. *See* NMDA receptor-regulated 2 gene
National Center for Biological Information (NCBI) maps, 165, 167

Necrotic cell death, 71
Negativism, 3, 5–8, 10–11, 13
Neurobiological syndromes. *See also* Autism; Catatonia
  criteria for, 153–156
Neurochemical etiologies for medical catatonia, 155
Neurochemistry, 155
Neurogenesis, 56, 71–72
Neuroleptic malignant syndrome, 68. *See also* Malignant catatonia
Neuroleptic malignant syndrome (NMS), 137
  use of ECT in, 216
Neuropathological changes, 196
Neuropsychiatric spectrum disorders, 202
NICE Guidelines for ECT, 219, 222
  aim of, 213
  recommendations of, 217
NMDA receptor-regulated 2 (*NARG2*) gene, 172
N-methyl-D-aspartate (NMDA) receptors, 172
NMS. *See* Neuroleptic malignant syndrome
Noradrenalin levels, 68
Nosology, 234

O

Obsessive-compulsive behaviors, worsening of, 198
Obsessive compulsive disorders (OCD), 110
  diagnosis of, 235
Oedipal phase, 86
Ohta Staging, 43
Opiates. *See* Illicit drugs
Opioids, 111–112

P

Pack therapy, 141
Parallel family-based studies in autism and catatonia, 159–160
Paralytic catatonia, 141
Paranoid schizophrenia, 160–161
Parental SZ-like psychosis, 166
Parkinsonism, 42, 44, 46, 50
Parkinson's disease, 137
Pathophysiological theory of autism, 156
Pathway dysfunction, 155
Patients
  ASD, 153, 156–157
  with frontal lobe syndromes, 158

PWS
    psychotic symptoms in, 121–126
    with UPD etiology, 126
    symptoms of autistic, 157
Patients' satisfaction and attitude to ECT, 224–225
PCP. *See* Illicit drugs
PECS. *See* Picture exchange communication system
Peer training, 185–186
Penicillin, 61
Periodic catatonia, 166, 216–217
Pervasive developmental disorders (PDD), 12, 23, 84, 86, 97, 110, 136, 197
    not otherwise specified, 56
Pharmacotherapy, 140
Photographic activity schedules, 189
Phylogenesis, 158
Picture exchange communication system (PECS), 183–184
Picture schedules in clinic setting, 190
Pimozide, 47
Pivotal response training (PRT)
    on using the child's motivation, 185–186
Practical intelligence. *See* Sensorimotor intelligence
Prader-Willi/Angelman syndrome loci, 167
Prader-Willi syndrome (PWS), 65, 282
    concluding remarks on, 127–128
    development and behavior, 120–121
    features of, 119–120
    imprinting in, 120
    patients
        psychotic symptoms in, 121–126
        with UPD etiology, 126
    psychomotor symptoms, 126–127
    psychotic disorders, 121–126
Pre-ECT consideration
    medication, 217–218
    Practice parameters, 217–218
Previous prematurity syndrome, 95
Primal cavity seat, of oral pleasure, 85
Priming
    applied to social-skills training methodology, 186–187
    effects of, 187
    examined using videotaped instruction, 186–187
Prolonged seizures, 221–222, 224
    and ECT, 223

Psilocybin. *See* Illicit drugs
Psychiatric symptomatology, 125–126
Psychodynamic therapy, 143
Psychological assessments, 249
    for catatonic-like deterioration, 254–255
    test of cognitive ability, 254
Psychological dysfunctions, 245
Psychological treatment and management, general principles of, 255
Psychomotor automatism, 131, 135, 144–145
    features of, 139
Psychomotor function, association of language acquisition and, in autistic children, 152
Psychomotor psychoses, 13
Psychomotor retardation, episodes of, 199
Psychomotor symptoms in PWS, 126–127
Psychopathological model for catatonia, 138–139, 144–145
Psychosis, 240
    atypical, 124
    cycloid, 124–126
    diagnosis of, 197
    paranoid, 124
    symptoms of, 126
Psychotherapy, 140–141
Psychotic symptomatology, 124
Psychotic symptoms in PWS patients, 121
    psychopathological symptoms, 124
PubMed and Medline search
    autism and motor signs, 199
    catatonia
        ASD, 199
        and autism, 199
        and children, 199
Purkinje cell survival, ECS effect on, 70–71
PWS. *See* Prader-Willi syndrome

Q

Quiet alert state, 85

R

Rapprochment phase, 86
Reifungsdissoziation, 94
Respiratory distress syndrome, 96
Response-cost system, 187
Rett's disorder, 12, 97
Risperidone, 235
    for psychosis, 200

S

Schizophrenia (SZ), 104–106. *See also* Catatonia in autism spectrum disorders
  with catatonia, 199
  catatonic, 160–161
  characteristic of, 203
  childhood, 133
  genomic scans for catatonia in, 165
    methods of, 166–167
    results of, 167–173
  paranoid, 160–161, 199
  unsystematic, 166
Schizophrenic negativism, 3, 10, 13
Seizures, effect of, on developing brain, 71
Self-injurious behavior (SIB), 45, 48, 111. *See also* Catatonia in austic spectrum disorders, etiology; Opioids
Self-management
  combined with video modeling, 190–191
  procedures
    autistic teenagers, 190
    to produce rapid behavior change, 189
    self-reinforcement component, 190
  use of
    to teach daily living skills, 189
    to teach social skills, 189
Self-reinforcement component, 190
Sensorimotor intelligence, 85
Sensory stimulation hypothesis for stereotypy, 112
Serotonin
  effect of, 111
  selective reuptake inhibitors, 272
  transporter gene, 166
Shah-Wing approach, 268, 273, 277
  limiting factor of, 274
Sign language training, 184
Signs
  advantages of using, 183–184
  disadvantages of using, 183–184
Skin-picking, 121, 126
SLC30A4. *See* Zinc transporter gene
SMD. *See* Stereotypic movement disorder
Social behaviors and autistic children, 185
Social initiations. *See also* Multiple baseline design
  level of, 186
Socialization process, in child, 91
Social-skills training methodology, 186
Social stories, use of
  to establish rule-governed behavior, 188
  to facilitate social initiations to typical peers, 188
  to improve greeting behavior, 187
  to reduce disruptive behaviors of students with autism, 187–188
  to teach children with autism, 187–189
Social stories intervention, 189
Sodium valproate, 69
Somatic disorders, 195
Spatial attention dysfunction, 159
Specific language disorder, 30
Speech
  acquisition, 183–184
  development, 198
Stereotypic movement disorder (SMD)
  assessment of, 113–114
  epidemiology, 110
  etiological hypothesis of, 110–113
  treatment, 114–115
Steroids, 272
Stroop Color-Word Test, 160
Stuporous catatonic patients, case study of, 142–143
Symptom autism, 11
SZ. *See* Schizophrenia

T

Tanaka-Binet scale of intelligence, 43
Tantrumous behavior, effects of video priming on, 187
Tardive seizures, 223
Teenagers with autism, self-management procedures of, 190
Tension insanity. *See* Catatonia
Theophylline, 224
Thioridazine, 140
Tics, 47, 63
Tourette syndrome, 42, 45, 47
Trail Making test, 160
Transient dystonia syndrome, 95–96
Trazodone, 224
TS. *See* Tourette syndrome
Typical catatonia, 5

U

UBE3A. *See* E6-AP ubiquitin-protein ligase gene
Uniparental disomy (UPD), 120–121
  etiology, 126
UPD. *See* Uniparental disomy

INDEX 295

V

Valproic acid, 47, 69, 119, 125, 127
Ventral prefrontal cortical function tests, 161
Verbal Fluency test, 160
Verbigeration, 4–5, 7
Video modeling
  to teach high-functioning autistic children, 190
Visual communication systems, for catatonic patients, 259
Voluntary human movements, cognitive model of, 138–139

W

Wisconsin Card Sorting test, 160

Wrist counter, 189
  use of, to record initiated compliments, 191
Writing type movement, 84

Y

Yale-Brown obsessive-compulsive scale, 44
Young orthopedic patients, impact of immobilization on, 88–89

Z

Zinc transporter gene, 165, 172
Ziprasidone, 235
Zolpidem, 66
  use of, in catatonic patients, 217, 226

CONTENTS OF RECENT VOLUMES

Volume 37

Section I: Selectionist Ideas and Neurobiology

Selectionist and Instructionist Ideas in Neuroscience
Olaf Sporns

Population Thinking and Neuronal Selection: Metaphors or Concepts?
Ernst Mayr

Selection and the Origin of Information
Manfred Eigen

Section II: Development and Neuronal Populations

Morphoregulatory Molecules and Selectional Dynamics during Development
Kathryn L. Crossin

Exploration and Selection in the Early Acquisition of Skill
Esther Thelen and Daniela Corbetta

Population Activity in the Control of Movement
Apostolos P. Georgopoulos

Section III: Functional Segregation and Integration in the Brain

Reentry and the Problem of Cortical Integration
Giulio Tononi

Coherence as an Organizing Principle of Cortical Functions
Wolf Singerl

Temporal Mechanisms in Perception
Ernst Pöppel

Section IV: Memory and Models

Selection versus Instruction: Use of Computer Models to Compare Brain Theories
George N. Reeke, Jr.

Memory and Forgetting: Long-Term and Gradual Changes in Memory Storage
Larry R. Squire

Implicit Knowledge: New Perspectives on Unconscious Processes
Daniel L. Schacter

Section V: Psychophysics, Psychoanalysis, and Neuropsychology

Phantom Limbs, Neglect Syndromes, Repressed Memories, and Freudian Psychology
V. S. Ramachandran

Neural Darwinism and a Conceptual Crisis in Psychoanalysis
Arnold H. Modell

A New Vision of the Mind
Oliver Sacks

INDEX

Volume 38

Regulation of GABA$_A$ Receptor Function and Gene Expression in the Central Nervous System
A. Leslie Morrow

Genetics and the Organization of the Basal Ganglia
Robert Hitzemann, Yeang Olan, Stephen Kanes, Katherine Dains, and Barbara Hitzemann

Structure and Pharmacology of Vertebrate GABA$_A$ Receptor Subtypes
Paul J. Whiting, Ruth M. McKernan, and Keith A. Wafford

Neurotransmitter Transporters: Molecular Biology, Function, and Regulation
Beth Borowsky and Beth J. Hoffman

Presynaptic Excitability
Meyer B. Jackson

Monoamine Neurotransmitters in Invertebrates and Vertebrates: An Examination of the Diverse Enzymatic Pathways Utilized to Synthesize and Inactivate Biogenic Amines
B. D. Sloley and A. V. Juorio

Neurotransmitter Systems in Schizophrenia
Gavin P. Reynolds

Physiology of Bergmann Glial Cells
Thomas Müller and Helmut Kettenmann

INDEX

Volume 39

Modulation of Amino Acid-Gated Ion Channels by Protein Phosphorylation
Stephen J. Moss and Trevor G. Smart

Use-Dependent Regulation of $GABA_A$ Receptors
Eugene M. Barnes, Jr.

Synaptic Transmission and Modulation in the Neostriatum
David M. Lovinger and Elizabeth Tyler

The Cytoskeleton and Neurotransmitter Receptors
Valerie J. Whatley and R. Adron Harris

Endogenous Opioid Regulation of Hippocampal Function
Michele L. Simmons and Charles Chavkin

Molecular Neurobiology of the Cannabinoid Receptor
Mary E. Abood and Billy R. Martin

Genetic Models in the Study of Anesthetic Drug Action
Victoria J. Simpson and Thomas E. Johnson

Neurochemical Bases of Locomotion and Ethanol Stimulant Effects
Tamara J. Phillips and Elaine H. Shen

Effects of Ethanol on Ion Channels
Fulton T. Crews, A. Leslie Morrow, Hugh Criswell, and George Breese

INDEX

Volume 40

Mechanisms of Nerve Cell Death: Apoptosis or Necrosis after Cerebral Ischemia
R. M. E. Chalmers-Redman, A. D. Fraser, W. Y. H. Ju, J. Wadia, N. A. Tatton, and W. G. Tatton

Changes in Ionic Fluxes during Cerebral Ischemia
Tibor Kristian and Bo K. Siesjo

Techniques for Examining Neuroprotective Drugs *In Vitro*
A. Richard Green and Alan J. Cross

Techniques for Examining Neuroprotective Drugs *In Vivo*
Mark P. Goldberg, Uta Strasser, and Laura L. Dugan

Calcium Antagonists: Their Role in Neuroprotection
A. Jacqueline Hunter

Sodium and Potassium Channel Modulators: Their Role in Neuroprotection
Tihomir P. Obrenovich

NMDA Antagonists: Their Role in Neuroprotection
Danial L. Small

Development of the NMDA Ion-Channel Blocker, Aptiganel Hydrochloride, as a Neuroprotective Agent for Acute CNS Injury
Robert N. McBurney

The Pharmacology of AMPA Antagonists and Their Role in Neuroprotection
Rammy Gill and David Lodge

GABA and Neuroprotection
Patrick D. Lyden

Adenosine and Neuroprotection
Bertil B. Fredholm

Interleukins and Cerebral Ischemia
Nancy J. Rothwell, Sarah A. Loddick, and Paul Stroemer

Nitrone-Based Free Radical Traps as Neuroprotective Agents in Cerebral Ischemia and Other Pathologies
Kenneth Hensley, John M. Carney, Charles A. Stewart, Tahera Tabatabaie, Quentin Pye, and Robert A. Floyd

Neurotoxic and Neuroprotective Roles of Nitric Oxide in Cerebral Ischemia
Turgay Dalkara and Michael A. Moskowitz

A Review of Earlier Clinical Studies on Neuroprotective Agents and Current Approaches
Nils-Gunnar Wahlgren

INDEX

Volume 41

Section I: Historical Overview

Rediscovery of an Early Concept
Jeremy D. Schmahmann

Section II: Anatomic Substrates

The Cerebrocerebellar System
Jeremy D. Schmahmann and Deepak N. Pandya

Cerebellar Output Channels
Frank A. Middleton and Peter L. Strick

Cerebellar-Hypothalamic Axis: Basic Circuits and Clinical Observations
Duane E. Haines, Espen Dietrichs, Gregory A. Mihailoff, and E. Frank McDonald

Section III. Physiological Observations

Amelioration of Aggression: Response to Selective Cerebellar Lesions in the Rhesus Monkey
Aaron J. Berman

Autonomic and Vasomotor Regulation
Donald J. Reis and Eugene V. Golanov

Associative Learning
Richard F. Thompson, Shaowen Bao, Lu Chen, Benjamin D. Cipriano, Jeffrey S. Grethe, Jeansok J. Kim, Judith K. Thompson, Jo Anne Tracy, Martha S. Weninger, and David J. Krupa

Visuospatial Abilities
Robert Lalonde

Spatial Event Processing
Marco Molinari, Laura Petrosini, and Liliana G. Grammaldo

Section IV: Functional Neuroimaging Studies

Linguistic Processing
Julie A. Fiez and Marcus E. Raichle

Sensory and Cognitive Functions
Lawrence M. Parsons and Peter T. Fox

Skill Learning
Julien Doyon

Section V: Clinical and Neuropsychological Observations

Executive Function and Motor Skill Learning
Mark Hallett and Jordon Grafman

Verbal Fluency and Agrammatism
Marco Molinari, Maria G. Leggio, and Maria C. Silveri

Classical Conditioning
Diana S. Woodruff-Pak

Early Infantile Autism
Margaret L. Bauman, Pauline A. Filipek, and Thomas L. Kemper

Olivopontocerebellar Atrophy and Friedreich's Ataxia: Neuropsychological Consequences of Bilateral versus Unilateral Cerebellar Lesions
Thérèse Botez-Marquard and Mihai I. Botez

Posterior Fossa Syndrome
Ian F. Pollack

Cerebellar Cognitive Affective Syndrome
Jeremy D. Schmahmann and Janet C. Sherman

Inherited Cerebellar Diseases
Claus W. Wallesch and Claudius Bartels

Neuropsychological Abnormalities in Cerebellar Syndromes—Fact or Fiction?
Irene Daum and Hermann Ackermann

Section VI: Theoretical Considerations

Cerebellar Microcomplexes
Masao Ito

Control of Sensory Data Acquisition
James M. Bower

Neural Representations of Moving Systems
Michael Paulin

How Fibers Subserve Computing Capabilities: Similarities Between Brains and Machines
Henrietta C. Leiner and Alan L. Leiner

Cerebellar Timing Systems
Richard Ivry

Attention Coordination and Anticipatory Control
Natacha A. Akshoomoff, Eric Courchesne, and Jeanne Townsend

Context-Response Linkage
W. Thomas Thach

Duality of Cerebellar Motor and Cognitive Functions
James R. Bloedel and Vlastislav Bracha

Section VII: Future Directions

Therapeutic and Research Implications
Jeremy D. Schmahmann

Volume 42

Alzheimer Disease
Mark A. Smith

Neurobiology of Stroke
W. Dalton Dietrich

Free Radicals, Calcium, and the Synaptic Plasticity-Cell Death Continuum: Emerging Roles of the Trascription Factor NFκB
Mark P. Mattson

AP-I Transcription Factors: Short- and Long-Term Modulators of Gene Expression in the Brain
Keith Pennypacker

Ion Channels in Epilepsy
Istvan Mody

Posttranslational Regulation of Ionotropic Glutamate Receptors and Synaptic Plasticity
Xiaoning Bi, Steve Standley, and Michel Baudry

Heritable Mutations in the Glycine, $GABA_A$, and Nicotinic Acetylcholine Receptors Provide New Insights into the Ligand-Gated Ion Channel Receptor Superfamily
Behnaz Vafa and Peter R. Schofield

INDEX

Volume 43

Early Development of the *Drosophila* Neuromuscular Junction: A Model for Studying Neuronal Networks in Development
Akira Chiba

Development of Larval Body Wall Muscles
Michael Bate, Matthias Landgraf, and Mar Ruiz Gómez Bate

Development of Electrical Properties and Synaptic Transmission at the Embryonic Neuromuscular Junction
Kendal S. Broadie

Ultrastructural Correlates of Neuromuscular Junction Development
Mary B. Rheuben, Motojiro Yoshihara, and Yoshiaki Kidokoro

Assembly and Maturation of the *Drosophila* Larval Neuromuscular Junction
L. Sian Gramates and Vivian Budnik

Second Messenger Systems Underlying Plasticity at the Neuromuscular Junction
Frances Hannan and Yi Zhong

Mechanisms of Neurotransmitter Release
J. Troy Littleton, Leo Pallanck, and Barry Ganetzky

Vesicle Recycling at the *Drosophila* Neuromuscular Junction
Daniel T. Stimson and Mani Ramaswami

Ionic Currents in Larval Muscles of *Drosophila*
Satpal Singh and Chun-Fang Wu

Development of the Adult Neuromuscular System
Joyce J. Fernandes and Haig Keshishian

Controlling the Motor Neuron
James R. Trimarchi, Ping Jin, and Rodney K. Murphey

Volume 44

Human Ego-Motion Perception
A. V. van den Berg

Optic Flow and Eye Movements
M. Lappe and K.-P. Hoffman

The Role of MST Neurons During Ocular Tracking in 3D Space
K. Kawano, U. Inoue, A. Takemura, Y. Kodaka, and F. A. Miles

Visual Navigation in Flying Insects
M. V. Srinivasan and S.-W. Zhang

Neuronal Matched Filters for Optic Flow Processing in Flying Insects
H. G. Krapp

A Common Frame of Reference for the Analysis of Optic Flow and Vestibular Information
B. J. Frost and D. R. W. Wylie

Optic Flow and the Visual Guidance of Locomotion in the Cat
H. Sherk and G. A. Fowler

Stages of Self-Motion Processing in Primate Posterior Parietal Cortex
F. Bremmer, J.-R. Duhamel, S. B. Hamed, and W. Graf

Optic Flow Analysis for Self-Movement Perception
C. J. Duffy

Neural Mechanisms for Self-Motion Perception in Area MST
R. A. Andersen, K. V. Shenoy, J. A. Crowell, and D. C. Bradley

Computational Mechanisms for Optic Flow Analysis in Primate Cortex
M. Lappe

Human Cortical Areas Underlying the Perception of Optic Flow: Brain Imaging Studies
M. W. Greenlee

What Neurological Patients Tell Us about the Use of Optic Flow
L. M. Vaina and S. K. Rushton

INDEX

Volume 45

Mechanisms of Brain Plasticity: From Normal Brain Function to Pathology
Philip. A. Schwartzkroin

Brain Development and Generation of Brain Pathologies
Gregory L. Holmes and Bridget McCabe

Maturation of Channels and Receptors: Consequences for Excitability
David F. Owens and Arnold R. Kriegstein

Neuronal Activity and the Establishment of Normal and Epileptic Circuits during Brain Development
John W. Swann, Karen L. Smith, and Chong L. Lee

The Effects of Seizures of the Hippocampus of the Immature Brain
Ellen F. Sperber and Solomon L. Moshe

Abnormal Development and Catastrophic Epilepsies: The Clinical Picture and Relation to Neuroimaging
Harry T. Chugani and Diane C. Chugani

Cortical Reorganization and Seizure Generation in Dysplastic Cortex
G. Avanzini, R. Preafico, S. Franceschetti, G. Sancini, G. Battaglia, and V. Scaioli

Rasmussen's Syndrome with Particular Reference to Cerebral Plasticity: A Tribute to Frank Morrell
Fredrick Andermann and Yvonne Hart

Structural Reorganization of Hippocampal Networks Caused by Seizure Activity
Daniel H. Lowenstein

Epilepsy-Associated Plasticity in gamma-Amniobutyric Acid Receptor Expression, Function and Inhibitory Synaptic Properties
Douglas A. Coulter

Synaptic Plasticity and Secondary Epileptogenesis
Timothy J. Teyler, Steven L. Morgan, Rebecca N. Russell, and Brian L. Woodside

Synaptic Plasticity in Epileptogenesis: Cellular Mechanisms Underlying Long-Lasting Synaptic Modifications that Require New Gene Expression
Oswald Steward, Christopher S. Wallace, and Paul F. Worley

Cellular Correlates of Behavior
Emma R. Wood, Paul A. Dudchenko, and Howard Eichenbaum

Mechanisms of Neuronal Conditioning
David A. T. King, David J. Krupa,
Michael R. Foy, and Richard F. Thompson

Plasticity in the Aging Central Nervous System
C. A. Barnes

Secondary Epileptogenesis, Kindling, and Intractable Epilepsy: A Reappraisal from the Perspective of Neuronal Plasticity
Thomas P. Sutula

Kindling and the Mirror Focus
Dan C. McIntyre and Michael O. Poulter

Partial Kindling and Behavioral Pathologies
Robert E. Adamec

The Mirror Focus and Secondary Epileptogenesis
B. J. Wilder

Hippocampal Lesions in Epilepsy: A Historical Review
Robert Naquet

Clinical Evidence for Secondary Epileptogensis
Hans O. Luders

Epilepsy as a Progressive (or Nonprogressive "Benign") Disorder
John A. Wada

Pathophysiological Aspects of Landau-Kleffner Syndrome: From the Active Epileptic Phase to Recovery
Marie-Noelle Metz-Lutz, Pierre Maquet,
Annd De Saint Martin, Gabrielle Rudolf,
Norma Wioland, Edouard Hirsch,
and Chriatian Marescaux

Local Pathways of Seizure Propagation in Neocortex
Barry W. Connors, David J. Pinto, and
Albert E. Telefeian

Multiple Subpial Transection: A Clinical Assessment
C. E. Polkey

The Legacy of Frank Morrell
Jerome Engel, Jr.

Volume 46

Neurosteroids: Beginning of the Story
Etienne E. Baulieu, P. Robel, and M. Schumacher

Biosynthesis of Neurosteroids and Regulation of Their Synthesis
Synthia H. Mellon and Hubert Vaudry

Neurosteroid 7-Hydroxylation Products in the Brain
Robert Morfin and Luboslav Stárka

Neurosteroid Analysis
Ahmed A. Alomary, Robert L. Fitzgerald, and
Robert H. Purdy

Role of the Peripheral-Type Benzodiazepine Receptor in Adrenal and Brain Steroidogenesis
Rachel C. Brown and Vassilios Papadopoulos

Formation and Effects of Neuroactive Steroids in the Central and Peripheral Nervous System
Roberto Cosimo Melcangi, Valerio Magnaghi,
Mariarita Galbiati, and Luciano Martini

Neurosteroid Modulation of Recombinant and Synaptic $GABA_A$ Receptors
Jeremy J. Lambert, Sarah C. Harney,
Delia Belelli, and John A. Peters

$GABA_A$-Receptor Plasticity during Long-Term Exposure to and Withdrawal from Progesterone
Giovanni Biggio, Paolo Follesa,
Enrico Sanna, Robert H. Purdy, and
Alessandra Concas

Stress and Neuroactive Steroids
Maria Luisa Barbaccia, Mariangela Serra,
Robert H. Purdy, and Giovanni Biggio

Neurosteroids in Learning and Memory Processes
Monique Vallée, Willy Mayo,
George F. Koob, and Michel Le Moal

Neurosteroids and Behavior
Sharon R. Engel and Kathleen A. Grant

Ethanol and Neurosteroid Interactions in the Brain
A. Leslie Morrow, Margaret J. VanDoren,
Rebekah Fleming, and Shannon Penland

Preclinical Development of Neurosteroids as Neuroprotective Agents for the Treatment of Neurodegenerative Diseases
Paul A. Lapchak and Dalia M. Araujo

Clinical Implications of Circulating Neurosteroids
Andrea R. Genazzani, Patrizia Monteleone, Massimo Stomati, Francesca Bernardi, Luigi Cobellis, Elena Casarosa, Michele Luisi, Stefano Luisi, and Felice Petraglia

Neuroactive Steroids and Central Nervous System Disorders
Mingde Wang, Torbjörn Bäckström, Inger Sundström, Göran Wahlström, Tommy Olsson, Di Zhu, Inga-Maj Johansson, Inger Björn, and Marie Bixo

Neuroactive Steroids in Neuropsychopharmacology
Rainer Rupprecht and Florian Holsboer

Current Perspectives on the Role of Neurosteroids in PMS and Depression
Lisa D. Griffin, Susan C. Conrad, and Synthia H. Mellon

INDEX

Volume 47

Introduction: Studying Gene Expression in Neural Tissues by *In Situ* Hybridization
W. Wisden and B. J. Morris

Part I: *In Situ* Hybridization with Radiolabelled Oligonucleotides

In Situ Hybridization with Oligonucleotide Probes
Wl. Wisden and B. J. Morris

Cryostat Sectioning of Brains
Victoria Revilla and Alison Jones

Processing Rodent Embryonic and Early Postnatal Tissue for *In Situ* Hybridization with Radiolabelled Oligonucleotides
David J. Laurie, Petra C. U. Schrotz, Hannah Monyer, and Ulla Amtmann

Processing of Retinal Tissue for *In Situ* Hybridization
Frank Müller

Processing the Spinal Cord for *In Situ* Hybridization with Radiolabelled Oligonucleotides
A. Berthele and T. R. Tölle

Processing Human Brain Tissue for *In Situ* Hybridization with Radiolabelled Oligonucleotides
Louise F. B. Nicholson

In Situ Hybridization of Astrocytes and Neurons Cultured *In Vitro*
L. A. Arizza-McNaughton, C. De Felipe, and S. P. Hunt

In Situ Hybridization on Organotypic Slice Cultures
A. Gerfin-Moser and H. Monyer

Quantitative Analysis of *In Situ* Hybridization Histochemistry
Andrew L. Gundlach and Ross D. O'Shea

Part II: Nonradioactive *In Situ* Hybridization

Nonradioactive *In Situ* Hybridization Using Alkaline Phosphatase-Labelled Oligonucleotides
S. J. Augood, E. M. McGowan, B. R. Finsen, B. Heppelmann, and P. C. Emson

Combining Nonradioactive *In Situ* Hybridization with Immunohistological and Anatomical Techniques
Petra Wahle

Nonradioactive *In Situ* Hybridization: Simplified Procedures for Use in Whole Mounts of Mouse and Chick Embryos
Linda Ariza-McNaughton and Robb Krumlauf

INDEX

Volume 48

Assembly and Intracellular Trafficking of $GABA_A$ Receptors Eugene Barnes

Subcellular Localization and Regulation of $GABA_A$ Receptors and Associated Proteins
Bernhard Lüscher and Jean-Marc Fritschy D_1 Dopamine Receptors
Richard Mailman

Molecular Modeling of Ligand-Gated Ion Channels: Progress and Challenges
Ed Bertaccini and James R. Trudel

Alzheimer's Disease: Its Diagnosis and Pathogenesis
Jillian J. Kril and Glenda M. Halliday

DNA Arrays and Functional Genomics in Neurobiology
Christelle Thibault, Long Wang, Li Zhang, and Michael F. Miles

INDEX

Volume 49

What Is West Syndrome?
Olivier Dulac, Christine Soufflet, Catherine Chiron, and Anna Kaminski

The Relationship Between Encephalopathy and Abnormal Neuronal Activity in the Developing Brain
Frances E. Jensen

Hypotheses from Functional Neuroimaging Studies
Csaba Juhász, Harry T. Chugani, Ouo Muzik, and Diane C. Chugani

Infantile Spasms: Unique Sydrome or General Age-Dependent Manifestation of a Diffuse Encephalopathy?
M. A. Koehn and M. Duchowny

Histopathology of Brain Tissue from Patients with Infantile Spasms
Harry V. Vinters

Generators of Ictal and Interictal Electroencephalograms Associated with Infantile Spasms: Intracellular Studies of Cortical and Thalamic Neurons
M. Steriade and I. Timofeev

Cortical and Subcortical Generators of Normal and Abnormal Rhythmicity
David A. McCormick

Role of Subcortical Structures in the Pathogenesis of Infantile Spasms: What Are Possible Subcortical Mediators?
F. A. Lado and S. L. Moshé

What Must We Know to Develop Better Therapies?
Jean Aicardi

The Treatment of Infantile Spasms: An Evidence-Based Approach
Mark Mackay, Shelly Weiss, and O. Carter Snead III

ACTH Treatment of Infantile Spasms: Mechanisms of Its Effects in Modulation of Neuronal Excitability
K. L. Brunson, S. Avishai-Eliner, and T. Z. Baram

Neurosteroids and Infantile Spasms: The Deoxycorticosterone Hypothesis
Michael A. Rogawski and Doodipala S. Reddy

Are there Specific Anatomical and/or Transmitter Systems (Cortical or Subcortical) That Should Be Targeted?
Phillip C. Jobe

Medical versus Surgical Treatment: Which Treatment When
W. Donald Shields

Developmental Outcome With and Without Successful Intervention
Rochelle Caplan, Prabha Siddarth, Gary Mathern, Harry Vinters, Susan Curtiss, Jennifer Levitt, Robert Asarnow, and W. Donald Shields

Infantile Spasms versus Myoclonus: Is There a Connection?
Michael R. Pranzatelli

Tuberous Sclerosis as an Underlying Basis for Infantile Spasm
Raymond S. Yeung

Brain Malformation, Epilepsy, and Infantile Spasms
M. Elizabeth Ross

Brain Maturational Aspects Relevant to Pathophysiology of Infantile Spasms
G. Auanzini, F. Panzica, and S. Franceschetti

Gene Expression Analysis as a Strategy to Understand the Molecular Pathogenesis of Infantile Spasms
Peter B. Crino

Infantile Spasms: Criteria for an Animal Model
Carl E. Stafstrom and Gregory L. Holmes

INDEX

Volume 50

Part I: Primary Mechanisms

How Does Glucose Generate Oxidative Stress in Peripheral Nerve?
Irina G. Obrosova

Glycation in Diabetic Neuropathy: Characteristics, Consequences, Causes, and Therapeutic Options
Paul J. Thornalley

Part II: Secondary Changes

Protein Kinase C Changes in Diabetes: Is the Concept Relevant to Neuropathy?
Joseph Eichberg

Are Mitogen-Activated Protein Kinases Glucose Transducers for Diabetic Neuropathies?
Tertia D. Purves and David R. Tomlinson

Neurofilaments in Diabetic Neuropathy
Paul Fernyhough and Robert E. Schmidt

Apoptosis in Diabetic Neuropathy
Aviva Tolkovsky

Nerve and Ganglion Blood Flow in Diabetes: An Appraisal
Douglas W. Zochodne

Part III: Manifestations

Potential Mechanisms of Neuropathic Pain in Diabetes
Nigel A. Calcutt

Electrophysiologic Measures of Diabetic Neuropathy: Mechanism and Meaning
Joseph C. Arezzo and Elena Zotova

Neuropathology and Pathogenesis of Diabetic Autonomic Neuropathy
Robert E. Schmidt

Role of the Schwann Cell in Diabetic Neuropathy
Luke Eckersley

Part IV: Potential Treatment

Polyol Pathway and Diabetic Peripheral Neuropathy
Peter J. Oates

Nerve Growth Factor for the Treatment of Diabetic Neuropathy: What Went Wrong, What Went Right, and What Does the Future Hold?
Stuart C. Apfel

Angiotensin-Converting Enzyme Inhibitors: Are there Credible Mechanisms for Beneficial Effects in Diabetic Neuropathy?
Rayaz A. Malik and David R. Tomlinson

Clinical Trials for Drugs Against Diabetic Neuropathy: Can We Combine Scientific Needs With Clinical Practicalities?
Dan Ziegler and Dieter Luft

INDEX

Volume 51

Energy Metabolism in the Brain
Leif Hertz and Gerald A. Dienel

The Cerebral Glucose-Fatty Acid Cycle: Evolutionary Roots, Regulation, and (Patho) physiological Importance
Kurt Heininger

Expression, Regulation, and Functional Role of Glucose Transporters (GLUTs) in Brain
Donard S. Dwyer, Susan J. Vannucci, and Ian A. Simpson

Insulin-Like Growth Factor-1 Promotes Neuronal Glucose Utilization During Brain Development and Repair Processes
Carolyn A. Bondy and Clara M. Cheng

CNS Sensing and Regulation of Peripheral Glucose Levels
Barry E. Levin, Ambrose A. Dunn-Meynell, and Vanessa H. Routh

Glucose Transporter Protein Syndromes
Darryl C. De Vivo, Dong Wang, Juan M. Pascual, and Yuan Yuan Ho

Glucose, Stress, and Hippocampal Neuronal Vulnerability
Lawrence P. Reagan

Glucose/Mitochondria in Neurological Conditions
John P. Blass

Energy Utilization in the Ischemic/Reperfused Brain
John W. Phillis and Michael H. O'Regan

Diabetes Mellitus and the Central Nervous System
Anthony L. McCall

Diabetes, the Brain, and Behavior: Is There a Biological Mechanism Underlying the Association between Diabetes and Depression?
A. M. Jacobson, J. A. Samson, K. Weinger, and C. M. Ryan

Schizophrenia and Diabetes
David C. Henderson and Elissa R. Ettinger

Psychoactive Drugs Affect Glucose Transport and the Regulation of Glucose Metabolism
Donard S. Dwyer, Timothy D. Ardizzone, and Ronald J. Bradley

Stress and Secretory Immunity
Jos A. Bosch, Christopher Ring, Eco J. C. de Geus, Enno C. I. Veerman, and Arie V. Nieuw Amerongen

Cytokines and Depression
Angela Clow

Immunity and Schizophrenia: Autoimmunity, Cytokines, and Immune Responses
Fiona Gaughran

Cerebral Lateralization and the Immune System
Pierre J. Neveu

Behavioral Conditioning of the Immune System
Frank Hucklebridge

Psychological and Neuroendocrine Correlates of Disease Progression
Julie M. Turner-Cobb

The Role of Psychological Intervention in Modulating Aspects of Immune Function in Relation to Health and Well-Being
J. H. Gruzelier

INDEX

INDEX

Volume 52

Neuroimmune Relationships in Perspective
Frank Hucklebridge and Angela Clow

Sympathetic Nervous System Interaction with the Immune System
Virginia M. Sanders and Adam P. Kohm

Mechanisms by Which Cytokines Signal the Brain
Adrian J. Dunn

Neuropeptides: Modulators of Immune Responses in Health and Disease
David S. Jessop

Brain–Immune Interactions in Sleep
Lisa Marshall and Jan Born

Neuroendocrinology of Autoimmunity
Michael Harbuz

Systemic Stress-Induced Th2 Shift and Its Clinical Implications
Ibia J. Elenkov

Neural Control of Salivary S-IgA Secretion
Gordon B. Proctor and Guy H. Carpenter

Volume 53

Section I: Mitochondrial Structure and Function

Mitochondrial DNA Structure and Function
Carlos T. Moraes, Sarika Srivastava, Ilias Kirkinezos, Jose Oca-Cossio, Corina van Waveren, Markus Woischnick, and Francisca Diaz

Oxidative Phosphorylation: Structure, Function, and Intermediary Metabolism
Simon J. R. Heales, Matthew E. Gegg, and John B. Clark

Import of Mitochondrial Proteins
Matthias F. Bauer, Sabine Hofmann, and Walter Neupert

Section II: Primary Respiratory Chain Disorders

Mitochondrial Disorders of the Nervous System: Clinical, Biochemical, and Molecular Genetic Features
Dominic Thyagarajan and Edward Byrne

Section III: Secondary Respiratory Chain Disorders

Friedreich's Ataxia
J. M. Cooper and J. L. Bradley

Wilson Disease
C. A. Davie and A. H. V. Schapira

Hereditary Spastic Paraplegia
Christopher J. McDermott and Pamela J. Shaw

Cytochrome c Oxidase Deficiency
Giacomo P. Comi, Sandra Strazzer, Sara Galbiati, and Nereo Bresolin

Section IV: Toxin Induced Mitochondrial Dysfunction

Toxin-Induced Mitochondrial Dysfunction
Susan E. Browne and M. Flint Beal

Section V: Neurodegenerative Disorders

Parkinson's Disease
L. V. P. Korlipara and A. H. V. Schapira

Huntington's Disease: The Mystery Unfolds?
Åsa Petersén and Patrik Brundin

Mitochondria in Alzheimer's Disease
Russell H. Swerdlow and Stephen J. Kish

Contributions of Mitochondrial Alterations, Resulting from Bad Genes and a Hostile Environment, to the Pathogenesis of Alzheimer's Disease
Mark P. Mattson

Mitochondria and Amyotrophic Lateral Sclerosis
Richard W. Orrell and Anthony H. V. Schapira

Section VI: Models of Mitochondrial Disease

Models of Mitochondrial Disease
Danae Liolitsa and Michael G. Hanna

Section VII: Defects of β Oxidation Including Carnitine Deficiency

Defects of β Oxidation Including Carnitine Deficiency
K. Bartlett and M. Pourfarzam

Section VIII: Mitochondrial Involvement in Aging

The Mitochondrial Theory of Aging: Involvement of Mitochondrial DNA Damage and Repair
Nadja C. de Souza-Pinto and Vilhelm A. Bohr

INDEX

Volume 54

Unique General Anesthetic Binding Sites Within Distinct Conformational States of the Nicotinic Acetylcholine Receptor
Hugo R. Ariaas, William, R. Kem, James R. Truddell, and Michael P. Blanton

Signaling Molecules and Receptor Transduction Cascades That Regulate NMDA Receptor-Mediated Synaptic Transmission
Suhas. A. Kotecha and John F. MacDonald

Behavioral Measures of Alcohol Self-Administration and Intake Control: Rodent Models
Herman H. Samson and Cristine L. Czachowski

Dopaminergic Mouse Mutants: Investigating the Roles of the Different Dopamine Receptor Subtypes and the Dopamine Transporter
Shirlee Tan, Bettina Hermann, and Emiliana Borrelli

Drosophila melanogaster, A Genetic Model System for Alcohol Research
Douglas J. Guarnieri and Ulrike Heberlein

INDEX

Volume 55

Section I: Virsu Vectors for Use in the Nervous System

Non-Neurotropic Adenovirus: A Vector for Gene Transfer to the Brain and Gene Therapy of Neurological Disorders
P. R. Lowenstein, D. Suwelack, J. Hu, X. Yuan, M. Jimenez-Dalmaroni, S. Goverdhama, and M. G. Castro

Adeno-Associated Virus Vectors
*E. Lehtonen and
L. Tenenbaum*

Problems in the Use of Herpes Simplex Virus as a Vector
L. T. Feldman

Lentiviral Vectors
*J. Jakobsson, C. Ericson,
N. Rosenquist, and C. Lundberg*

Retroviral Vectors for Gene Delivery to Neural Precursor Cells
K. Kageyama, H. Hirata, and J. Hatakeyama

Section II: Gene Therapy with Virus Vectors for Specific Disease of the Nervous System

The Principles of Molecular Therapies for Glioblastoma
G. Karpati and J. Nalbatonglu

Oncolytic Herpes Simplex Virus
J. C. C. Hu and R. S. Coffin

Recombinant Retrovirus Vectors for Treatment of Brain Tumors
N. G. Rainov and C. M. Kramm

Adeno-Associated Viral Vectors for Parkinson's Disease
I. Muramatsu, L. Wang, K. Ikeguchi, K-i Fujimoto, T. Okada, H. Mizukami, Y. Hanazono, A. Kume, I. Nakano, and K. Ozawa

HSV Vectors for Parkinson's Disease
D. S. Latchman

Gene Therapy for Stroke
K. Abe and W. R. Zhang

Gene Therapy for Mucopolysaccharidosis
A. Bosch and J. M. Heard

INDEX

Volume 56

Behavioral Mechanisms and the Neurobiology of Conditioned Sexual Responding
Mark Krause

NMDA Receptors in Alcoholism
Paula L. Hoffman

Processing and Representation of Species-Specific Communication Calls in the Auditory System of Bats
George D. Pollak, Achim Klug, and Eric E. Bauer

Central Nervous System Control of Micturition
Gert Holstege and Leonora J. Mouton

The Structure and Physiology of the Rat Auditory System: An Overview
Manuel Malmierca

Neurobiology of Cat and Human Sexual Behavior
Gert Holstege and J. R. Georgiadis

INDEX

Volume 57

Cumulative Subject Index of Volumes 1–25

Volume 58

Cumulative Subject Index of Volumes 26–50

Volume 59

Loss of Spines and Neuropil
Liesl B. Jones

Schizophrenia as a Disorder of Neuroplasticity
Robert E. McCullumsmith, Sarah M. Clinton, and James H. Meador-Woodruff

The Synaptic Pathology of Schizophrenia: Is Aberrant Neurodevelopment and Plasticity to Blame?
Sharon L. Eastwood

Neurochemical Basis for an Epigenetic Vision of Synaptic Organization
E. Costa, D. R. Grayson, M. Veldic, and A. Guidotti

Muscarinic Receptors in Schizophrenia: Is There a Role for Synaptic Plasticity?
Thomas J. Raedler

Serotonin and Brain Development
Monsheel S. K. Sodhi and Elaine Sanders-Bush

Presynaptic Proteins and Schizophrenia
William G. Honer and Clint E. Young

Mitogen-Activated Protein Kinase Signaling
Svetlana V. Kyosseva

Postsynaptic Density Scaffolding Proteins at Excitatory Synapse and Disorders of Synaptic Plasticity: Implications for Human Behavior Pathologies
Andrea de Bartolomeis and Germano Fiore

Prostaglandin-Mediated Signaling in Schizophrenia
S. Smesny

Mitochondria, Synaptic Plasticity, and Schizophrenia
Dorit Ben-Shachar and Daphna Laifenfeld

Membrane Phospholipids and Cytokine Interaction in Schizophrenia
Jeffrey K. Yao and Daniel P. van Kammen

Neurotensin, Schizophrenia, and Antipsychotic Drug Action
Becky Kinkead and Charles B. Nemeroff

Schizophrenia, Vitamin D, and Brain Development
Alan Mackay-Sim, François Féron, Darryl Eyles, Thomas Burne, and John McGrath

Possible Contributions of Myelin and Oligodendrocyte Dysfunction to Schizophrenia
Daniel G. Stewart and Kenneth L. Davis

Brain-Derived Neurotrophic Factor and the Plasticity of the Mesolimbic Dopamine Pathway
Oliver Guillin, Nathalie Griffon, Jorge Diaz, Bernard Le Foll, Erwan Bezard, Christian Gross, Chris Lammers, Holger Stark, Patrick Carroll, Jean-Charles Schwartz, and Pierre Sokoloff

S100B in Schizophrenic Psychosis
Matthias Rothermundt, Gerald Ponath, and Volker Arolt

Oct-6 Transcription Factor
Maria Ilia

NMDA Receptor Function, Neuroplasticity, and the Pathophysiology of Schizophrenia
Joseph T. Coyle and Guochuan Tsai

INDEX

Volume 60

Microarray Platforms: Introduction and Application to Neurobiology
Stanislav L. Karsten, Lili C. Kudo, and Daniel H. Geschwind

Experimental Design and Low-Level Analysis of Microarray Data
B. M. Bolstad, F. Collin, K. M. Simpson, R. A. Irizarry, and T. P. Speed

Brain Gene Expression: Genomics and Genetics
Elissa J. Chesler and Robert W. Williams

DNA Microarrays and Animal Models of Learning and Memory
Sebastiano Cavallaro

Microarray Analysis of Human Nervous System Gene Expression in Neurological Disease
Steven A. Greenberg

DNA Microarray Analysis of Postmortem Brain Tissue
Károly Mirnics, Pat Levitt, and David A. Lewis

INDEX

Volume 61

Section I: High-Throughput Technologies

Biomarker Discovery Using Molecular Profiling Approaches
Stephen J. Walker and Arron Xu

Proteomic Analysis of Mitochondrial Proteins
Mary F. Lopez, Simon Melov, Felicity Johnson, Nicole Nagulko, Eva Golenko, Scott Kuzdzal, Suzanne Ackloo, and Alvydas Mikulskis

Section II: Proteomic Applications

NMDA Receptors, Neural Pathways, and Protein Interaction Databases
Holger Husi

Dopamine Transporter Network and Pathways
Rajani Maiya and R. Dayne Mayfield

Proteomic Approaches in Drug Discovery and Development
Holly D. Soares, Stephen A. Williams,

Peter J. Snyder, Feng Gao, Tom Stiger,
Christian Rohlff, Athula Herath, Trey Sunderland,
Karen Putnam, and W. Frost White

Section III: Informatics

Proteomic Informatics
Steven Russell, William Old, Katheryn Resing, and
Lawrence Hunter

Section IV: Changes in the Proteome by Disease

Proteomics Analysis in Alzheimer's Disease: New Insights into Mechanisms of Neurodegeneration
D. Allan Butterfield and Debra Boyd-Kimball

Proteomics and Alcoholism
Frank A. Witzmann and Wendy N. Strother

Proteomics Studies of Traumatic Brain Injury
Kevin K. W. Wang, Andrew Ottens,
William Haskins, Ming Cheng Liu,
Firas Kobeissy, Nancy Denslow,
SuShing Chen, and Ronald L. Hayes

Influence of Huntington's Disease on the Human and Mouse Proteome
Claus Zabel and Joachim Klose

Section V: Overview of the Neuroproteome

Proteomics—Application to the Brain
Katrin Marcus, Oliver Schmidt, Heike Schaefer,
Michael Hamacher, AndrÅ van Hall, and Helmut E. Meyer

INDEX

Volume 62

$GABA_A$ Receptor Structure–Function Studies: A Reexamination in Light of New Acetylcholine Receptor Structures
Myles H. Akabas

Dopamine Mechanisms and Cocaine Reward
Aiko Ikegami and Christine L. Duvauchelle

Proteolytic Dysfunction in Neurodegenerative Disorders
Kevin St. P. McNaught

Neuroimaging Studies in Bipolar Children and Adolescents

Rene L. Olvera, David C. Glahn, Sheila C. Caetano,
Steven R. Pliszka, and Jair C. Soares

Chemosensory G-Protein-Coupled Receptor Signaling in the Brain
Geoffrey E. Woodard

Disturbances of Emotion Regulation after Focal Brain Lesions
Antoine Bechara

The Use of *Caenorhabditis elegans* in Molecular Neuropharmacology
Jill C. Bettinger, Lucinda Carnell, Andrew G. Davies, and Steven L. McIntire

INDEX

Volume 63

Mapping Neuroreceptors at work: On the Definition and Interpretation of Binding Potentials after 20 years of Progress
Albert Gjedde, Dean F. Wong, Pedro Rosa-Neto, and Paul Cumming

Mitochondrial Dysfunction in Bipolar Disorder: From ^{31}P-Magnetic Resonance Spectroscopic Findings to Their Molecular Mechanisms
Tadafumi Kato

Large-Scale Microarray Studies of Gene Expression in Multiple Regions of the Brain in Schizophrenia and Alzheimer's Disease
Pavel L. Katsel, Kenneth L. Davis, and Vahram Haroutunian

Regulation of Serotonin 2C Receptor PRE-mRNA Editing By Serotonin
Claudia Schmauss

The Dopamine Hypothesis of Drug Addiction: Hypodopaminergic State
Miriam Melis, Saturnino Spiga, and Marco Diana

Human and Animal Spongiform Encephalopathies are Autoimmune Diseases: A Novel Theory and Its supporting Evidence
Bao Ting Zhu

Adenosine and Brain Function
Bertil B. Fredholm, Jiang-Fan Chen, Rodrigo A. Cunha, Per Svenningsson, and Jean-Marie Vaugeois

INDEX

Volume 64

Section I. The Cholinergic System
John Smythies

Section II. The Dopamine System
John Symythies

Section III. The Norepinephrine System
John Smythies

Section IV. The Adrenaline System
John Smythies

Section V. Serotonin System
John Smythies

INDEX

Volume 65

Insulin Resistance: Causes and Consequences
Zachary T. Bloomgarden

Antidepressant-Induced Manic Conversion: A Developmentally Informed Synthesis of the Literature
Christine J. Lim, James F. Leckman, Christopher Young, and Andrés Martin

Sites of Alcohol and Volatile Anesthetic Action on Glycine Receptors
Ingrid A. Lobo and R. Adron Harris

Role of the Orbitofrontal Cortex in Reinforcement Processing and Inhibitory Control: Evidence from Functional Magnetic Resonance Imaging Studies in Healthy Human Subjects
Rebecca Elliott and Bill Deakin

Common Substrates of Dysphoria in Stimulant Drug Abuse and Primary Depression: Therapeutic Targets
Kate Baicy, Carrie E. Bearden, John Monterosso, Arthur L. Brody, Andrew J. Isaacson, and Edythe D. London

The Role of cAMP Response Element–Binding Proteins in Mediating Stress-Induced Vulnerability to Drug Abuse
Arati Sadalge Kreibich and Julie A. Blendy

G-Protein–Coupled Receptor Deorphanizations
Yumiko Saito and Olivier Civelli

Mechanistic Connections Between Glucose/Lipid Disturbances and Weight Gain Induced by Antipsychotic Drugs
Donard S. Dwyer, Dallas Donohoe, Xiao-Hong Lu, and Eric J. Aamodt

Serotonin Firing Activity as a Marker for Mood Disorders: Lessons from Knockout Mice
Gabriella Gobbi

INDEX

Volume 66

Brain Atlases of Normal and Diseased Populations
Arthur W. Toga and Paul M. Thompson

Neuroimaging Databases as a Resource for Scientific Discovery
John Darrell Van Horn, John Wolfe, Autumn Agnoli, Jeffrey Woodward, Michael Schmitt, James Dobson, Sarene Schumacher, and Bennet Vance

Modeling Brain Responses
Karl J. Friston, William Penny, and Olivier David

Voxel-Based Morphometric Analysis Using Shape Transformations
Christos Davatzikos

The Cutting Edge of fMRI and High-Field fMRI
Dae-Shik Kim

Quantification of White Matter Using Diffusion-Tensor Imaging
Hae-Jeong Park

Perfusion fMRI for Functional Neuroimaging
Geoffrey K. Aguirre, John A. Detre, and Jiongjiong Wang

Functional Near-Infrared Spectroscopy: Potential and Limitations in Neuroimaging Studies
Yoko Hoshi

Neural Modeling and Functional Brain Imaging: The Interplay Between the Data-Fitting and Simulation Approaches
Barry Horwitz and Michael F. Glabus

Combined EEG and fMRI Studies of Human Brain Function
V. Menon and S. Crottaz-Herbette

INDEX

Volume 67

Distinguishing Neural Substrates of Heterogeneity Among Anxiety Disorders
Jack B. Nitschke and Wendy Heller

Neuroimaging in Dementia
K. P. Ebmeier, C. Donaghey, and N. J. Dougall

Prefrontal and Anterior Cingulate Contributions to Volition in Depression
Jack B. Nitschke and Kristen L. Mackiewicz

Functional Imaging Research in Schizophrenia
H. Tost, G. Ende, M. Ruf, F. A. Henn, and A. Meyer-Lindenberg

Neuroimaging in Functional Somatic Syndromes
Patrick B. Wood

Neuroimaging in Multiple Sclerosis
Alireza Minagar, Eduardo Gonzalez-Toledo, James Pinkston, and Stephen L. Jaffe

Stroke
Roger E. Kelley and Eduardo Gonzalez-Toledo

Functional MRI in Pediatric Neurobehavioral Disorders
Michael Seyffert and F. Xavier Castellanos

Structural MRI and Brain Development
Paul M. Thompson, Elizabeth R. Sowell, Nitin Gogtay, Jay N. Giedd, Christine N. Vidal, Kiralee M. Hayashi, Alex Leow, Rob Nicolson, Judith L. Rapoport, and Arthur W. Toga

Neuroimaging and Human Genetics
Georg Winterer, Ahmad R. Hariri, David Goldman, and Daniel R. Weinberger

Neuroreceptor Imaging in Psychiatry: Theory and Applications

W. Gordon Frankle, Mark Slifstein, Peter S. Talbot, and Marc Laruelle

INDEX

Volume 68

Fetal Magnetoencephalography: Viewing the Developing Brain In Utero
Hubert Preissl, Curtis L. Lowery, and Hari Eswaran

Magnetoencephalography in Studies of Infants and Children
Minna Huotilainen

Let's Talk Together: Memory Traces Revealed by Cooperative Activation in the Cerebral Cortex
Jochen Kaiser, Susanne Leiberg, and Werner Lutzenberger

Human Communication Investigated With Magnetoencephalography: Speech, Music, and Gestures
Thomas R. Knösche, Burkhard Maess, Akinori Nakamura, and Angela D. Friederici

Combining Magnetoencephalography and Functional Magnetic Resonance Imaging
Klaus Mathiak and Andreas J. Fallgatter

Beamformer Analysis of MEG Data
Arjan Hillebrand and Gareth R. Barnes

Functional Connectivity Analysis in Magnetoencephalography
Alfons Schnitzler and Joachim Gross

Human Visual Processing as Revealed by Magnetoencephalographys
Yoshiki Kaneoke, Shoko Watanabe, and Ryusuke Kakigi

A Review of Clinical Applications of Magnetoencephalography
Andrew C. Papanicolaou, Eduardo M. Castillo, Rebecca Billingsley-Marshall, Ekaterina Pataraia, and Panagiotis G. Simos

INDEX

Volume 69

Nematode Neurons: Anatomy and Anatomical Methods in *Caenorhabditis elegans*
David H. Hall, Robyn Lints, and Zeynep Altun

Investigations of Learning and Memory in *Caenorhabditis elegans*
Andrew C. Giles, Jacqueline K. Rose, and Catharine H. Rankin

Neural Specification and Differentiation
Eric Aamodt and Stephanie Aamodt

Sexual Behavior of the *Caenorhabditis elegans* Male
Scott W. Emmons

The Motor Circuit
Stephen E. Von Stetina, Millet Treinin, and David M. Miller III

Mechanosensation in *Caenorhabditis elegans*
Robert O'Hagan and Martin Chalfie

INDEX

Volume 70

Spectral Processing by the Peripheral Auditory System Facts and Models
Enrique A. Lopez-Poveda

Basic Psychophysics of Human Spectral Processing
Brian C. J. Moore

Across-Channel Spectral Processing
John H. Grose, Joseph W. Hall III, and Emily Buss

Speech and Music Have Different Requirements for Spectral Resolution
Robert V. Shannon

Non-Linearities and the Representation of Auditory Spectra
Eric D. Young, Jane J. Yu, and Lina A. J. Reiss

Spectral Processing in the Inferior Colliculus
Kevin A. Davis

Neural Mechanisms for Spectral Analysis in the Auditory Midbrain, Thalamus, and Cortex
Monty A. Escabí and Heather L. Read

Spectral Processing in the Auditory Cortex
Mitchell L. Sutter

Processing of Dynamic Spectral Properties of Sounds
Adrian Rees and Manuel S. Malmierca

Representations of Spectral Coding in the Human Brain
Deborah A. Hall, PhD

Spectral Processing and Sound Source Determination
Donal G. Sinex

Spectral Information in Sound Localization
Simon Carlile, Russell Martin, and Ken McAnally

Plasticity of Spectral Processing
Dexter R. F. Irvine and Beverly A. Wright

Spectral Processing In Cochlear Implants
Colette M. McKay

INDEX

Volume 71

Autism: Neuropathology, Alterations of the GABAergic System, and Animal Models
Christoph Schmitz, Imke A. J. van Kooten, Patrick R. Hof, Herman van Engeland, Paul H. Patterson, and Harry W. M. Steinbusch

The Role of GABA in the Early Neuronal Development
Marta Jelitai and Emília Madarasz

GABAergic Signaling in the Developing Cerebellum
Chitoshi Takayama

Insights into GABA Functions in the Developing Cerebellum
Mónica L. Fiszman

Role of GABA in the Mechanism of the Onset of Puberty in Non-Human Primates
Ei Terasawa

Rett Syndrome: A Rosetta Stone for Understanding the Molecular Pathogenesis of Autism
Janine M. LaSalle, Amber Hogart, and Karen N. Thatcher

GABAergic Cerebellar System in Autism: A Neuropathological and Developmental Perspective
Gene J. Blatt

Reelin Glycoprotein in Autism and Schizophrenia
S. Hossein Fatemi

Is There A Connection Between Autism, Prader-Willi Syndrome, Catatonia, and GABA?
Dirk M. Dhossche, Yaru Song, and Yiming Liu

Alcohol, GABA Receptors, and Neurodevelopmental Disorders
Ujjwal K. Rout

Effects of Secretin on Extracellular GABA and Other Amino Acid Concentrations in the Rat Hippocampus
Hans-Willi Clement, Alexander Pschibul, and Eberhard Schulz

Predicted Role of Secretin and Oxytocin in the Treatment of Behavioral and Developmental Disorders: Implications for Autism
Martha G. Welch and David A. Ruggiero

Immunological Findings in Autism
Hari Har Parshad Cohly and Asit Panja

Correlates of Psychomotor Symptoms in Autism
Laura Stoppelbein, Sara Sytsma-Jordan, and Leilani Greening

GABRB3 Gene Deficient Mice: A Potential Model of Autism Spectrum Disorder
Timothy M. DeLorey

The Reeler Mouse: Anatomy of a Mutant
Gabriella D'Arcangelo

Shared Chromosomal Susceptibility Regions Between Autism and Other Mental Disorders
Yvon C. Chagnon

INDEX